Progress in Inflammation Research

Series Editor

Prof. Dr. Michael J. Parnham
Senior Scientific Advisor
PLIVA dd
Prilaz baruna Filipovica 25
HR-10000 Zagreb
Croatia

Advisory Board

Forthcoming titles:
Pharmacotherapy of Gastrointestinal Inflammation, A. Guglietta (Editor), 2003
Arachidonate Remodeling and Inflammation, A.N. Fonteh, R.L. Wykle (Editor), 2003
Inflammatory Processes and Cancer,
D.W. Morgan, M. Nakada, U. Forssmann (Editors), 2003
Recent Advances in Pathophysiology of COPD, T.T. Hansel, P.J. Barnes (Editors), 2003
Anti-Inflammatory or Anti-Rheumatic Drugs,
R.O. Day, D.E. Furst, P.L.C.M. van Riel (Editors), 2004
Cytokines and Joint Injury, P. Miossec, W.B. van den Berg (Editors), 2004
Antibiotics as Immunomodulatory Agents, B. Rubin, J. Tamaoki (Editors), 2004

(Already published titles see last page.)

Heat Shock Proteins and Inflammation

W. van Eden

Editor

Springer Basel AG

Editor

Willem van Eden
Division of Immunology
Department of Infectious Diseases and Immunology
Faculty of Veterinary Medicine
University of Utrecht
Yalelaan 1
3584CL Utrecht
The Netherlands

Library of Congress Cataloging-in-Publication Data
Heat shock proteins and inflammation / W. van Eden, editor.
p. cm. -- (Progress in inflammation research)
Includes bibliographical references and index.
ISBN 978-3-0348-9410-4 ISBN 978-3-0348-8028-2 (eBook)
DOI 10.1007/978-3-0348-8028-2
1. Heat shock proteins. 2. Inflammation--Mediators. I. Eden, Willem van, 1953- II. PIR (Series)
QP552.H43H427 2003
616'.0473--dc22 2003052375

Bibliographic information published by Die Deutsche Bibliothek
Die Deutsche Bibliothek lists this publication in the Deutsche Nationalbibliografie;
detailed bibliographic data is available in the internet at <http://dnb.ddb.de>

ISBN 978-3-0348-9410-4

Originally published by Birkhäuser Verlag in 2003
Softcover reprint of the hardcover 1st edition 2003
Printed on acid-free paper produced from chlorine-free pulp. TCF ∞
Cover design: Markus Etterich, Basel
Cover illustration: Expression of HSP60 in a rejected human allograft with cytoplasmic immunoreactivity in the tubular epithelium, glomerulas and blood vessels (see page 225.)

ISBN 978-3-0348-9410-4

9 8 7 6 5 4 3 2 1 www.birkhäuser-science.com

Contents

List of contributors

Salvatore Albani, Departments of Medicine and Pediatrics, University of California San Diego, 9500 Gilman Drive, La Jolla, CA 92093-0663, USA; e-mail: salbani@ucsd.edu; Androclus Therapeutics, 4204 Sorrento Valley Blvd., Ste A-C, San Diego, CA 92121, USA; and IACOPO Institute for Translational Medicine, 9500 Gilman Drive, La Jolla, CA 92093-0663, USA

Sreyashi Basu, Center for Immunotherapy of Cancer and Infectious Diseases, University of Connecticut School of Medicine, MC1601, Farmington, CT 06030-1601, USA; e-mail: basu@up.uchc.edu

Minka Breloer, Bernhard-Nocht-Institute for Tropical Medicine, Bernhard-Nocht-str. 74, D-20359 Hamburg, Germany; e-mail: mbreloer@aol.com

Rebecca J. Brownlie, Department of Life Sciences, King's College London, Franklin-Wilkins Building, 150 Stamford Street, London SE1 9NN, UK;
e-mail: Rebecca.brownlie@kcl.ac.uk

Irun R. Cohen, Department of Immunology, The Weizmann Institute of Science, Rehovot, 76100; Israel; e-mail: irun.cohen@weizmann.ac.il

Michal Cohen, Department of Immunology, The Weizmann Institute of Science, Rehovot, 76100; Israel; e-mail: michal.cohen@weizmann.ac.il

Valerie M. Corrigall, Department of Rheumatology, GKT School of Medicine, King's College London, Guy's Hospital, London SE1 9RT, UK;
e-mail: valerie.corrigall@kcl.ac.uk

Clarissa U.I. Prazeres da Costa, Institute for Medical Microbiology, Immunology and Hygiene, Technical University of Munich, Trogerstr. 9, D-81675 Munich, Germany; e-mail: clarissa.dacosta@lrz.tum.de

Ismé M. de Kleer, University Medical Center Utrecht, Wilhelmina Children's Hospital, Department of Pediatric Immunology, P.O. Box 85090, NL-3508 AB Utrecht, The Netherlands; e-mail: i.m.dekleer@azu.nl

Richard C. Duggleby, Department of Medicine, University of Cambridge School of Clinical Medicine, Box 157, Addenbrooke's Hospital, Hills Road, Cambridge CB2 2QQ, UK; e-mail: rcd27@cam.ac.uk

J.S. Hill Gaston, Department of Medicine, University of Cambridge School of Clinical Medicine, Box 157, Addenbrooke's Hospital, Hills Road, Cambridge CB2 2QQ, UK; e-mail: jshg2@medschl.cam.ac.uk

Jane C. Goodall, Department of Medicine, University of Cambridge School of Clinical Medicine, Box 157, Addenbrooke's Hospital, Hills Road, Cambridge CB2 2QQ, UK; e-mail: jcg23@medschl.cam.ac.uk

Brian Henderson, Cellular Microbiology Research Group, Eastman Dental Institute, University College London, 256 Gray's Inn Road, London WC1X 8LD, UK; e-mail: b.henderson@eastman.ucl.ac.uk

Charles Kelly, Department of Oral Medicine and Pathology; Guy's, King's & St. Thomas' Medical Schools, King's College, London SE1 9RT, UK; e-mail: charles.kelly@kcl.ac.uk

Michael Knoflach, Institute of Pathophysiology, University of Innsbruck, Medical School, Fritz-Pregl-Str. 3, A-6020 Innsbruck, Austria; e-mail: Michael.Knoflach@uibk.ac.at

Ad P. Koets, Department of Farm Animal Health and Division of Immunology, Department of Infectious Diseases and Immunology, Faculty of Veterinary Medicine, Utrecht University, P.O. Box 80.165, NL-3508 TD Utrecht, The Netherlands; e-mail: a.p.koets@vet.uu.nl

Wietse Kuis, University Medical Center Utrecht, Wilhelmina Children's Hospital, Department of Pediatric Immunology, P.O. Box 85090, NL-3508 AB Utrecht, The Netherlands; e-mail: w.kuis@wkz.azu.nl

Tho D. Le, Departments of Medicine and Pediatrics, University of California San Diego, 9500 Gilman Drive, La Jolla, CA 92093-0663, USA; e-mail: tdle60@hotmail.com

Thomas Lehner, Department of Immunobiology; Guy's, King's & St. Thomas' Medical Schools, King's College, London SE1 9RT, UK; e-mail: thomas.lehner@kcl.ac.uk

Ofer Lider, Department of Immunology, The Weizmann Institute of Science, Rehovot, 76100, Israel; e-mail: ofer.lider@weizmann.ac.il

Mark S. Lillicrap, Department of Medicine, University of Cambridge School of Clinical Medicine, Box 157, Addenbrooke's Hospital, Hills Road, Cambridge CB2 2QQ, UK

Bruno Mayrl, Institute of Pathophysiology, University of Innsbruck, Medical School, Fritz-Pregl-Str. 3, A-6020 Innsbruck, Austria

Thomas C. Miethke, Institute for Medical Microbiology, Immunology and Hygiene, Technical University of Munich, Trogerstr. 9, D-81675 Munich, Germany; e-mail: Thomas.Miethke@lrz.tu-muenchen.de

Solveig H. Moré, Bernhard-Nocht-Institute for Tropical Medicine, Bernhard-Nocht-str. 74, D-20359 Hamburg, Germany; e-mail: more@folk.de

Gabriel Nussbaum, Department of Immunology, The Weizmann Institute of Science, Rehovot, 76100, Israel; e-mail: nussbaum@md.huji.ac.il

Gabriel S. Panayi, Department of Rheumatology, GKT School of Medicine, King's College London, Guy's Hospital, London SE1 9RT, UK; e-mail: Gabriel.Panayi@kcl.ac.uk

Liesbeth Paul, Biomedical Primate Centre, Lange Kleiweg 139, NL-2288 GJ Rijswijk, The Netherlands; e-mail: paul@bprc.nl

Berent P. Prakken, University Medical Center Utrecht, Wilhelmina Children's Hospital, Department of Pediatric Immunology, P.O. Box 85090, 3508 AB Utrecht, The Netherlands; e-mail: b.prakken@wkz.azu.nl

Gisella L. Puga Yung, Departments of Medicine and Pediatrics, University of California San Diego, 9500 Gilman Drive, La Jolla, CA 92093-0663, USA; e-mail: gpugayung@ucsd.edu; and Instituto de Ciencias Biomédicas (ICBM), Facultad de Medicina Norte, Universidad de Chile, Av. Independencia 1027, Casilla 13898, Santiago 653499, Chile

Francisco J. Quintana, Department of Immunology, The Weizmann Institute of Science, Rehovot, 76100, Israel; e-mail: francisco.quintana@weizmann.ac.il

Roberto Raggiaschi, Department of Medicine, University of Cambridge School of Clinical Medicine, Box 157, Addenbrooke's Hospital, Hills Road, Cambridge CB2 2QQ, UK

Sarah Roord, University Medical Center Utrecht, Wilhelmina Children's Hospital, P.O. Box 85090, NL-3508 AB Utrecht, The Netherlands;
e-mail: sarahroord@hotmail.com

Mahavir Singh, Lionex Diagnostics and Therapeutics GmbH, Mascheroder Weg 1B, D-38124 Braunschweig, Germany; e-mail: MSi@lionex.de

Kalle Söderström, In Silico R&D, Entelos Inc., 110 Marsh Drive, Foster City, CA 94404, USA; soderstrom@entelos.com

Pramod Srivastava, Center for Immunotherapy of Cancer and Infectious Diseases, University of Connecticut School of Medicine, MC1601, Farmington, CT 06030-1601, USA; e-mail: srivastava@nso2.uchc.edu

Stephen J. Thompson, Department of Life Sciences, King's College London, Franklin-Wilkins Building, 150 Stamford Street, London SE1 9NN;
e-mail: steve.thompson@kcl.ac.uk

Klemens Trieb, Department of Orthopedics, University of Vienna, Währingergürtel 18-20, A-1090 Vienna, Austria; e-mail: Klemens.trieb@akh-wien.ac.at

Ruurd van der Zee, Division of Immunology, Department of Infectious Diseases and Immunology, Faculty of Veterinary Medicine, University of Utrecht, Yalelaan 1, 3584CL Utrecht, The Netherlands; e-mail: r.zee@vet.uu.nl

Willem van Eden, Division of Immunology, Department of Infectious Diseases and Immunology, Faculty of Veterinary Medicine, University of Utrecht, Yalelaan 1, 3584CL Utrecht, The Netherlands; e-mail: w.eden@vet.uu.nl

Johannes M. van Noort, Division of Immunological and Infectious Diseases, TNO Prevention and Health, P.O. Box 2215, NL-2301 CE Leiden, The Netherlands;
e-mail: jm.vannoort@pg.tno.nl

Arne von Bonin, Bernhard-Nocht-Institute for Tropical Medicine, Bernhard-Nocht-str. 74, D-20359 Hamburg, Germany; e-mail: Arne_von_Bonin@magicvillage.de

Hermann Wagner, Institute for Medical Microbiology, Immunology and Hygiene, Technical University of Munich, Trogerstr. 9, D-81675 Munich, Germany; e-mail: h.wagner@lrz.tum.de

Yufei Wang, Department of Immunobiology; Guy's, King's & St. Thomas' Medical Schools, King's College, London SE1 9RT, UK

Georg Wick, Institute for Biomedical Aging Research, Austrian Academy of Sciences, Rennweg 10; 6020 Innsbruck, Austria; e-mail: Georg.Wick@oeaw.ac.at; and Institute of Pathophysiology, University of Innsbruck, Medical School, Fritz-Pregl-Str. 3, A-6020 Innsbruck, Austria

Alexandra Zanin-Zhorov, Department of Immunology, The Weizmann Institute of Science, Rehovot, 76100, Israel; e-mail: alexandra.zanin@weizmann.ac.il

Preface

Heat shock proteins or stress proteins are very immunogenic and abundant intracellular proteins. Their synthesis is up-regulated by a multitude of stressors, such as raised temperature, glucose deprivation, toxic compounds and inflammation.

The original grouping of heat shock proteins into families was based on their molecular weights. Members within a family have high levels of sequence homology, even between eukaryotic and prokaryotic members, and no homologies exist between different families.

The connection of heat shock proteins to inflammation was established when T-cells reactive to HSP60 of mycobacteria were found to have a crucial role in the induction and, more interestingly, regulation of experimental arthritis. Since then, the presence of immunity to HSP in virtually all clinical conditions of inflammation, including autoimmune diseases, transplant rejection and atherosclerosis, has emphasised the critical significance of immunity to heat shock proteins for inflammatory diseases.

Based on the initial observations it now seems that their abundant presence, especially under conditions of inflammatory stress, may have turned heat shock proteins into most dependable targets for the immune system, to trigger inflammation promoting responses or to trigger regulation in order to control inflammation.

Recently, interest in the immunology of heat shock proteins has increased to a great extent. Firstly, the characterisation of regulatory T-cells being central to mechanisms of peripheral tolerance to avoid harmful auto-aggression has aroused interest in the definition of the antigens that drive such regulatory T-cells: Heat shock proteins being good candidates as such. Secondly, the capacity of heat shock proteins to interact with receptors of the innate immune system, such as Toll-like receptors, has generated the awareness of their potential crucial impact on the initiation of immune responses and subsequently on the development of adaptive immunity.

Experience with clinical application of heat shock proteins as anti-inflammatory agents has started to emanate. Clinical trials with heat shock protein derived synthetic peptides have been undertaken in patients with diabetes and arthritis, with promising results. In this book the emphasis is on the underpinning research, which has generated the basic concepts of the role of heat shock proteins in inflammatory diseases.

W. van Eden

Preface

HSP60 and the regulation of inflammation: Physiological and pathological

Irun R. Cohen, Francisco J. Quintana, Gabriel Nussbaum, Michal Cohen, Alexandra Zanin-Zhorov and Ofer Lider

Department of Immunology, The Weizmann Institute of Science, Rehovot 76100, Israel

Introduction

This chapter positions HSP60 at the center of inflammation and body maintenance. We shall discuss the following topics:
- Inflammation: Physiological
- Inflammation: Pathological
- HSP60: Autoimmune target
- HSP60: Regulator signal
- HSP60: Innate ligand
- HSP60 model
- Signal fidelity and HSP60

Inflammation: Physiological

Inflammation has come to have a bad name. We talk about inflammatory diseases – diseases apparently caused by the inflammatory process. The pharmaceutical industry abets inflammation's ill repute and works hard to develop "anti-inflammatory" drugs, which are widely prescribed by physicians and even sold over the counter to the public.

But inflammation has not always been disparaged. The bio-medical scientists who developed the concept of inflammation through the first half of the 20th Century were aware of the beneficial aspects of inflammation [1]. In his book *General Pathology* [2], Lord Florey defines inflammation by citing Ebert: "Inflammation is a process which begins following a sub-lethal injury and ends with complete healing" [3].

Defined so, inflammation is physiological. From the moment of birth, the body must be maintained in the face of constant exposure to sub-lethal injury; the response to injury is inflammation and repair. The physiological system responsible for regulating inflammation is the immune system. The cytokines, chemokines,

Heat Shock Proteins and Inflammation, edited by Willem van Eden

adhesion molecules, and other molecules produced by the immune system's adaptive and innate agents are required for angiogenesis, wound healing, tissue remodeling and regeneration, connective tissue formation, phagocytosis, apoptosis, and other processes needed for body maintenance. Even recognition of specific antigens is involved in the regulation of inflammation. A telling example is the phenomenon of neuroprotection: It appears that the preservation and recovery of function following trauma to the central nervous system is enhanced by activated autoimmune T-cells that recognize myelin antigens [4]. The point is that the adaptive arm of the immune system also takes part in the physiology of inflammation: antibodies, B-cells and T-cells. We shall discuss below how T-cells that recognize heat shock protein 60 (HSP60) aid the regulation of inflammation. HSP60, as a ligand for innate Toll-like receptors (TLR), helps connect innate and adaptive immunity into one integrated system. Defense against infectious agents is just one aspect of the immune maintenance of a healthy body; here too, both the innate and the adaptive arms of the immune system play critical roles [5].

To maintain the body, the immune system has to diagnose the need for inflammation at any particular site and at all times, and to respond dynamically with the exact mix of inflammatory molecules, in the degree needed to repair the damage. The inflammatory response needs to be turned on, fine tuned, and turned off dynamically as the healing process progresses [6]. The physiological regulation of inflammation by the immune system involves a dynamic dialog between the immune cells and the damaged tissue. The immune system responds to molecules from the tissue that signals the state of the tissue. As we shall discuss below, the expression of HSP60 is a reliable signal. The immune system, in turn, produces molecules (cytokines, chemokines, angiogenic factors, growth factors, apoptotic factors, and so forth) that induce changes in the target tissue that, properly orchestrated, lead to healing.

Inflammation: Pathological

If the inflammatory process is not properly regulated, or not terminated, or activated at the wrong place, at the wrong time, or to an inappropriate degree, then the inflammatory process itself can become the cause of significant damage [7]. Indeed, infectious agents bent on damaging the host, usually do so by triggering inappropriate inflammation through their toxins; the host is made sick by his or her own inflammatory reaction to pathogenic stimuli that trigger TNF-α, IFN-γ and other strong pro-inflammatory mediators [8]. Autoimmune diseases are the classic example of inappropriate inflammation. Chronic inflammation plays a role in diseases such as atherosclerosis, which bear autoimmune stigmata [9]. Allergies too are the expression of inappropriate inflammation [10]. Even agents of chemical or biological warfare have been designed to activate pathological inflammation [11].

Clearly, pathological inflammation is a feature of many diseases. However, pathologic inflammation is only physiologic inflammation gone wrong. The pathophysiology of inflammation emerges from the physiology of inflammation. The immune system normally deploys the inflammatory reaction so that it maintains and repairs the body; occasionally, however, the inflammatory process runs wild and can become a pathologic reaction [12]. Indeed, inappropriate healing can be as damaging as inappropriate destruction: the pannus of rheumatoid arthritis is scar tissue [13]; scleroderma too is caused by the unregulated formation of connective tissue [14]; angiogenesis in the retina is a major cause of blindness [15, 16]. The immune system needs to receive reliable signals if it is to dispense beneficial inflammation while avoiding pathological inflammation.

HSP60: Autoimmune target

HSP60 was first discovered to function as a molecular chaperone inside cells. The HSP60 molecule is required to assist the folding of polypeptides into mature proteins in routine protein synthesis, in normal transport of proteins across membranes and in response to protein denaturation during cell stress [17, 18]. It could be said that HSP60, like the other stress proteins, performs an important function in intracellular maintenance. Intra-cellular maintenance was a subject for biochemists.

Later, and in parallel, HSP60 was unknowingly being studied as a dominant antigen in the host response to infectious bacteria. It was noted that the immune response to different bacteria tended to focus on a "common bacterial antigen" [19]. This common antigen was discovered to be the variants of HSP60 expressed by different bacteria [20–23]. The fact that HSP60 was such a dominant antigen was not explained by these studies.

The dominance of HSP60 as a T-cell antigen came to light in the study of adjuvant arthritis (AA), an autoimmune disease inducible in rats by immunization to killed mycobacteria [24]. It was discovered that a T-cell clone cross-reactive with cartilage and mycobacteria could mediate arthritis in irradiated rats [25, 26]. The mycobacterial antigen was later identified to be the HSP60 (HSP65) molecule [27]. The idea was that mycobacterial HSP65 bore a peptide epitope cross-reactive with a self-epitope in the rat joint [26]. This seminal finding aroused interest in HSP60 as a target in an autoimmune disease, albeit in an autoimmune disease induced by bacterial immunization.

The connection of HSP60 to autoimmune disease was confirmed in another system when it was discovered that HSP60, both mouse and human, was a target antigen in the Type 1 diabetes developing spontaneously in NOD mice [28, 29]. HSP60 autoimmunity was functional: immunization to human HSP60 could accelerate or abort the diabetes, an anti-HSP60 T-cell clone could mediate disease in NOD mice [30], and immunization with an HSP60 target peptide (p277) could activate transient insulitis and hyperglycemia in standard strains of mice, provided that the peptide was conjugated to an immunogenic carrier (ovalbumin) [31]. Open questions were how autoimmunity to HSP60 could be involved in diverse diseases such as AA and NOD diabetes, and how might a ubiquitous molecule like HSP60 be a tissue-specific target [32]. Autoimmunity to HSP60 was soon discovered to characterize a variety of inflammatory and autoimmune conditions such as human type 1 diabetes [28], atherosclerosis [33], Bechet's disease [34], lupus [35], and others. These findings only compounded the questions of the role of HSP60 autoimmunity in inflammation.

HSP60: Regulator signal

In direct contrast to HSP60 autoimmunity as a target in pathologic inflammation, HSP60 was also noted to down-regulate pathological inflammation. Mycobacterial HSP65 or its 180–188 peptide were found early on to vaccinate rats against adjuvant arthritis [27, 36, 37]. Work with HSP60 as a regulator of Type 1 diabetes followed.

Vaccination of NOD mice with the p277 peptide of HSP60 arrested the development of diabetes [30] and even induced remission of overt hyperglycemia [38]. Successful p277 treatment was associated with the down-regulation of spontaneous T-cell reactivity to p277 and with the induction of antibodies to p277 displaying Th2-like isotypes IgG1 and IgG2b [39]. Other peptides of HSP60 could also inhibit the development of spontaneous diabetes in NOD mice [40].

NOD mice can also develop a more robust form of diabetes induced by the administration of cyclophosphamide – cyclophosphamide-accelerated diabetes (CAD) [41]. We used DNA vaccination with constructs encoding human HSP60 (pHSP60) or mycobacterial HSP65 (pHSP65) to explore the regulatory role of HSP65 [42]. Vaccination with pHSP60 protected NOD mice from CAD. In contrast, vaccination with pHSP65, with an empty vector or with a CpG-positive oligonucleotide was not effective, suggesting that the efficacy of the pHSP60 construct might be based on regulatory HSP60 epitopes not shared with its mycobacterial counterpart, HSP65. Vaccination with pHSP60 modulated the T-cell responses to HSP60, and also to the glutamic acid decarboxilase (GAD) and insulin autoantigens: T-cell proliferative responses were significantly reduced and the pattern of cytokine secretion to HSP60, GAD and insulin showed an increase in IL-10 and IL-5 secretion and a decrease in IFN-γ secretion, compatible with a shift from a Th1-like towards a Th2-like autoimmune response. Thus, immunoregulatory networks activated by vaccination with pHSP60 or p277 can spread to other β-cell antigens like insulin and GAD and control NOD diabetes. To understand the role of HSP60 in immune signaling, we shall have to understand how HSP60 can affect autoimmunity to other molecules.

Type 1 diabetes in humans was also found to be susceptible to immunomodulation by p277 therapy. A double-blind, Phase II clinical trial was designed to study the effects of p277 therapy on newly diagnosed patients [43]. The administration of p277 after the onset of clinical diabetes preserved the endogenous levels of C-peptide (which fell in the placebo group) and was associated with lower requirements for exogenous insulin, revealing the arrest of inflammatory β-cell destruction. Treatment with p277 was associated with an enhanced Th2 response to HSP60 and p277. Taken together, these results suggest that treatment with HSP60 or its p277 peptide can lead to the induction of HSP60-specific regulators that can control the collective of pathogenic reactivities involved in the progression of autoimmune diabetes.

The administration of HSP60 or some of its peptides could also prolong the survival of skin allografts in mice [44]. Thus the regulatory effects of HSP60 were not limited to autoimmune disease.

HSP60 can also regulate AA. Vaccination of rats with HSP65 or some of its T-cell epitopes was found to prevent AA [27, 36, 37, 45]. The mechanism of protection was thought to involve cross-reactivity with the self-60 KDa heat shock protein (HSP60) [46]. We studied the roles of HSP60 and HSP65 in modulating AA [47], and identified regulatory epitopes within the HSP60 protein using DNA vaccines (Quintana et al., submitted). Susceptible rats were immunized with DNA vaccines encoding human HSP60 (pHSP60) or HSP65 (pHSP65) and AA was induced. Both pHSP60 and pHSP65 protected against AA. However pHSP60 was significantly more effective. We identified immunoregulatory regions within HSP60 using HSP60 DNA fragments and HSP60-derived overlapping peptides. A regulatory HSP60 peptide (Hu3, aa 31-50) that was specifically recognized by the T-cells of rats protected from AA by DNA-vaccination. Vaccination with Hu3, or transfer of splenocytes from Hu3-vaccinated rats, prevented the development of AA. Vaccination with the mycobacterial homologue of Hu3 had no effect. Effective DNA or peptide vaccination was associated with enhanced T-cell proliferation to a variety of disease-associated antigens, along with a Th2/3-like shift (down-regulation of IFN-γ secretion and concomitant enhanced production of IL-10 and TGF-β1) in the response to peptide Mt176-190 (the 180–188 epitope of HSP65). The regulatory response to HSP60 or its Hu3 epitope included both Th1 (IFN-γ and Th2/3 (IL10/TGF-β1) secretors. These results showed that HSP60-specific regulation can control AA and be activated by immunization with relevant HSP60-derived epitopes, administered as peptides or as DNA vaccines.

Along with the perplexing association of HSP60 autoimmunity in different autoimmune diseases, we are confronted by the effect of HSP60 as a modulator of the immune response phenotype and a terminator of autoimmune damage and allograft rejection.

HSP60: Innate immune signal

Innate immune system receptors have been shown to control the cytokine phenotype of the adaptive immune response [48]. Headway in unraveling the pleiotropic effects of HSP60 has been the discovery that HSP60 can signal macrophages and other cells through an innate signaling pathway, dependent on functional CD14 plus TLR-4 and/or TLR-2 [49–51]. The TLR-4 molecule does not seem to bind HSP60 directly, but TLR-4 is required to transduce the signal [50, 52]. Binding to the cell surface may be mediated by more than one receptor, as HSP60 molecules derived from various sources do not all compete for binding of labeled human HSP60 [53]. Macrophages exposed to soluble HSP60 secrete pro-inflammatory mediators such as TNF-α, IL-6, IL-12, and nitric oxide [49, 50, 54]. The pro-inflammatory effects of HSP60 can explain how HSP60 can be associated with up-regulation of inflammation in so many conditions. Early on, we observed that soluble HSP60 was present in the circulation of NOD mice developing type 1 diabetes and peaked before the onset of disease [55]; this increased HSP60 might accelerate islet inflammation through TLR-4 signaling. However, how can the same HSP60 molecule block inflammation?

Some explanation may be found in the discovery that HSP60 can directly activate anti-inflammatory effects in T-cells by way of an innate receptor. We find that HSP60 and its fragments can regulate the physiology of inflammation itself by acting as ligands for TLR-2 in T-cells [56]. HSP60 activated human T-cell adhesion to fibronectin, to a degree similar to other activators: IL-2, SDF-1α and RANTES. T-cell type and state of activation was important; non-activated $CD45RA^+$ and IL-2-activated $CD45RO^+$ T-cells responded optimally at low concentrations (0.1–1 ng/ml), but non-activated $CD45RO^+$ T-cells required higher concentrations (>1 μg/ml) of HSP60. T-cell HSP60 signaling was inhibited specifically by a mAb to TLR-2, but not by a mAb to TLR-4. The human T-cell response to soluble HSP60 depended on PI-3 kinase and PKC signaling, and involved the phosphorylation of Pyk-2. Soluble HSP60 also inhibited actin polymerization and T-cell chemotaxis through ECM-like gels towards the chemokines SDF-1α or ELC. Exposure to HSP60 could also down-regulate the expression of chemokine receptors CXCR4 and CCR7. Most importantly, HSP60 prevented the secretion of IFN-γ by activated T-cells (unpublished observations). These results suggest that soluble HSP60 (and its fragments), through TLR-2-dependent interactions, can down-regulate T-cell behavior and control inflammation.

To examine further the contribution of innate immune signaling to autoimmunity, we inserted a TLR-4 mutation into NOD mice. Mutated TLR-4 appears to markedly increase susceptibility to autoimmune Type I diabetes (unpublished observations). Apparently TLR-4 signaling, whether by endogenous ligands such as HSP60 or foreign ligands such as LPS, can educate the immune system to avoid pathogenic autoimmunity. Further studies will examine the requirement of TLR-4 signaling for the therapeutic effects of HSP60- and HSP60-based therapies. The importance of innate receptor signaling for the down-regulation of inflammation was also confirmed in studies showing that CpG, a ligand for TLR-9 [57], can inhibit the spontaneous development of Type I diabetes in NOD mice [58]. It would appear that CpG can actually up-regulate the expression of HSP60 and enhance HSP60 regulators (unpublished observations).

Thus, HSP60 can have both pro-inflammatory and anti-inflammatory effects on various cell types. HSP60 works as a ligand both for antigen receptors on T-cells and B-cells (and auto-antibodies) and for innate receptors TLR-4 and TLR-2 on various cells types.

HSP60 model

Figure 1 summarizes our current views of inflammation (physiologic and pathologic), body maintenance and HSP60. Infection, trauma and noxious agents cause stress (damage) to cells and tissues. Unless repaired, stress can be lethal. Stress of any kind induces up-regulation of HSP60 and other stress molecules. The chaperone function of HSP60 and its allies inside stressed cells protects the cells as a type of intra-cellular maintenance. However, there is also an extra-cellular maintenance system – the immune system. The HSP60 molecule, as it performs its chaperone function in the stressed cells, also functions as a molecular signal to the wandering cells of the immune system. Macrophages, dendritic cells, endothelial cells, and others recognize HSP60 epitopes *via* innate receptors. T-cells and B-cells recognize HSP60 both *via* their adaptive antigen receptors and their innate receptors. Healthy individuals are born with a high frequency of T-cells that have been positively selected to see HSP60 epitopes; HSP60 is a member of the set of self-molecules for which there exists natural autoimmunity [6, 59].

The fine balance of the amounts of HSP60 and other molecules expressed by the damaged/healing tissues and the responding immune cells is integrated into the dynamic process we call physiologic inflammation. Physiologic inflammation results in beneficial immune maintenance. Thus the processes set into motion by HSP60 and the other stress molecules responding to cellular damage lead to both intra-cellular maintenance (chaperone function) and extra-cellular immune maintenance (signal function).

Malfunction of the inflammatory response, however, can produce pathologic inflammation and, rather than heal, compound the damage.

This model, at our present state of knowledge, is mostly words. HSP60 research has to fill in the picture in a precise and quantitative way. Which cells recognize HSP60? Which cells and which conditions lead to the secretion of soluble HSP60? What are the functional receptors and epitopes? What are the varied responses? How do the amounts and concentrations of various molecules orchestrate the

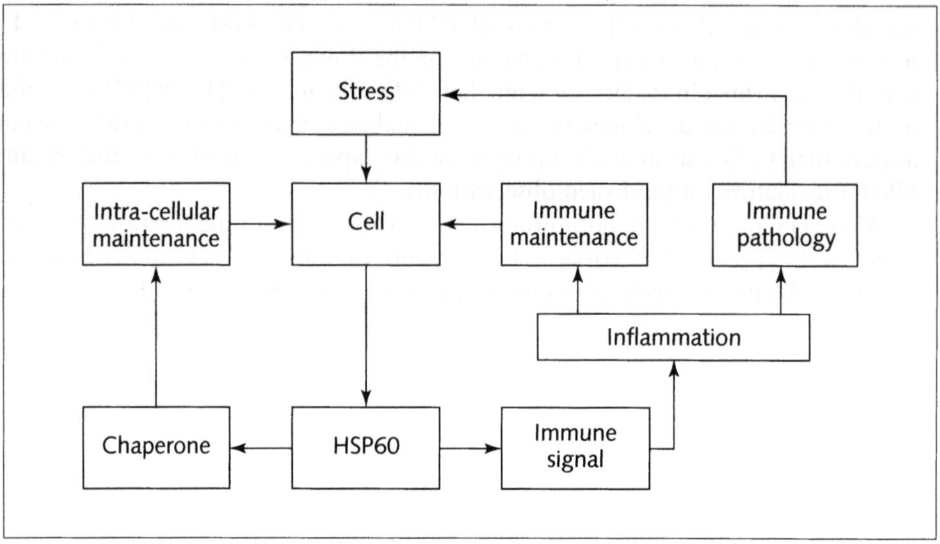

Figure 1
HSP60, as a chaperone, mediates intra-cellular maintenance; as an immune signal, HSP60 regulates inflammation. See text.

inflammatory response? How does healing occur? How does pathologic inflammation emerge? How can we control it and restore health? The program is very heavy, but HSP60 has given us a powerful tool for manipulating the inflammatory response and not only for exploring it.

Signal fidelity and HSP60

In closing, let us take note of the central position of HSP60. Why should the same HSP60 molecule function intra-cellularly as a chaperone and extracellularly as a signal? Would it not be more efficient to divide the functions and have two different molecules do the job? A chaperone should be a chaperone and a signal molecule should specialize in signaling. Is it not confusing for the system to load one HSP60 molecule with more than one important function?

The paradox of pleiotropism tells us something fundamental about signaling and about evolution. An important aspect of signaling is the fidelity of the signal. A reliable signal is a signal that never tells a lie [60]. What signal could be a more faithful sign of stress than a stress protein with a chaperone function? HSP60 signaling is foolproof.

The pleiotropism of HSP60 is yet another example of the way evolution uses old information for new purposes. HSP60 was invented at the onset of cellular life – at least in prokaryotes [61]. Moreover, the homologues of the TLR molecules involved in the transduction of the HSP60 signal appeared with multicellular organisms, but associated with development and not only with innate immunity [62]. However, like other ancient molecules, HSP60 gained a new function in higher multicellular organisms and immune systems. That's the way it goes; it takes old information to make new information [63].

Acknowledgements

Irun R. Cohen is the Mauerberger Professor of Immunology at The Weizmann Institute of Science and Director of the Center for the Study of Emerging Diseases, Jerusalem. Ofer Lider is the incumbent of the Weizmann League Career Development Chair in Children's Diseases. Parts of the research described here were supported by the European Union 5th Framework Program, by the Center for the Study of Emerging Diseases, and by the Juvenile Diabetes Foundation International.

References

1 Parnes O (2003) "Trouble from within": Allergy, autoimmunity and pathology in the first half of the twentieth century. *Studies in History and Philosophy of Biological and Biomedical Sciences; in press*

2 Florey L (1970) *General pathology*. 4th ed, Lloyd-Luke medical books, London

3 Ebert RH (1965) The inflammatory process. In: BW Zweifach, L Grant, RT McCluskey (eds): *The Inflammatory Process*. Academic Press, New York, 1–7

4 Schwartz M, Cohen IR (2000) Autoimmunity can benefit self-maintenance. *Immunol Today* 21: 265–268

5 Cohen IR (2000) Discrimination and dialogue in the immune system. *Semin Immunol* 12: 215–219; discussion 257–344

6 Cohen IR (2000) *Tending Adam's garden: Evolving the cognitive immune self*. Academic Press, London

7 Cohen IR, Efroni S (2003) Inflammation and vaccination: Cause and cure for Type 1 diabetes. In: I Raz, JS Skyler, E Shafir (eds): *Diabetes: From research to diagnosis and treatment*. Martin Dunitz, London, 223–233

8 Calandra T, Bochud PY, Heumann D (2002) Cytokines in septic shock. *Curr Clin Top Infect Dis* 22: 1–23

9 Wick G, Perschinka H, Millonig G (2001) Atherosclerosis as an autoimmune disease: An update. *Trends Immunol* 22: 665–669

10 Abbas AK, Lichtman AH, Pober JS (1994) *Cellular and molecular immunology*. 2nd ed. WB Saunders Company, Philadelphia

11 Sidell FR, Franz DR (1997) Overview: defense against the effects of chemical and biological warfare agents. In: FR Sidell, ET Takafuji, DR Franz (eds): *Medical Aspects of Chemical and Biological Warfare*. Office of The Surgeon General, Washington, D.C., 2–7

12 Efroni S, Cohen IR (2002) Simplicity belies a complex system: A response to the minimal model of immunity of Langman and Cohn. *Cell Immunol* 216: 23–30

13 Bresnihan B (1999) Pathogenesis of joint damage in rheumatoid arthritis. *J Rheumatol* 26: 717–719

14 Bertinotti L, Miniati I, Cerinic MM (2002) Angioedema and systemic sclerosis. A review of the literature. *Scand J Rheumatol* 31: 178–180

15 Feman SS (1997) New discoveries in diabetes-and thyroid-related eye disease. *Curr Opin Ophthalmol* 8: 61–65

16 Jackson JR, Seed MP, Kircher CH, Willoughby DA, Winkler JD (1997) The codependence of angiogenesis and chronic inflammation. *FASEB J* 11: 457–465

17 Hartl FU, Hayer-Hartl M (2002) Molecular chaperones in the cytosol: From nascent chain to folded protein. *Science* 295: 1852–1858

18 Zinsmaier KE, Bronk P (2001) Molecular chaperones and the regulation of neurotransmitter exocytosis. *Biochem Pharmacol* 62: 1–11

19 Young DB, Ivanyi J, Cox JH, Lamb JR (1987) The 65 kDa antigen of mycobacteria – A common bacterial protein? *Immunol Today* 8: 215–219

20 Hansen K, Bangsborg JM, Fjordvang H, Pedersen NS, Hindersson P (1988) Immunochemical characterization of and isolation of the gene for a *Borrelia burgdorferi* immunodominant 60-kilodalton antigen common to a wide range of bacteria. *Infect Immun* 56: 2047–2053

21 Morrison RP, Belland RJ, Lyng K, Caldwell HD (1989) Chlamydial disease pathogenesis. The 57 kD chlamydial hypersensitivity antigen is a stress response protein. *J Exp Med* 170: 1271–1283

22 Lamb JR, Young DB (1990) T-cell recognition of stress proteins. A link between infectious and autoimmune disease. *Mol Biol Med* 7: 311–321

23 Lamb JR, Bal V, Rothbard JB, Mehlert A, Mendez-Samperio P, Young DB (1989) The mycobacterial GroEL stress protein: A common target of T-cell recognition in infection and autoimmunity. *J Autoimmun* 2 (Suppl): 93–100

24 Wauben MHM, Wagenaar-Hilbers JPA, van Eden W (1994) Adjuvant arthritis. In: IR Cohen, A Miller (eds): *Autoimmune disease models*. Academic Press Inc., San Diego, California, USA

25 Holoshitz J, Matitiau A, Cohen IR (1984) Arthritis induced in rats by cloned T lymphocytes responsive to mycobacteria but not to collagen type II. *J Clin Invest* 73: 211–215

26 van Eden W, Holoshitz J, Nevo Z, Frenkel A, Klajman A, Cohen IR (1985) Arthritis induced by a T-lymphocyte clone that responds to Mycobacterium tuberculosis and to cartilage proteoglycans. *Proc Natl Acad Sci USA* 82: 5117–5120

27 van Eden W, Thole JE, van der Zee R, Noordzij A, van Embden JD, Hensen EJ, Cohen

IR (1988) Cloning of the mycobacterial epitope recognized by T lymphocytes in adjuvant arthritis. *Nature* 331: 171–173

28 Abulafia-Lapid R, Elias D, Raz I, Keren-Zur Y, Atlan H, Cohen IR (1999) T-cell proliferative responses of type 1 diabetes patients and healthy individuals to human hsp60 and its peptides. *J Autoimmun* 12: 121–129

29 Birk OS, Elias D, Weiss AS, Rosen A, van-der Zee R, Walker MD, Cohen IR (1996) NOD mouse diabetes: The ubiquitous mouse hsp60 is a beta-cell target antigen of autoimmune T-cells. *J Autoimmun* 9: 159–166

30 Elias D, Reshef T, Birk OS, van der Zee R, Walker MD, Cohen IR (1991) Vaccination against autoimmune mouse diabetes with a T-cell epitope of the human 65-kDa heat shock protein. *Proc Natl Acad Sci USA* 88: 3088–3091

31 Elias D, Marcus H, Reshef T, Ablamunits V, Cohen IR (1995) Induction of diabetes in standard mice by immunization with the p277 peptide of a 60-kDa heat shock protein. *Eur J Immunol* 25: 2851–2857

32 Cohen IR (1991) Autoimmunity to chaperonins in the pathogenesis of arthritis and diabetes. *Annu Rev Immunol* 9: 567–589

33 Wick G (2000) Atherosclerosis – an autoimmune disease due to an immune reaction against heat-shock protein 60. *Herz* 25: 87–90

34 Lehner T (1997) The role of heat shock protein, microbial and autoimmune agents in the aetiology of Behcet's disease. *Int Rev Immunol* 14: 21–32

35 Dhillon V, Latchman D, Isenberg D (1991) Heat shock proteins and systemic lupus erythematosus. *Lupus* 1: 3–8

36 Billingham ME, Carney S, Butler R, Colston MJ (1990) A mycobacterial 65-kD heat shock protein induces antigen-specific suppression of adjuvant arthritis, but is not itself arthritogenic. *J Exp Med* 171: 339–344

37 Yang XD, Gasser J, Feige U (1992) Prevention of adjuvant arthritis in rats by a nonapeptide from the 65-kD mycobacterial heat shock protein: Specificity and mechanism. *Clin Exp Immunol* 87: 99–104

38 Elias D, Cohen IR (1994) Peptide therapy for diabetes in NOD mice. *Lancet* 343: 704–706

39 Elias D, Meilin A, Ablamunits V, Birk OS, Carmi P, Konen-Waisman S, Cohen IR (1997) Hsp60 peptide therapy of NOD mouse diabetes induces a Th2 cytokine burst and downregulates autoimmunity to various beta-cell antigens. *Diabetes* 46: 758–764

40 Bockova J, Elias D, Cohen IR (1997) Treatment of NOD diabetes with a novel peptide of the hsp60 molecule induces Th2-type antibodies. *J Autoimmun* 10: 323–329

41 Yasunami R, Bach JF (1988) Anti-suppressor effect of cyclophosphamide on the development of spontaneous diabetes in NOD mice. *Eur J Immunol* 18: 481–484

42 Quintana FJ, Carmi P, Cohen IR (2002) DNA vaccination with heat shock protein 60 inhibits cyclophosphamide-accelerated diabetes. *J Immunol* 169: 6030–6035

43 Raz I, Elias D, Avron A, Tamir M, Metzger M, Cohen IR (2001) Beta-cell function in new-onset type 1 diabetes and immunomodulation with a heat-shock protein peptide (DiaPep277): A randomised, double-blind, phase II trial. *Lancet* 358: 1749–1753

44 Birk OS, Gur SL, Elias D, Margalit R, Mor F, Carmi P, Bockova J, Altmann DM, Cohen IR (1999) The 60-kDa heat shock protein modulates allograft rejection. *Proc Natl Acad Sci USA* 96: 5159–5163

45 Hogervorst EJ, Schouls L, Wagenaar JP, Boog CJ, Spaan WJ, van Embden JD, van Eden W (1991) Modulation of experimental autoimmunity: Treatment of adjuvant arthritis by immunization with a recombinant Vaccinia virus. *Infect Immun* 59: 2029–2035

46 van Eden W, Wendling U, Paul L, Prakken B, van Kooten P, van der Zee R (2000) Arthritis protective regulatory potential of self-heat shock protein cross-reactive T-cells. *Cell Stress Chaperones* 5: 452–457

47 Quintana FJ, Carmi P, Mor F, Cohen IR (2002) Inhibition of adjuvant arthritis by a DNA vaccine encoding human heat shock protein 60. *J Immunol* 169: 3422–3428

48 Akira S, Takeda K, Kaisho T (2001) Toll-like receptors: Critical proteins linking innate and acquired immunity. *Nat Immunol* 2: 675–680

49 Kol A, Lichtman AH, Finberg RW, Libby P, Kurt-Jones EA (2000) Cutting edge: Heat shock protein (HSP) 60 activates the innate immune response: CD14 is an essential receptor for HSP60 activation of mononuclear cells. *J Immunol* 164: 13–17

50 Ohashi K, Burkart V, Flohe S, Kolb H (2000) Cutting edge: Heat shock protein 60 is a putative endogenous ligand of the toll-like receptor-4 complex. *J Immunol* 164: 558–561

51 Vabulas RM, Ahmad-Nejad P, da Costa C, Miethke T, Kirschning CJ, Hacker H, Wagner H (2001) Endocytosed HSP60s use toll-like receptor 2 (TLR2) and TLR4 to activate the toll/interleukin-1 receptor signaling pathway in innate immune cells. *J Biol Chem* 276: 31332–31339

52 Habich C, Baumgart K, Kolb H, Burkart V (2002) The receptor for heat shock protein 60 on macrophages is saturable, specific, and distinct from receptors for other heat shock proteins. *J Immunol* 168: 569–576

53 Habich C, Kempe K, van der Zee R, Burkart V, Kolb H (2003) Different heat shock protein 60 species share pro-inflammatory activity but not binding sites on macrophages. *FEBS Lett* 533: 105–109

54 Flohe SB, Bruggemann J, Lendemans S, Nikulina M, Meierhoff G, Flohe S, Kolb H (2003) Human heat shock protein 60 induces maturation of dendritic cells *versus* a Th1-promoting phenotype. *J Immunol* 170: 2340–2348

55 Elias D, Markovits D, Reshef T, van der Zee R, Cohen IR (1990) Induction and therapy of autoimmune diabetes in the non-obese diabetic (NOD/Lt) mouse by a 65-kDa heat shock protein. *Proc Natl Acad Sci USA* 87: 1576–1580

56 Zanin-Zhorov A, Nussbaum G, Franitza S, Cohen IR, Lider O (2003) T-cells respond to heat Shock Protein 60 *via* TLR-2: Activation of adhesion and inhibition of chemokine receptors. *FASEB J* 11: 1567–1569

57 Hemmi H, Takeuchi O, Kawai T, Kaisho T, Sato S, Sanjo H, Matsumoto M, Hoshino K, Wagner H, Takeda K et al (2000) A Toll-like receptor recognizes bacterial DNA. *Nature* 408: 740–745

58 Quintana FJ, Rotem A, Carmi P, Cohen IR (2000) Vaccination with empty plasmid

DNA or CpG oligonucleotide inhibits diabetes in non-obese diabetic mice: Modulation of spontaneous 60-kDa heat shock protein autoimmunity. *J Immunol* 165: 6148–6155

59 Cohen IR (1992) The cognitive paradigm and the immunological homunculus. *Immunol Today* 13: 490–494

60 Zehavi A (1997) *The handicap principle: A missing piece of Darwin's puzzle.* Oxford University Press, London

61 Macario AJ, Lange M, Ahring BK, De Macario EC (1999) Stress genes and proteins in the archaea. *Microbiol Mol Biol Rev* 63: 923–967

62 Imler JL, Hoffmann JA (2002) Toll receptors in Drosophila: A family of molecules regulating development and immunity. *Curr Top Microbiol Immunol* 270: 63–79

63 Atlan H, Cohen IR (1998) Immune information, self-organization and meaning. *Int Immunol* 10: 711–717

Heat shock proteins and suppression of inflammation

Willem van Eden[1], *Liesbeth Paul*[2] *and Ruurd van der Zee*[1]

[1]Division of Immunology, Department of Infectious Diseases and Immunology, Faculty of Veterinary Medicine, University of Utrecht, Yalelaan 1, 3584 CL Utrecht, The Netherlands;
[2]Biomedical Primate Centre, Lange Kleiweg 139, 2288 GJ Rijswijk, The Netherlands

Introduction

The analysis of T-cell responses in the classical model of mycobacteria in oil induced arthritis in rats, which was discovered by Pearson et al. [1], through the observation that immunisation with complete Freund's adjuvant inadvertently led to induction of disease, has led to the first findings on the relationship between immunity to HSP and inflammatory diseases. This chapter aims to summarise these first findings in the model and to describe studies that have led to a broad concept of immunity to HSP as a critical contributor to mechanisms of peripheral tolerance. From the experimental data discussed, attractive possibilities seem to emanate for preventive and therapeutic interventions in inflammatory diseases.

HSP60 discovered as a critical antigen in adjuvant arthritis

Mycobacteria in oil induced arthritis in rats was known as a disease that was transferable into naïve syngeneic recipient rats with lymphocytes in the absence of added mycobacterial antigens, which characterised the model as an autoimmune disease model [2]. These transfer studies were refined by Holoshitz et al. when they succeeded in transferring disease in irradiated Lewis rats with a T-cell clone, called A2b, obtained from mycobacteria immunised animals, having a single specificity for an unknown *M. tuberculosis*-derived antigen [3]. Subsequent experimentation revealed the possible self-antigen involved to be related to an antigen associated with the proteoglycan fraction of joint cartilage [4]. In mycobacteria the antigen was found to reside in an acetone precipitable fraction. The exact nature of the antigen, however, was discovered when we tested a supposedly 65 kDa recombinant mycobacterial antigen, serendipitously cloned by van Embden [5]. A2b showed a vigorous proliferative response in the presence of this antigen, which even exceeded the level of responsiveness towards its original antigen, crude mycobacteria. Analysis of the gene sequence coding for this antigen revealed its nature as being one of the first of newly discovered HSP60 family of bacterial heat shock proteins, formerly known as "com-

Heat Shock Proteins and Inflammation, edited by Willem van Eden

mon antigen of gram-negatives". This finding of a conserved bacterial antigen as the antigen of arthritis producing T-cells in the AA model, led to a wide search in other disease models and in human autoimmune diseases for further associations of reactivity to HSP with disease. As described in several other chapters of this book, the latter search has opened a fruitful area of research, which has led to the perception that immunity to HSP is a common feature in virtually every inflammatory condition.

The search for mimicry epitopes in arthritis leads to matrixmetalloproteinases

After having defined *Mycobacterium tuberculosis* HSP60 (mtHSP60) as the antigen of arthritogenic T-cell A2b in the AA model, the epitope of A2b was uncovered by screening proteins produced by deletion mutants of the HSP60 gene, and by testing synthetic peptides. The epitope turned out to reside in a non-conserved area of the molecule, with the sequence at positions 180–188 representing the minimal sequence for inducing A2b responses [6]. Searches for the postulated mimicked epitope, associated with cartilage proteoglycans, were not leading to a definite answer. The non-conserved nature of the epitope however, did rule out the possibility of the rat tissue HSP60 harbouring the target self-epitope. Also, when tested no responses were seen in the presence of the cloned rat HSP60 [7], however followed an innovative search strategy, where information on MHC and A2b TcR contact residues of the epitope was used as the basis of a computer search profile. With this profile, a search was carried out in the Swiss-Prot database for putatively arthritis-associated T-cell epitopes. Of the 12,000 + hits, 51 were selected based on the presence of these sequences in proteins associated with cartilage, joints and/or arthritis. Peptides were synthesised on the basis of the selected sequences and were tested for their capacity to elicit T-cell responses in spleen or lymph node-derived cells obtained from animals at various stages of AA development. 14 peptides were initially found to elicit such T-cell responses. Of these 14 peptides, six were found to have the capacity to induce arthritis in Lewis rats when emulsified in the adjuvant DDA. Interestingly, three of these six peptides were derived from matrixmetalloproteinases (MMP). Furthermore, transfer experiments using T-cells from peptide immunised animals revealed that only T-cells obtained from MMP peptide immunised animals transferred disease, whereas none of the non-MMP peptides had this capacity. Despite the fact that these MMP peptides were not found to directly stimulate T-cell A2b, this A2b based search for arthritis-associated sequences revealed epitopes with arthritogenic potential in proteins with an already established impact in arthritis. Whether, therefore, MMPs are the actual targets of arthritogenic T-cells in AA remains questionable. Nonetheless, the testing of T-cell responses to these MMP peptides in RA patients has indicated that also in human arthritis, responses to these peptides do develop. It is possible that administration of these peptides under tolerising conditions may yield avenues for antigen specific immunotherapies specifically targeted to joint inflammation.

Arthritis suppressive effect of mtHSP60

Despite the fact that mtHSP60 was the mycobacterial antigen that stimulated the arthritis producing T-cell A2b in AA, when the mtHSP60 molecule was isolated from the context of the entire mycobacterium and used to immunise rats, no disease was seen to develop. Interestingly however, induction of AA with mycobacteria in oil following prior mtHSP60 immunisation, however, was found to be impossible [5]. Apparently, mtHSP60 immunisation led to the production of resistance to subsequent disease induction. This was also found in a similar model in Lewis rats, which does not involve any microbial agent, produced by administration of avridine or CP20961 [8]. Subsequent experiments carried out by various groups using various experimental autoimmune models have now substantiated the disease inhibitory effect of mtHSP60 immunisation [9–13]. A comprehensive analysis of HSP60 T-cell epitopes in adjuvant arthritis indicated the induction of self-HSP (host HSP) cross-recognition as an underlying mechanistic principle of the arthritis suppressive potential of mtHSP60. By testing an overlapping set of 15-mer peptides spanning the complete mtHSP60 sequence, nine distinct dominant T-cell epitopes were detected. Subsequent adoptive transfer studies, using T-cell lines generated to all epitopes, revealed that only T-cells directed to a very conserved 256–265 sequence transferred protection [12]. The same T-cells were shown to recognise the tissue or (mammalian) maHSP60 homologous peptide and also heat-shocked autologous cells. Furthermore, active immunisation with the conserved peptide protected against the induction of both mycobacteria-induced adjuvant arthritis and avridine arthritis. All other (non-conserved) peptides failed to produce such protection.

Recently the mtHSP60 mode of action in suppressing arthritis has been reproduced very similarly for mtHSP70 by others [14] and ourselves [13]. Also, in these latter cases T-cells recognising the very conserved mtHSP70 peptides were found to produce protection.

Given the multitude of models involving different triggering substances or antigens, where single molecules such as mtHSP60 and 70 were found to be protective, the induction of regulatory self-HSP cross-reactive T-cell responses may well explain the observation made so far.

T-cell responses to stressed antigen presenting cells (APC)

It is assumed that the regulatory potential of self-HSP reactive T-cells is exerted at the site of inflammation, where under the influence of the stressed environment, caused by toxic mediators of inflammation, cells do express enhanced levels of HSP. The potential of these T-cells to recognise endogenously produced self-HSP peptides, captured in the MHC Class II molecules of stressed APC, was demonstrated in a variety of experiments.

Self HSP60 cross-reactive T-cells were produced by repeated re-stimulations with the conserved core epitope (256–265) of MtHSP60. Subsequent testing of these T-cell lines for their proliferative responses in the presence of either normal or heat shocked (30 minutes culture at 43°C and a recovery period of four hours at 37°C) spleen cells, showed moderate responses in the presence of normal APC and showed high proliferative responses in the presence of the heat shocked APC [15, 16]. Thus, T-cells with specificity for self-HSP60 were responsive to endogenously-produced self HSP, as presented by cells in the absence of added antigens. These responses in the absence of added antigens were shown to be MHC Class II restricted, as these responses were fully inhibited by adding an antibody specific for RT1-B (OX6) to the culture. An antibody specific for RT1-D (OX17) had no inhibitory effect for the T-cell lines tested so far. In some cases, the self-HSP60 reactive T-cell line became auto-proliferative and expanded without the need of re-stimulations with antigens. These lines were also seen to produce IL-4, IL-10 and INF-γ. Upon transfer they were shown to have arthritis suppressive potential.

In co-culture experiments the suppressive potential of self-HSP specific T-cells was analysed. T-cell lines were generated by immunising with the rat homologous sequence of MtHSP60 256–265 (R256–265). For control purposes, the same protocol was followed for the generation of T-cell lines with specificity for OVA 323–339. Following isolation of the draining lymph node T-cells, the cells were re-stimulated *in vitro* with normal or heat-shocked (protocol as above) spleen cells. Following an expansion phase on IL-2, these T-cells were co-cultured with responder cells A2b in the presence of its antigen MtHSP60 180–188. It turned out that R256–265 specific T-cells, which had been re-stimulated with stressed spleen cells, had a very significant suppressive effect on the level of A2b proliferation. In other words, self-HSP reactive T-cells are not only responsive to stressed APC, but also they seem to develop a regulatory phenotype upon recognition of stressed APC (Fig. 1.).

Possible mechanisms of HSP mediated suppression of inflammation

How self-HSP cross-reactive T-cells become regulators

Recent studies of peripheral immunological tolerance have amply substantiated the potential of self-reactive T-cells as regulatory T-cells that have the capacity to control autoimmune diseases. These T-cells have regulatory activities based on either

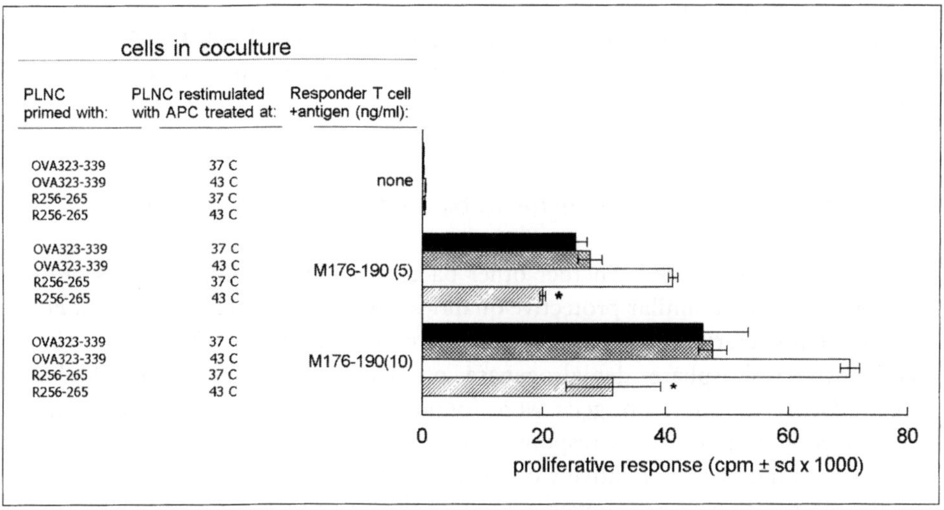

Figure 1

Short-term T-cell lines generated against a self-HSP60 epitope inhibit the proliferative responses of the arthritogenic T-cell clone A2b.

Short-term T-cell lines were generated by re-stimulation of OVA323-339 primed LNC with normal (solid bars) or heat stressed APC (cross-hatched bars) or by re-stimulation of R256-265 (the rat homologous sequence of MtHSP60 256–265) primed LNC with normal (open bars) or heat stressed APC (hatched bars). After an expansion-phase of seven days, T-cells were co-cultured with T-cell clone A2b in a ratio of 1:1 in the presence of normal APC and proliferative responses to peptide M176-190 were determined.

** $p < 0.05$ as compared to co-cultures with OVA323-339 short-term T-cell lines and as compared to the co-culture with the R256–265 short-term T-cell line generated by re-stimulation with normal APC.*

skewed regulatory cytokine production (Th1 *versus* Th2, Tr1 [13, 17, 18]) or a diverted functional status such as anergy [19]. In the case of anergy, T-cells may produce suppressive IL10 [20] or modulate APC functional activity [21–23].

Given the particular nature of heat-shock proteins, self-HSP60 reactive T-cells can be regulatory through grossly two additive mechanisms [24, 25].

Firstly, presence of constitutive self-HSP (low levels) in parenchymal cells and microbial homologues in the GALT may maintain tolerance in the peripheral self HSP specific T-cell repertoire through induction of anergy (lack of co-stimulation or cross-tolerance induction by professional APC) and induction of regulatory cytokine profiles (mucosal tolerance). Secondly, self-HSP60 (due to the uniquely subtle amino acid sequence variations between self and microbial homologues) does provide natural altered peptide ligands (APL) or partial agonists for the microbial

HSP60 oriented T-cell repertoire and therefore sets the self-HSP cross-reactive repertoire in the regulatory mode. The thus actively tolerised self-HSP60 repertoire, exerts its regulatory activity in the context of stress up-regulated self-HSP60 (maHSP60) at the site of inflammation.

Protection against arthritis seems to be a unique quality of HSP

Recently, we have tested whether other bacterially-derived immunogens of a conserved nature have similar protective qualities in experimental arthritis [26]. For this we selected bacterial antigens such as superoxide dismutase (SOD) of *E. coli*, glyceraldehyde-3-phosphate dehydrogenase of *Bacillus* (G3PDH) and aldolase of *Staphylococcus*. These antigens were found to be immunogenic as they induced readily proliferative T-cells responses and delayed type hypersensitivity reactions. All three antigens were relatively conserved and have homologous enzymes present in the mammalian hosts. However, none of the antigens were seen to affect arthritis induction upon immunisation both in adjuvant and avridine arthritis.

Why would HSP, including minor sub-families such as HSP10, protect against arthritis and other inflammatory processes, whereas other conserved and immunogenic bacterial proteins do not?

One major difference between ordinary bacterial proteins and HSP is the stress protein nature of HSP. Inflammation leads to the locally up-regulated expression of HSP, which was documented by various studies to occur in the synovium in the case of arthritis [27]. Therefore, the up-regulated presence of the target for regulatory T-cell activity can focus the regulatory control towards the inflammatory process.

In addition, it is possible that in comparison with other bacterial antigens, HSP have specialised receptors (such as CD14 and Toll-like receptors [28, 29] for their entry into for instance macrophages, as suggested by the evidence that HSP have a unique capacity to signal "danger". For gut flora associated HSP this could possibly mean that HSP have a relatively easy entry into the gut associated lymphoid tissues. In this case, the immune system would have a more continuous and intense relationship with HSP than with other microbial antigens.

Commensal bacterial environment and promotion of tolerance through HSP

The significance of gut flora for resistance to arthritis has been widely documented. In the majority of cases presence of gut flora was seen to produce a relative resistance to arthritis induction. Classical experiments by Kohashi et al. [30] have shown in Fisher rats that germ-free animals were susceptible to adjuvant arthritis whereas conventional animals were not. Reconstitution of the germ-free with the original flora or with *E. coli* was shown to reproduce resistance.

Of interest in this respect seems to be a slowly acting orally-administered anti-rheumatic drug called Subreum or OM89 (Laboratoires OM, Geneva) [31]. Subreum consists of an extract of selected *E. coli* strains and contains *E. coli* HSP, mainly HSP70. Upon intragastric administration in Lewis rats the material was found to inhibit adjuvant arthritis and to induce proliferative T-cell responses specific for HSP60 and HSP70 [32]. It is possible that the mode of action of this material is due to the induction of regulatory T-cell activity directed to HSP. And indeed, recent experiments by Cobelens et al. [33] have provided the evidence that oral HSP can directly mediate an arthritis therapeutic effect. Mycobacterial HSP60 was administered; starting at the time adjuvant arthritis was manifest, orally in combination with soybean trypsin inhibitor to avoid small-intestinal disintegration. This led to an immediate reduction of arthritis severity in the treated animals. Cobelens et al. have shown more recently that β_2 agonists, such as Salbutamol [34], have the capacity to promote production of IL10 and TGF-β in intestinal cells. When given orally in combination with HSP60, a suppressive effect on the development of adjuvant arthritis was noted. A separate series of experiments using a classical protocol of oral OVA induced tolerance in combination with Salbutamol showed a remarkably long lasting (several months) tolerance promoting effect of Salbutamol.

More indirect evidence was collected by Nieuwenhuis et al. [35] when they showed that an antibiotic regimen that led to a demonstrable predominance of *E. coli* bacteria in the gut produced a strong resistance to adjuvant arthritis induction. Interestingly enough, the same regimen produced resistance to induction of experimental autoimmune encephalomyelitis (EAE).

The Sercarz laboratory obtained interesting observations that relate microbial environment-induced mtHSP60 specific T-cell responses to arthritis resistance [36]. Two groups of Fischer (F344) rats were compared: one group housed in a barrier facility and one group housed in a conventional facility. The animals held in the conventional facility were seen to develop better resistance to AA induction as compared to the barrier kept animals. Furthermore, the conventional facility-reared animals were shown to have increased their repertoire and responses of T-cells directed to a number of carboxy-terminally located mtHSP60 epitopes that had been described to induce arthritis protective T-cell responses. Also, transfer of T-cells from conventional animals and not from barrier-kept animals, stimulated by the latter epitopes was shown to protect naïve recipients from active disease induction. Thus, apparently, environmental bacteria can modulate arthritis through the spontaneous induction of T-cell responses to regulatory determinants of mtHSP60.

Self-HSP cross-reactive T-cells produce IL-10

Immunisation with mycobacterial HSP70 in rats was found to induce production of IL-10. First draining lymph node cells were analysed by RT-PCR. In samples obtained after HSP70 immunisation message for IL-10 was detected, whereas in samples obtained after immunisation with other control bacterial antigens (as mentioned above) it was not detected [13]. Follow-up experiments, using intracellular staining for cytokines in selected $CD4^+$ T-cells, showed that HSP70 and HSP60 specific T-cells were producing IL-10 and IL- 4 to a lesser extent. Control T-cells specific for aldolase were found to produce IFN-γ and TNF-α and no IL-10 or IL-4. In the same assay system, a T-cell line raised by immunisation with a conserved HSP70 epitope was also found to produce IL-10. In accordance with that, the same epitope in the form of a synthetic peptide was shown to be arthritis protective upon nasal administration.

In a recent study by van Halteren et al. [37] the role of HSP60 specific T-cells in mouse NOD diabetes was studied. From the analysis comparing the diabetes susceptible NOD with the diabetes resistant transgenic strain expressing the resistant MHC type $I\text{-}A^{g7asp}$ on the NOD background it appeared that HSP60 reactive T-cells were triggered by insulitis, although HSP60 was not a primary autoantigen in NOD diabetes. Furthermore, the analysis of cytokines produced suggested that IL-10 producing HSP60 specific T-cells were associated with the control of insulitis. A drop in IL-10 producing HSP60 reactive T-cells was found to precede clinical expression of diabetes.

Other studies have also indicated a propensity of HSP to induce production of IL-10. For example in the mouse it was shown that chlamydial HSP60 primarily induced pro-inflammatory cytokines [38]. However, in a similar set-up the mouse HSP60 protein was found to induce primarily IL-10 and in the combination of chlamydial and mouse HSP60 both pro-inflammatory cytokines and IL-10 were produced. Therefore, it was postulated that in the natural situation of chlamydial infection both cytokines that drive inflammation and cytokines that control inflammation are produced. A similar situation was reported for experimental Listeria infections in the rat [39]. In that case *Listeria* HSP70 specific T-cell were found to have a regulatory role during listeriosis through production of TGF-β and IL-10. It was suggested that HSP reactive T-cells were involved to terminate the Th1 cell mediated excessive inflammation after the "battle against *Listeria monocytogenes* has been won".

Other forms of inflammation, apart from that resulting from infection, have also been seen to be associated with production of IL-10. In fact it appears as if cellular injuries of diverse origin lead to production of IL-10. Therefore, Stordeur and Goldman have called IL-10 a "stress cytokine" [40].

Altogether, it is possible that IL-10 production by stressed cells and the propensity of self HSP (cross-) reactive T-cells to produce IL-10 are both a reflection of the regulatory arm of the immune system meant to control excessive or non-productive inflammation (Fig. 2).

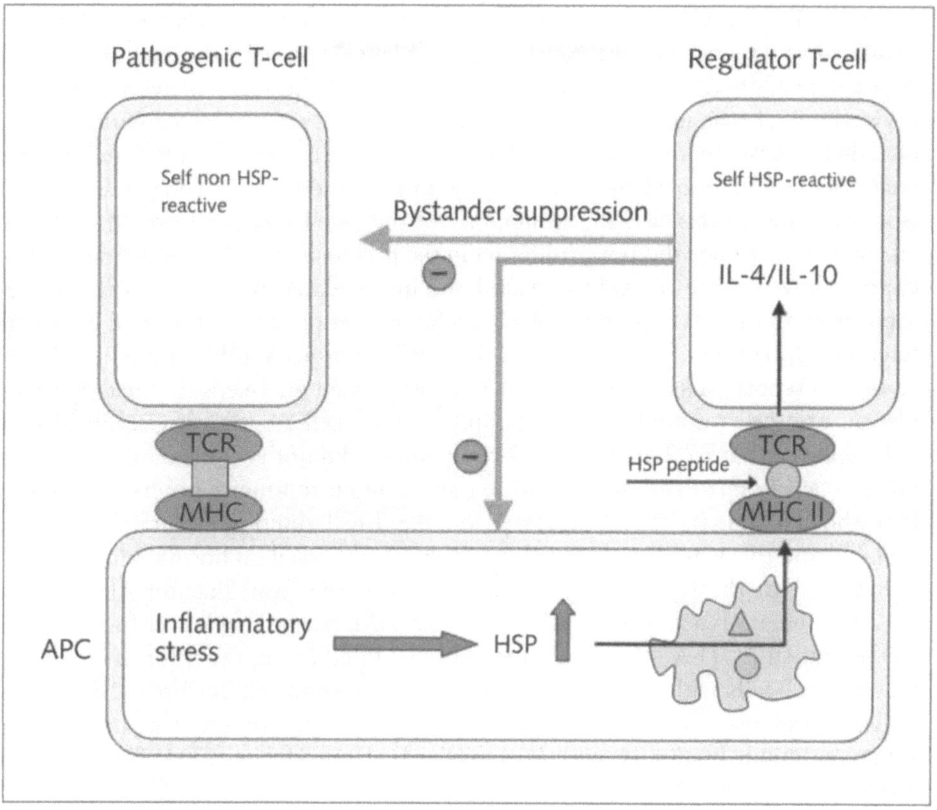

Figure 2
Self HSP-reactive T-cells suppress pathogenic T-cells present in the same T-cell-APC cluster, either through bystander suppression or by down-modulating the T-cell activating capacity of APC, by the secretion of IL-10 and/or IL-4.

B7.2 expression on activated self-HSP cross-reactive T-cells

Apart from their propensity to produce IL-10 some distinct phenotypic markers were found to become up-regulated on self-HSP reactive T-cells. Recent experiments in rat adjuvant arthritis have shown the selective up-regulation of B7.2 on HSP60 reactive T-cells when stimulated by the self (rat) HSP60 sequence [41]. T-cell lines were raised against the 256–270 conserved mycobacterial HSP60 peptide. Earlier, such lines had been found to induce arthritis protection [12]. Upon re-stimulation with the mycobacterial peptide (256–265) analysis by FACS showed that the cell was up-regulating activation markers such as IL-2R, ICAM-1 and OX40. Furthermore, the TCR was down-modulated and the co-stimulatory molecule B7.1 was up-regulated at a higher peptide concentration. Stimulation with the homologous self-peptide (rat 256–265) did lead to up-regulation of the same activation markers. However, B7.1 was not up-regulated and there was a strong up-regulation of B7.2 already at a low peptide concentration. The same selective B7.2 up-regulation was seen after stimulation of this T-cell line with heat-shocked antigen presenting cells (spleen). Thus, in this case the stimulation of the T-cell specific for the mycobacterial conserved epitope did not proliferate in the presence of the corresponding rat self epitope, however the cell was activated and up-regulated B7.2. By this the self-epitope acted as a partial agonist or APL. So far, no data has been reported on the differential expression of B7.1 and B7.2 on T-cells after activation with APL. However, several reports suggested a role for B7.2, expressed on T-cells, in the down-modulation of T-cell responses, including anti-T-cell T-cell responses. Greenfield et al. [42] reported that B7.2 transfected T-cell tumours did not provide co-stimulation to other T-cells *in vitro* and, in fact, inhibit anti-tumour immunity *in vivo*. In addition they showed that the B7.2 expressed on the T-cell tumour preferentially bound CTLA-4 and this was also observed for the B7.2 expressed on normal murine T-cells indicating that the B7.2 expressed on T-cells differed from that on APC. Indeed, B7.2 on human T-cells was shown to have a different glycosylation form with no CD28, but still CTLA-4 binding capacity [43]. In addition, the interaction of B7.2 on murine T-cells with CTLA-4 was shown to be responsible for down-modulation of T-cell responses *in vitro* [44]. Recently, B7.2 on T-cell tumours was reported to suppress tumour immunity, with IL-4 and/or IL-10 producing $CD4^+$ T-cells playing a critical role in the suppression.

Thus, the up-regulation of B7.2 on T-cells, induced upon recognition of self-HSP on APC, may well provide such T-cells with a mechanism to control pathogenic T-cells *in vivo*. As pathogenic activated rat T-cells do express MHC Class II antigens [45] and enhanced levels of self HSP60 [46], they may present the self-HSP60 peptide to T-cells that are induced to express B7.2. Also in a more direct manner, this T-T interaction involving B7.2 would lead to a negative signal through CTLA-4 on the activated pathogenic T-cells, leading to a suppressive form of regulation (Fig. 3). And indeed a self-HSP specific highly autoreactive T-cell line expressing B7.2 after culture with just APC and no added antigen was seen to transfer protection in rat adjuvant arthritis.

Anti-inflammatory activity in infection

Risk of inducing non-specific immune suppression by HSP immunisation

Theoretically, enhancement of self-HSP directed regulation could lead to an inhibited immune responsiveness causing a relatively immuno-deficient state. In such a

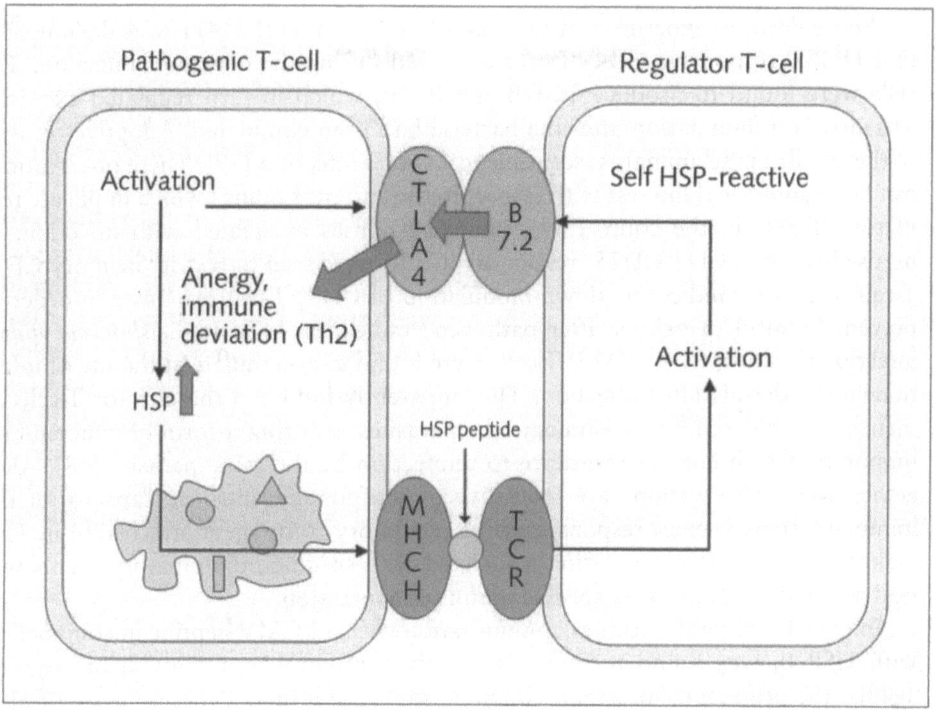

Figure 3
Self HSP-reactive T-cells down-modulate HSP-presenting pathogenic T-cells directly through inhibitory T-cell interactions involving B7.2 (CD86) and its ligand CTLA-4.

case the undesired side effect of HSP interventions would be the increased susceptibility to infection. However, from existing experience with HSP immunised animals no evidence for a raised incidence of infection was documented. In addition, responses to other non-HSP antigens were never found to be suppressed in such animals. Apparently, the localised over-expression of self-HSP at sites of inflammation confines the regulatory activity of HSP responsive T-cells to those areas where inflammation persists to some extent, allowing the regulation to develop.

In the case of infection it is the kinetics of HSP expression that dictates the course of events to develop into a safely controlled anti-infection inflammatory response. Infection produced by bacteria or other cellular organisms may lead first to expression of microbial non-self HSP. The resulting microbe HSP specific, possibly high affinity, pro-inflammatory T-cell response is held to contribute to the elimination of the microbial invader. Subsequent over-expression of self-HSP by the stressed cells at the site of infection may attract and activate the regulatory T-cells that control the inflammation which then subsides.

And indeed, in experimental *Listeria* infection in rats (F344) it was documented that HSP70 responsive CD4+ T-cells appeared during the course of infection. The cells were found to produce TGF-β and IL-10, which in turn regulated excessive Th1 driven inflammation after the bacteria had been eliminated. Adoptive transfer of these cells made animals susceptible to *Listeria* infection [39]. These observations can be of additional interest with respect to the current findings which implicate regulatory T-cells in the control of immune responses associated with infection. In mouse listeriosis CD4+CD25+ regulatory T-cells were seen to restrict memory CD8+ T-cell responses, indicating down-modulation of CD8+ T-cell responses in order to prevent harmful overshoot after pathogen eradication [47]. In *Leishmania major* infections in mice, CD4+CD25+ T-cells were found to accumulate at the site of infection in the dermis after infection. This apparently inhibited the effector T-cells in their efforts to completely eliminate the parasites, securing a form of concomitant immunity, which confers resistance to reinfection by the same parasite [48]. Altogether, these observations are indicative of the possible intrinsic capacity of the immune system to exert responses with a regulatory component attached to it. This makes it unlikely that regulation associated with specific inflammatory events will lead to a risky form of non-specific immune suppression.

In the case of viral infection, immunisation with a LCMV peptide in conjunction with HSP70, was found to induce protective anti-viral immunity, again arguing against the production of general immune suppression under the influence of HSP [49].

Use of HSP in anti-infection therapy

Further evidence that HSP immunity contributes to anti-infectious host defensive strategies, despite its regulatory potential, comes from recent successes of inhibiting tuberculosis in mice using an HSP60 DNA vaccine [50]. Whereas earlier experiments had shown prevention to be possible using HSP60 DNA vaccines, now an established (eight weeks) infection with virulent *M. tuberculosis* H37Rv was dramatically reduced following four doses (two week) intervals of a plasmid expressing mycobacterial HSP60. Interestingly, in combination with antimicrobial chemotherapy the approach led in some mice to the complete elimination of residual bacteria.

A very similar approach using mycobacterial HSP60 DNA vaccines inhibited adjuvant arthritis in Lewis rats [51].

Use of HSP for vaccines

The immunological dominance of HSP has led to testing their potential as carrier molecules for providing T-cell help to less immunogenic vaccine components [52].

This has resulted in observations where mycobacterial HSP60 and HSP70 were found to be excellent carrier molecules in BCG primed animals. Furthermore, the undesired carrier-induced suppression, as known to exist for other carrier molecules such as tetanus toxoid, was not observed for HSP.

The principle was shown to be valid for peptide conjugates as well, where defined microbial HSP60 T-cell epitopes were used to induce pneumococcal polysaccharide specific antibodies [53]. More recently, a self-HSP60 derived peptide was used to produce an effective vaccine in mice when conjugated to pneumococcal polysaccharide [54].

And interestingly, for HSP70, BCG priming turned out to be not a prerequisite, possibly reflecting a continuous and spontaneous state of immune priming for this antigen [55]. Therefore, HSP70 may well be an ideal "adjuvant" when used as a carrier molecule. This has been shown to work for an adjuvant-free *M. tuberculosis* HSP70-HIVp24 (Gag) fusion protein.

Sub-unit vaccines based on HSP's have been developed and tested for various infectious organisms. In *Legionella* HSP60 was found to induce immunity cross-protective across multiple *Legionella* spp. against aerosol challenge. For *H. pylori* HSP60 or HSP10 were found to induce protection against orogastric challenge [56]. HSP60 of *Yersinia* was found protective against infection with *Y. enterocolitica* [57].

In human candidiasis antibodies against a fragment of *Candida* HSP90 were found to correlate with patient recovery from systemic *Candida* infection [58]. Passive transfer of patient sera or monoclonal antibodies to an HSP90 epitope protected mice against systemic *Candida* infections.

Acknowledgements

We thank Mrs Anita Beijer for expert editorial assistance. Part of the work was supported by the EC FP5: HSP60 as a novel therapeutic target for diabetes and rheumatoid arthritis (QLRT-2001-01287).

References

1 Pearson C (1956) Development of arthritis, periarthritis and periostitis in rats given adjuvant. *Proc Soc Exp Biol Med* 91: 95–101

2 Whitehouse DJ, Whitehouse WM, Pearson CM (1969) Passive transfer of adjuvant-induced arthritis and allergic encephalomyelitis in rats using thoracic duct lymphocytes. *Nature* 224: 1322

3 Holoshitz J, Naparstek Y, Ben-Nun A, Cohen IR (1983) Lines of T lymphocytes induce or vaccinate against autoimmune arthritis. *Science* 219: 56–58

4 van Eden W, Holoshitz J, Nevo Z, Frenkel A, Klajman A, Cohen IR (1985) Arthritis

induced by a T-lymphocyte clone that responds to *Mycobacterium tuberculosis* and to cartilage proteoglycans. *Proc Natl Acad Sci USA* 82: 5117–5120

5 van Eden W, Thole JE, van der Zee R et al (1988) Cloning of the mycobacterial epitope recognized by T lymphocytes in adjuvant arthritis. *Nature* 331: 171–173

6 Van der Zee R, Van Eden W, Meloen RH, Noordzij A, Van Embden JD (1989) Efficient mapping and characterization of a T cell epitope by the simultaneous synthesis of multiple peptides. *Eur J Immunol* 19: 43–47

7 van Bilsen JH, Wagenaar-Hilbers JP, van der Cammen MJ, van Dijk ME, van Eden W, Wauben MH (2002) Successful immunotherapy with matrix metalloproteinase-derived peptides in adjuvant arthritis depends on the timing of peptide administration. *Arthritis Res* 4: R2

8 Kingston AE, Hicks CA, Colston MJ, Billingham ME (1996) A 71-kD heat shock protein (hsp) from *Mycobacterium tuberculosis* has modulatory effects on experimental rat arthritis. *Clin Exp Immunol* 103: 77–82

9 van Eden W (1991) Heat-shock proteins as immunogenic bacterial antigens with the potential to induce and regulate autoimmune arthritis. *Immunol Rev* 121: 5–28

10 van den Broek MF, Hogervorst EJ, Van Bruggen MC, Van Eden W, van der Zee R, van den Berg WB (1989) Protection against streptococcal cell wall-induced arthritis by pretreatment with the 65-kD mycobacterial heat shock protein. *J Exp Med* 170: 449–466

11 Thompson SJ, Francis JN, Siew LK, Webb GR, Jenner PJ, Colston MJ, Elson CJ (1998) An immunodominant epitope from mycobacterial 65-kDa heat shock protein protects against pristane-induced arthritis. *J Immunol* 160: 4628–4634

12 Anderton SM, van der Zee R, Prakken B, Noordzij A, van Eden W (1995) Activation of T cells recognizing self 60-kD heat shock protein can protect against experimental arthritis. *J Exp Med* 181: 943–952

13 Wendling U, Paul L, van Der Zee R, Prakken B, Singh M, van Eden W (2000) A conserved mycobacterial heat shock protein (hsp) 70 sequence prevents adjuvant arthritis upon nasal administration and induces IL-10-producing T cells that cross-react with the mammalian self-hsp70 homologue. *J Immunol* 164: 2711–2717

14 Tanaka S, Kimura Y, Mitani A et al (1999) Activation of T cells recognizing an epitope of heat-shock protein 70 can protect against rat adjuvant arthritis. *J Immunol* 163: 5560–5565

15 Paul AG, van Kooten PJ, van Eden W, van der Zee R (2000) Highly autoproliferative T cells specific for 60-kDa heat shock protein produce IL-4/IL-10 and IFN-gamma and are protective in adjuvant arthritis. *J Immunol* 165: 7270–7277

16 van Eden W, Wendling U, Paul L, Prakken B, van Kooten P, van der Zee R (2000) Arthritis protective regulatory potential of self-heat shock protein cross-reactive T cells. *Cell Stress Chaperones* 5: 452–457

17 Groux H, O'Garra A, Bigler M et al (1997) A CD4+ T-cell subset inhibits antigen-specific T-cell responses and prevents colitis. *Nature* 389: 737–742

18 Sakaguchi S, Sakaguchi N, Asano M, Itoh M, Toda M (1995) Immunologic self-tolerance maintained by activated T cells expressing IL-2 receptor alpha-chains (CD25)

Breakdown of a single mechanism of self-tolerance causes various autoimmune diseases. *J Immunol* 155: 1151–1164
19 Sakaguchi S (2000) Regulatory T cells: Key controllers of immunologic self-tolerance. *Cell* 101: 455–458
20 Buer J, Lanoue A, Franzke A, Garcia C, von Boehmer H, Sarukhan A (1998) Interleukin 10 secretion and impaired effector function of major histocompatibility complex class II-restricted T cells anergized *in vivo. J Exp Med* 187: 177–183
21 Jordan MS, Riley MP, von Boehmer H, Caton AJ (2000) Anergy and suppression regulate CD4(+) T cell responses to a self peptide. *Eur J Immunol* 30: 136–144
22 Chai JG, Bartok I, Chandler P, Vendetti S, Antoniou A, Dyson J, Lechler R (1999) Anergic T cells act as suppressor cells *in vitro* and *in vivo. Eur J Immunol* 29: 686–692
23 Taams LS, van Rensen AJ, Poelen MC, van Els CA, Besseling AC, Wagenaar JP, van Eden W, Wauben MH (1998) Anergic T cells actively suppress T cell responses *via* the antigen-presenting cell. *Eur J Immunol* 28: 2902–2912
24 van Eden W, van der Zee R, Paul AG, Prakken BJ, Wendling U, Anderton SM, Wauben MH (1998) Do heat shock proteins control the balance of T-cell regulation in inflammatory diseases? *Immunol Today* 19: 303–307
25 van Eden W, van der Zee R, Taams LS, Prakken AB, van Roon J, Wauben MH (1998) Heat-shock protein T-cell epitopes trigger a spreading regulatory control in a diversified arthritogenic T-cell response. *Immunol Rev* 164: 169–174
26 Prakken BJ, Wendling U, van der Zee R, Rutten VP, Kuis W, van Eden W (2001) Induction of IL-10 and inhibition of experimental arthritis are specific features of microbial heat shock proteins that are absent for other evolutionarily conserved immunodominant proteins. *J Immunol* 167: 4147–4153
27 Boog CJ, de Graeff-Meeder ER, Lucassen MA, van der Zee R, Voorhorst-Ogink MM, van Kooten PJ, Geuze HJ, van Eden W (1992) Two monoclonal antibodies generated against human hsp60 show reactivity with synovial membranes of patients with juvenile chronic arthritis. *J Exp Med* 175: 1805–1810
28 Ohashi K, Burkart V, Flohe S, Kolb H (2000) Cutting edge: Heat shock protein 60 is a putative endogenous ligand of the toll-like receptor-4 complex. *J Immunol* 164: 558–561
29 Habich C, Kempe K, van der Zee R, Burkart V, Kolb H (2003) Different heat shock protein 60 species share pro-inflammatory activity but not binding sites on macrophages. *FEBS Lett* 533: 105–109
30 Kohashi O, Kohashi Y, Takahashi T, Ozawa A, Shigematsu N (1986) Suppressive effect of *Escherichia coli* on adjuvant-induced arthritis in germ-free rats. *Arthritis Rheum* 29: 547–553
31 Vischer TL, Van Eden W (1994) Oral desensitisation in rheumatoid arthritis. *Ann Rheum Dis* 53: 708–710
32 Bloemendal A, Van der Zee R, Rutten VP, van Kooten PJ, Farine JC, van Eden W (1997) Experimental immunization with anti-rheumatic bacterial extract OM-89 induces T cell responses to heat shock protein (hsp)60 and hsp70; modulation of peripheral immuno-

logical tolerance as its possible mode of action in the treatment of rheumatoid arthritis (RA). *Clin Exp Immunol* 110: 72–78

33 Cobelens PM, Heijnen CJ, Nieuwenhuis EE, Kramer PP, van der Zee R, van Eden W, Kavelaars A (2000) Treatment of adjuvant-induced arthritis by oral administration of mycobacterial Hsp65 during disease. *Arthritis Rheum* 43: 2694–2702

34 Cobelens PM, Kavelaars A, Vroon A, Ringeling M, van der Zee R, van Eden W, Heijnen CJ (2002) The beta 2-adrenergic agonist salbutamol potentiates oral induction of tolerance, suppressing adjuvant arthritis and antigen-specific immunity. *J Immunol* 169: 5028–5035

35 Nieuwenhuis EE, Visser MR, Kavelaars A, Cobelens PM, Fleer A, Harmsen W, Verhoef J, Akkermans LM, Heijnen CJ (2000) Oral antibiotics as a novel therapy for arthritis: evidence for a beneficial effect of intestinal *Escherichia coli*. *Arthritis Rheum* 43:2583–2589

36 Moudgil KD, Kim E, Yun OJ, Chi HH, Brahn E, Sercarz EE (2001) Environmental modulation of autoimmune arthritis involves the spontaneous microbial induction of T cell responses to regulatory determinants within heat shock protein 65. *J Immunol* 166: 4237–4243

37 van Halteren AG, Mosselman B, Roep BO, van Eden W, Cooke A, Kraal G, Wauben MH (2000) T cell reactivity to heat shock protein 60 in diabetes-susceptible and genetically protected non-obese diabetic mice is associated with a protective cytokine profile. *J Immunol* 165: 5544–5551

38 Yi Y, Yang X, Brunham RC (1997) Autoimmunity to heat shock protein 60 and antigen-specific production of interleukin-10. *Infect Immun* 65: 1669–1674

39 Kimura Y, Yamada K, Sakai T, Mishima K, Nishimura H, Matsumoto Y, Singh M, Yoshikai Y (1998) The regulatory role of heat shock protein 70-reactive $CD4^+$ T cells during rat listeriosis. *Int Immunol* 10: 117–130

40 Stordeur P, Goldman M (1998) Interleukin-10 as a regulatory cytokine induced by cellular stress: Molecular aspects. *Int Rev Immunol* 16: 501–522

41 Paul AG, van Der Zee R, Taams LS, van Eden W (2000) A self-hsp60 peptide acts as a partial agonist inducing expression of B7-2 on mycobacterial hsp60-specific T cells: A possible mechanism for inhibitory T cell regulation of adjuvant arthritis? *Int Immunol* 12: 1041–1050

42 Greenfield EA, Howard E, Paradis T, Nguyen K, Benazzo F, McLean P, Hollsberg P, Davis G, Hafler DA, Sharpe AH et al (1997) B7.2 expressed by T cells does not induce CD28-mediated costimulatory activity but retains CTLA4 binding: Implications for induction of antitumor immunity to T cell tumors. *J Immunol* 158: 2025–2034

43 Hollsberg P, Scholz C, Anderson DE, Greenfield EA, Kuchroo VK, Freeman GJ, Hafler DA (1997) Expression of a hypoglycosylated form of CD86 (B7-2) on human T cells with altered binding properties to CD28 and CTLA-4. *J Immunol* 159: 4799–4805

44 Stremmel C, Greenfield EA, Howard E, Freeman GJ, Kuchroo VK (1999) B7-2 expressed on EL4 lymphoma suppresses antitumor immunity by an interleukin 4-dependent mechanism. *J Exp Med* 189: 919–930

45 Broeren CP, Wauben MH, Lucassen MA, Van Meurs M, Van Kooten PJ, Boog CJ, Claassen E, Van Eden W (1995) Activated rat T cells synthesize and express functional major histocompatibility class II antigens. *Immunology* 84: 193–201
46 Ferris DK, Harel-Bellan A, Morimoto RI, Welch WJ, Farrar WL (1988) Mitogen and lymphokine stimulation of heat shock proteins in T lymphocytes. *Proc Natl Acad Sci USA* 85: 3850–3854
47 Kursar M, Bonhagen K, Fensterle J, Kohler A, Hurwitz R, Kamradt T, Kaufmann SH, Mittrucker HW (2002) Regulatory CD4+CD25+ T cells restrict memory $CD8^+$ T cell responses. *J Exp Med* 196: 1585–1592
48 Belkaid Y, Piccirillo CA, Mendez S, Shevach EM, Sacks DL (2002) $CD4^+CD25^+$ regulatory T cells control Leishmania major persistence and immunity. *Nature* 420: 502–507
49 Ciupitu AM, Petersson M, O'Donnell CL, Williams K, Jindal S, Kiessling R, Welsh RM (1998) Immunization with a lymphocytic choriomeningitis virus peptide mixed with heat shock protein 70 results in protective antiviral immunity and specific cytotoxic T lymphocytes. *J Exp Med* 187: 685–691
50 Lowrie DB, Tascon RE, Bonato VL, Lima VM, Faccioli LH, Stavropoulos E, Colston MJ, Hewinson RG, Moelling K, Silva CL (1999) Therapy of tuberculosis in mice by DNA vaccination. *Nature* 400: 269–271
51 Ragno S, Colston MJ, Lowrie DB, Winrow VR, Blake DR, Tascon R (1997) Protection of rats from adjuvant arthritis by immunization with naked DNA encoding for mycobacterial heat shock protein 65. *Arthritis Rheum* 40: 277–283
52 Barrios C, Georgopoulos C, Lambert PH, Del Giudice G (1994) Heat shock proteins as carrier molecules: *in vivo* helper effect mediated by *Escherichia coli* GroEL and DnaK proteins requires cross-linking with antigen. *Clin Exp Immunol* 98: 229–233
53 de Velasco EA, Merkus D, Anderton S, Verheul AF, Lizzio EF, Van der Zee R, Van Eden W, Hoffman T, Verhoef J, Snippe H (1995) Synthetic peptides representing T-cell epitopes act as carriers in pneumococcal polysaccharide conjugate vaccines. *Infect Immun* 63: 961–968
54 Konen-Waisman S, Fridkin M, Cohen IR (1995) Self and foreign 60-kilodalton heat shock protein T cell epitope peptides serve as immunogenic carriers for a T cell-independent sugar antigen. *J Immunol* 154: 5977–5985
55 Perraut R, Lussow AR, Gavoille S, Garraud O, Matile H, Tougne C, van Embden J, van der Zee R, Lambert PH, Gysin J et al (1993) Successful primate immunization with peptides conjugated to purified protein derivative or mycobacterial heat shock proteins in the absence of adjuvants. *Clin Exp Immunol* 93: 382–386
56 Yamaguchi H, Osaki T, Kai M, Taguchi H, Kamiya S (2000) Immune response against a cross-reactive epitope on the heat shock protein 60 homologue of *Helicobacter pylori*. *Infect Immun* 68: 3448–3454
57 Noll A, Bucheler N, Bohn E, Schirmbeck R, Reimann J, Autenrieth IB (1999) DNA immunization confers systemic, but not mucosal, protection against entero-invasive bacteria. *Eur J Immunol* 29: 986–996
58 Matthews R, Burnie J (1996) Antibodies against candida: potential therapeutics? *Trends Microbiol* 4: 354–358

Heat shock proteins in immune response

Sreyashi Basu and Pramod Srivastava

Center for Immunotherapy of Cancer and Infectious Diseases, University of Connecticut School of Medicine, MC1601, Farmington, CT 06030-1601, USA

Heat shock proteins, the natural adjuvants

The connection between HSPs and tumor rejection antigens became apparent through tumor rejection studies in mice and rats. In search for individually distinct tumor rejection antigens, tumor cell lysates were fractionated biochemically and each fraction was tested for its ability to immunize *in vivo* against the tumor. The tumor rejection antigens thus found were mostly HSPs (Tab. 1). Preparations of HSPs, e.g., HSP70, HSP90, gp96 and calreticulin (CRT) from Meth A fibrosarcoma (tumor induced in BALB/c mice by methylcholanthrene) when used to immunize BALB/c mice, rendered the mice immune to subsequent challenge with live Meth A tumor cells in a classical tumor rejection assay [1–3]. However, immunization of mice with HSP preparations from normal tissues [2] or from antigenically-distinct tumor cells [1, 4] did not protect the animals against tumor challenge.

HSP preparations chaperone peptides

HSP preparations from a given cancer cell were observed to mount a protective immune response to the specific cancer. HSPs purified from normal tissue did not elicit protective immunity against any cancer tested. As HSPs are chaperones and nonpolymorphic, it was speculated that HSPs chaperone low molecular weight peptides which are responsible for the specific immunogenicity of HSP preparations [5, 6]. Structural evidence for peptides being associated with HSPs came from a number of different studies and can be summarized as follows:

- Homogenous preparation of HSP70 and gp96 was found to be associated with peptides, as in the case of Meth A [2], E.G7 tumor expressing ovalbumin protein [7] and mammalian liver [8].
- Gp96 purified from vesicular stomatitis virus infected cells was found to be associated with a known Kb restricted viral epitope [9].

Heat Shock Proteins and Inflammation, edited by Willem van Eden

Table 1 - Cancer-derived proteins which have been shown to elicit protective immunity to cancers

Cancer	Induced by	Immunogenicity	Host	Molecules	Assay	References
Zajdela hepatocarcinoma	Chemical	++	Rat	gp100 (gp96)	Prophylaxis	[4]
MethA fibrosarcoma	Chemical	++	BALB/c mice	gp96	Prophylaxis	[1]
				gp96	Therapy	[15]
				HSP90	Prophylaxis	[51]
				HSP70	Prophylaxis	[2]
				CRT	Prophylaxis	[3]
CMS5	Chemical	+	BALB/c mice	gp96	Prophylaxis	[1]
CMS13	Chemical	++	BALB/c mice	gp96	Prophylaxis	[52]
Lewis lung carcinoma	Spontaneous	–	C57BL/6 mice	gp96	Therapy	[15]
				HSP70	Therapy	[15]
B16 melanoma	Spontaneous	–	C57BL/6 mice	gp96	Therapy	[15]
				gp96	Therapy	[53]
CT26 Colon carcinoma	Chemical	++	BALB/c mice	gp96	Therapy	[15]
UV 6138	UV	+++	C3H mice	gp96	Prophylaxis	[15]
UV 139J	UV	++	C3H mice	gp96	Prophylaxis	[54]
				gp96	Therapy	[15]

- Cytotoxic T lymphocyte (CTL) epitope of a mouse leukemia RLmale1 and its precursors were found to be associated with HSP90 and HSP70 in the cytosol and gp96 in the lumen of the endoplasmic reticulum. HSP70 was associated only with the final sized 8mer epitope, whereas HSP90 was found to be associated with the octamer and two other precursor peptides. Gp96 was associated with the octamer and one of the two precursor peptides [10].
- The crystal structure of a peptide complex with the substrate-binding unit of DnaK (a member of the HSP70 family protein) has been determined. The structure consists of a β-sandwich sub-domain followed by α-helical segments. The peptide is bound to DnaK in an extended conformation in the β-sandwich. The α-helical domain stabilizes the complex, but does not contact the peptide directly [11].
- When peptide transport studies from cytosol to endoplasmic reticulum (ER) were carried out, one cytoplasmic protein p100 [12] and several ER resident proteins such as protein disulfide isomerase (PDI), calreticulin (CRT), calnexin, ERp72, gp96, and grp170 were found to bind to the transported peptides [13, 14].

Functional evidence for the peptide-binding property came from several independent immunological observations:

- Immunization of mice with tumor derived HSP70 preparations depleted of peptides by treating the HSP with ATP rendered the HSP ineffective in protecting the animals against tumor challenge [2, 15].
- Immunization with gp96 preparations from cells expressing certain minor antigens elicited cytotoxic T lymphocytes (CTLs) specific against the minor antigen [16].
- Immunization with gp96 preparations from cells expressing β-galactosidase protein could generate CTLs against the Ld-restricted β-galactosidase epitope [16].
- Immunization with gp96 preparations from cells expressing β-galactosidase was found dependent on the status of TAP in the cells from which gp96 was derived. Gp96 from TAP $^{+/+}$ cells was able to immunize while preparations from $^{-/-}$ cells were unable to immunize [17].
- Gp96 purified from vesicular stomatitis virus infected cells was found to elicit CTLs against a known Kb restricted viral epitope [18].
- HSP70 and gp96 has been shown to bind peptides *in vitro* and these *in vitro* generated HSP-peptide complexes were found to be immunogenic with respect to generation of CTLs specific for the peptide complexed to the HSP but not to the HSP itself [19].
- Gp96 isolated from cells infected with the intracellular bacteria *Listeria monocytogenes* and *Mycobacterium tuberculosis* were shown to induce a protective immunity against the specific infections in mice [20].
- Isolation of peptide from purified gp96 and microsequencing showed that a virus-

specific peptide is bound to gp96 derived from liver tissues of patients with hepatitis B virus (HBV)-induced hepatocellular carcinoma [21].
- HSP70 purified from human melanoma was shown to activate T-cells recognizing melanoma differentiation antigens in an antigen- and Class I-dependent fashion [22].
- HSP70-peptide complex purified from tyrosinase-positive but not from tyrosinase-negative melanoma cells were found to deliver the tyrosinase Ag to immature DCs for MHC Class I-restricted T cell recognition [23].

These observations indicate that HSPs are not antigenic *per se* but are carriers of antigenic peptides.

Mechanisms of immunogenicity of HSP-peptide complex

The mechanism that leads to tumor resistance upon immunization of mice with tumor-derived gp96 preparations has been elucidated to a great extent. Earlier work [24] has shown that HSP immunization is dependent on the presence of functional phagocytic cells and $CD8^+$ T-cells. Depletion of $CD8^+$ T cell and phagocytic cells in this study showed that the phagocytic cells are essential for the priming of $CD8^+$ T-cells by gp96-peptide complex without the requirement of $CD4^+$ T-cell help (Tab. 2). This study suggested for the first time that antigen-presenting cells (APCs) are necessary to take up the HSP-peptide complex (HSP-PC) and then present the HSP-chaperoned peptides to the $CD8^+$ T-cell.

Subsequent studies on the mechanism of re-presentation of HSP-chaperoned peptides by APCs confirmed that APCs such as $CD11b^+$ cells but not B-cells or fibroblast cells can take up HSP-PC. The peptides are then processed and re-presented on the surface of the APC in association with their MHC I molecule which then stimulate antigen-specific $CD8^+$ T lymphocytes.

Re-presentation of HSP-PC is mediated *via* an HSP receptor

Immunization of mice with extremely small quantity (~1 μg) of tumor-derived gp96-peptide complex intradermally can protect mice from live tumor challenge. This small quantity of HSPs chaperoning even smaller quantities of antigenic peptides, are able to immunize against tumors presenting those antigenic peptides [2, 8]. As immunogenicity of HSP-peptide complexes is dependent on the presence of functional professional antigen-presenting cells [24] it was hypothesised that HSPs are taken up by APCs through HSP-receptors [25]. As HSP-PCs are targeted to the HSP specific receptors on APCs the small quantity of antigenic peptides targeted to the APC will be large enough to mount a strong immune response specific to the anti-

Table 2 - Cell types required for tumor immunity elicited by immunization with intact MethA cells or MethA derived gp96 [24]

Immunization	$CD4^+$ cells	$CD8^+$ cells	Macrophages
Priming phase			
Intact Meth A cells	R	NR	NR
MethA derived gp96	NR	R	R
Effector phase			
Intact Meth A cells	R	R	NR
MethA derived gp96	R	R	R

R, required; NR, not required.

genic peptide [19]. Evidence for the specific binding of HSPs on APCs came from a number of studies [26–28]. Later studies identified a number of receptors on APCs responsible for binding and uptake of HSP-PC such as CD91, CD40, LOX-1 [29–31]. HSP-PC are internalized through its receptor on the surface of macrophages and dendritic cells (DCs) and re-presented on their surface in association with MHCI and MHCII molecules [29, 31–33]. This re-presentation process is dependant on functional proteasome transporter associated with antigen presentation (TAP) and also sensitive to Brefeldin A (BFA) but not to chloroquine. The sensitivity of re-presentation of gp96-chaperoned peptides to proteasome inhibitor, lactacystin and to BFA but not to chloroquine suggests that peptides chaperoned by HSPs are channeled from an endosomal (non-acidic) compartment to the cytosol and then through the TAP from cytosol (or alternatively depending on the peptide in a TAP independent pathway) to the ER to Golgi for antigen presentation by MHC I molecules of the APC [18, 33–35].

Activation of APCs by HSPs

The HSP-APC interaction lies at the center of the many properties of HSPs as immunogens. The first insights into this interaction revealed that it was a central event in adaptive immune response to HSP-chaperoned peptides, whether naturally-derived, or artificially reconstituted. Subsequently, it appears that the HSP-APC interaction results in an even more fundamental chain of events. HSPs purified from normal tissue when injected into mice activate resident dendritic cells, which eventually migrate to the draining lymph nodes to mount a specific immune response ([36], and R. Wang, P.K. Srivastava and S. Basu, submitted). Interaction of HSP with

APC induce activation and translocation of the NF-κB complex into the nuclei of APCs, where it leads to a chain of transcriptional events including secretion of inflammatory cytokines by APCs e.g., TNF-α, IL-1β, IL-12 and GM-CSF [37, 38], induction of inducible nitric oxide synthase (iNOS), production of nitric oxide by macrophages and dendritic cells [39] and maturation of dendritic cells as measured by enhanced expression of MHC Class II, CD86 and CD40 [37, 40, 41].

Re-presentation of HSP-PC by APC requires endocytosis of the HSP-PC; on the other hand the innate immune consequences of HSP-APC interaction, requires signaling. We have shown that CD91 molecule on APCs to be the common receptor for gp96, HSP90, HSP70 and CRT [33]. The readout in these studies was re-presentation of HSP-chaperoned peptides by the MHC I of the APC. Recent evidence from other systems shows CD91 to be a signaling receptor, in addition to being an endocytosing receptor [41–43]. Thus, CD91 could be involved in both phenomena. A number of receptors on APCs have been implicated in interaction with HSPs but the exact identity of the receptor for activation of APCs by HSPs is still perplexing. Although some of the innate functions of HSPs mimic the characteristics of LPS, the activity of HSPs is also distinct from that of LPS in significant ways. The activation receptors thus far identified for HSPs are the same as those for endotoxin e.g., TLR4 and TLR2 have been suggested to be receptors for HSP60 [44, 45], for HSP70 [46] and GP96 [47]. On the other hand CD40, a TNF receptor family member, has been shown to be involved in the interaction of APCs with mycobacterial but not human HSP70 [48] and data from our work suggests that CD36, a member of the superfamily of scavenger receptors – to be a signaling receptor only specific for gp96 [49]. It is important to note here that effect of endotoxin contamination in HSP preparations can supersede the innate effect of HSP per se and can mislead us in identifying the HSP specific receptors [50].

The HSP-APC interaction thus results in activation of, not only the adaptive but also the innate component of immune response. This might be helpful in explaining the extraordinary ability of HSP-PC to act as a tumor rejection vaccine.

References

1 Srivastava PK, DeLeo AB, Old LJ (1986) Tumor rejection antigens of chemically induced sarcomas of inbred mice. *Proc Natl Acad Sci USA* 83: 3407–3411

2 Udono H, Srivastava PK (1993) Heat shock protein 70-associated peptides elicit specific cancer immunity. *J Exp Med* 178: 1391–1396

3 Basu S, Srivastava PK (1999).Calreticulin, a peptide-binding chaperone of the endoplasmic reticulum, elicits tumor- and peptide-specific immunity. *J Exp Med* 189: 797–802

4 Srivastava PK, Das MR (1984) The serologically unique cell surface antigen of Zajdela

ascitic hepatoma is also its tumor-associated transplantation antigen. *Int J Cancer* 33: 417–419

5 Srivastava PK, Heike M (1991) Tumor-specific immunogenicity of stress-induced proteins: Convergence of two evolutionary pathways of antigen presentation? *Semin Immunol* 3: 57–64

6 Srivastava PK, Maki RG (1991) Stress-induced proteins in immune response to cancer. *Curr Top Microbiol Immunol* 167: 109–123

7 Breloer M, Marti T, Fleischer B, von Bonin A (1998) Isolation of processed, H-2Kb-binding ovalbumin-derived peptides associated with the stress proteins HSP70 and gp96. *Eur J Immunol* 28: 1016–1021

8 Li Z, Srivastava PK (1993) Tumor rejection antigen gp96/grp94 is an ATPase: Implications for protein folding and antigen presentation. *EMBO J* 12: 3143–3151

9 Nieland TJF, Tan MCA, Monnee-van Muijen M, Koning F, Kruisbeek AM, Van Bleek GM (1996) Isolation of an immunodominant viral peptide that is endogenously bound to the stress protein gp96/grp94. *Proc Natl Acad Sci USA* 93: 6135–6139

10 Ishii T, Udono H, Yamano T, Ohta H, Uenaka A, Ono T, Hizuta A, Tanaka N, Srivastava PK, Nakayama E (1999) Isolation of MHC class I-restricted tumor antigen peptide and its precursors associated with heat shock proteins hsp70, hsp90, and gp96. *J Immunol* 162: 1303–1309

11 Zhu X, Zhao X, Burkholder WF, Gragerov A, Ogata CM, Gottesman ME, Hendrickson WA (1996) Structural analysis of substrate binding by the molecular chaperone DnaK. *Science* 272: 1606–1614

12 Marusina K, Reid G, Gabathuler R, Jefferies W, Monaco JJ (1997) Novel peptide-binding proteins and peptide transport in normal and TAP-deficient microsomes. *Biochemistry* 36: 856–863

13 Lammert E, Arnold D, Nijenhuis M, Momburg F, Hammerling GJ, Brunner J, Stevanovic S, Rammensee HG, Schild H (1997) The endoplasmic reticulum-resident stress protein gp96 binds peptides translocated by TAP. *Eur J Immunol* 27: 923–927

14 Spee P, Neefjes J (1997) TAP-translocated peptides specifically bind proteins in the endoplasmic reticulum, including gp96, protein disulfide isomerase and calreticulin. *Eur J Immunol* 27: 2441–2449

15 TamuraY, Peng P, Liu K, Daou M, Srivastava PK (1997) Immunotherapy of tumors with autologous tumor-derived heat shock protein preparations. *Science* 278: 117–120

16 Arnold D, Faath S, Rammensee H, Schild H (1995) Cross-priming of minor histocompatibility antigen-specific cytotoxic T cells upon immunization with the heat shock protein gp96. *J Exp Med* 172: 885–889

17 Arnold D, Wahl C, Faath S, Rammensee H-G, Schild H (1997) Influences of transporter associated with antigen processing (TAP) on the repertoire of peptides associated with the endoplasmic reticulum-resident stress protein gp96. *J Exp Med* 176: 461–466

18 Suto R, Srivastava PK (1995) A mechanism for the specific immunogenicity of heat shock protein-chaperoned peptides. *Science* 269: 1585–1588

19 Blachere NE, Li Z, Chandawarkar RY, Suto R, Jaikaria NS, Basu S, Udono H, Srivasta-

va PK (1997) Heat shock protein-peptide complexes, reconstituted *in vitro*, elicit peptide-specific cytotoxic T lymphocyte response and tumor immunity. *J Exp Med* 176: 1315–1319

20 Zugel U, Sponaas AM, Neckermann J, Schoel B, Kaufmann SH (2001) gp96-Peptide vaccination of mice against intracellular bacteria. *Infect Immun* 69: 4164–4167

21 Meng SD, Gao T, Gao GF, Tien P (2001) HBV-specific peptide associated with heat-shock protein gp96. *Lancet* 17: 528–529

22 Castelli C, Ciupitu AM, Rini F, Rivoltini L, Mazzocchi A, Kiessling R, Parmiani G (2001) Human heat shock protein 70 peptide complexes specifically activate antimelanoma T cells. *Cancer Res* 61: 222–227

23 Noessner E, Gastpar R, Milani V, Brandl A, Hutzler PJ, Kuppner MC, Roos M, Kremmer E, Asea A, Calderwood SK et al (2002) Tumor-derived heat shock protein 70 peptide complexes are cross-presented by human dendritic cells. *J Immunol* 169: 5424–5432

24 Udono H, Levey D, Srivastava PK (1994) Cellular requirements for tumor-specific immunity elicited by heat shock proteins: Tumor rejection antigen gp96 primes $CD8^+$ T cells *in vivo*. *Proc Natl Acad Sci USA* 91: 3077–3081

25 Srivastava PK, Udono H, Blachere N, Li Z (1994) Heat shock proteins transfer peptides during antigen processing and CTL priming. *Immunogenetics* 39: 93–98

26 Wassenberg JJ, Dezfulian C, Nicchitta CV (1999) Receptor mediated and fluid phase pathways for internalization of the ER Hsp90 chaperone GRP94 in murine macrophages. *J Cell Sci* 112: 2167–2175

27 Arnold-Schild D, Hanau D, Spehner D, Schmid C, Rammensee HG, de la Salle H, Schild H (1999) Receptor-mediated endocytosis of heat shock proteins by professional antigen-presenting cells. *J Immunol* 162: 3757–3760

28 Binder RJ, Harris M, Menoret A, Srivastava PK (2000) Saturation, competition and specificity in interaction of heat shock proteins (hsp) gp96, hsp90, and hsp70 with CD11b+ cells. *J Immunol* 165: 2582–2587

29 Binder R, Han DK, Srivastava PK (2000). CD91: A receptor for heat shock protein gp96. *Nature Immunol* 2: 151–155

30 Becker T, Hartl FU, Wieland F (2002) CD40: An extracellular receptor for binding and uptake of Hsp70-peptide complexes. *J Cell Biol* 158:1277–1285

31 Delneste Y, Magistrelli G, Gauchat J, Haeuw J, Aubry J, Nakamura K, Kawakami-Honda N, Goetsch L, Sawamura T, Bonnefoy J et al (2002) Involvement of LOX-1 in dendritic cell-mediated antigen cross-presentation. *Immunity* 17: 353–362

32 Matsutake T, Srivastava PK. (2000) CD91 is involved in MHC Class II Presentation of gp96 – chaperoned peptides. *Cell Stress and Chaperones* 5: 378

33 Basu S, Binder RJ, Ramalingam T, Srivastava PK (2001) CD91 is a common receptor for heat shock proteins gp96, hsp90, hsp70, and calreticulin. *Immunity* 14: 303–313

34 Singh-Jasuja H, Toes RE, Spee P, Munz C, Hilf N, Schoenberger SP, Ricciardi-Castagnoli P, Neefjes J, Rammensee HG, Arnold-Schild D et al (2000) Cross-presentation of

glycoprotein 96-associated antigens on major histocompatibility complex class I molecules requires receptor-mediated endocytosis. *J Exp Med* 191: 1965–1974

35 Castellino F, Boucher PE, Eichelberg K, Mayhew M, Rothman JE, Houghton AN, Germain RN (2000) Receptor-mediated uptake of antigen/heat shock protein complex results in major histocompatibility complex class I antigen presentation *via* two distinct processing pathways. *J Exp Med* 191: 1957–1964

36 Binder RJ, Anderson KM, Basu S, Srivastava PK (2000) Cutting edge: Heat shock protein gp96 induces maturation and migration of CD11c[+] cells *in vivo*. *J Immunol* 165: 6029–6035

37 Basu S, Binder RJ, Suto R, Anderson KM, Srivastava PK (2000) Necrotic but not apoptotic cell death releases heat shock proteins, which deliver a partial maturation signal to dendritic cells and activate the NF-kappa B pathway. *Int Immunol* 12: 1539–1546

38 Asea A, Kraeft SK, Kurt-Jones EA, Stevenson MA, Chen LB, Finberg RW, Koo GC, Calderwood SK (2000) HSP70 stimulates cytokine production through a CD14-dependant pathway, demonstrating its dual role as a chaperone and cytokine. *Nat Med* 6: 435–442

39 Panjwani NN, Popova L, Srivastava PK (2002) Heat shock proteins gp96 and hsp70 activate release of nitric oxide by antigen presenting cells. *J Immunol* 168: 2997–3003

40 Singh-Jasuja H, Scherer HU, Hilf N, Arnold-Schild D, Rammensee HG, Toes RE, Schild H (2000) The heat shock protein gp96 induces maturation of dendritic cells and down-regulation of its receptor. *Eur J Immunol* 30: 2211–2215

41 Somersan S, Larsson M, Fonteneau JF, Basu S, Srivastava P, Bhardwaj N (2001) Primary tumor tissue lysates are enriched in heat shock proteins and induce the maturation of human dendritic cells. *J Immunol* 167: 4844–4852

42 Su HP, Nakada-Tsukui K, Tosello-Trampont AC, Li Y, Bu G, Henson PM, Ravichandran K (2001) Interaction of CED-6/GULP, an adapter protein involved in engulfment of apoptotic cells, with CED-1 and CD91/LRP. *J Biol Chem* 277: 11772–11779

43 Misra UK, Pizzo SV (2001) Induction of cyclooxygenase-2 synthesis by ligation of the macrophage alpha (2)-macroglobulin signalling receptor. *Cell Signal* 13: 801–808

44 Ohashi K, Burkart V, Flohe S, Kolb H (2000) Cutting edge: Heat shock protein 60 is a putative endogenous ligand of the toll-like receptor-4 complex. *J Immunol* 164: 558–561

45 Vabulas RM, Ahmad-Nejad P, da Costa C, Miethke T, Kirschning CJ, Hacker H, Wagner H (2001) Endocytosed HSP60s use toll-like receptor 2 (TLR2) and TLR4 to activate the toll/interleukin-1 receptor signaling pathway in innate immune cells. *J Biol Chem* 276: 31332–31339

46 Asea A, Rehli M, Kabingu E, Boch JA, Bare O, Auron PE, Stevenson MA, Calderwood SK. (2002) Novel signal transduction pathway utilized by extracellular HSP70: Role of toll-like receptor (TLR) 2 and TLR4. *J Biol Chem* 277: 15028–15034

47 Vabulas RM, Braedel S, Hilf N, Singh-Jasuja H, Herter S, Ahmad-Nejad P, Kirschning CJ, Da Costa C, Rammensee HG, Wagner H et al (2002) The endoplasmic reticulum-

resident heat shock protein Gp96 activates dendritic cells *via* the toll-like receptor 2/4 pathway. *J Biol Chem* 277: 20847–20853
48 Wang Y, Kelly CG, Singh M, McGowan EG, Carrara AS, Bergmeier LA, Lehner T (2002) Stimulation of Th1-polarizing cytokines, C-C chemokines, maturation of dendritic cells, and adjuvant function by the peptide binding fragment of heat shock protein 70. *J Immunol* 169: 2422–2429
49 Panjwani NN, Popova L, Febbraio F, Srivastava PK (2000) The CD36 scavenger receptor as a receptor for gp96. *Cell Stress and Chaperones* 5: 391
50 Gao B, Tsan MF (2002 Endotoxin contamination in recombinant human heat shock protein 70 (Hsp70) preparation is responsible for the induction of tumor necrosis factor alpha release by murine macrophages. J Biol Chem 278: 174–179
51 Ullrich, SJ, Robinson EA, Law LW, Willingham M, Appella E (1986) A mouse tumor-specific transplantation antigen is a heat shock-related protein. *Proc Natl Acad Sci USA* 83: 3118–3125
52 Palladino MA, Srivastava PK, Oettgen HF, DeLeo AB (1987) Expression of a shared tumor-specific antigen by two chemically induced BALB/c sarcomas. *Cancer Res* 47: 5074–5079
53 Nicchitta CV (1998) Biochemical, cell biological and immunological issues surrounding the endoplasmic reticulum chaperone GRP94/gp96. *Curr Opin Immunol* 10: 103–109
54 Janetzki S, Blachere NE, Srivastava PK (1998) Generation of tumor-specific cytotoxic T lymphocytes and memory T cells by immunization with tumor-derived heat shock protein gp96. *J Immunother* 21: 269–276

Heat shock protein-mediated activation of innate immune cells

Clarissa U.I. Prazeres da Costa, Hermann Wagner and Thomas C. Miethke

Institute for Medical Microbiology, Immunology and Hygiene, Technical University of Munich, Trogerstr. 9, 81675 Munich, Germany

Introduction

Cells of the innate immune system are equipped with germline encoded receptors allowing the recognition of structurally-conserved pathogen-associated microbial patterns (PAMP). A subgroup of those receptors are members of a recently defined family of proteins called the Toll-like receptors (TLR). The prototypic and name defining gene "Toll" was cloned in *Drosophila melanogaster* and was found to be not only responsible for dorso-ventral patterning of the developing larvae but also to play a central role in the defense against fungal and gram-positive infections. It soon became clear that mammals adopted this defense system, since at least ten different functional Toll-like receptors have been discovered in the genome of humans and nine in mice. These receptors are involved in the recognition of PAMPs like endotoxin, peptidoglycan, bacterial DNA, double-stranded RNA, flagellin, lipopeptides, etc. Remarkably, individual receptors are specialized to recognize certain PAMPs, e.g., TLR4 senses endotoxin and TLR9 bacterial DNA. Until recently the TLR system was thought to have evolved for the recognition of conserved and vital foreign structures of various micro-organisms, i.e., structures not present in the host. However, this principle was challenged by the discovery that HSPs of bacterial and mammalian origin were also able to stimulate innate immune cells *via* the TLR-system. Mammalian and bacterial HSPs share a high degree of homology, probably to maintain their essential function in mammalian and bacterial cells, i.e., the proper folding of a variety of proteins. The unexpected observation that HSPs are able to activate cells of the innate immune system is the central topic of this review. We will focus on the cell types activated by HSPs, in particular dendritic cells and macrophages, the molecular mechanisms of its binding by cell surface receptors and its recognition by TLRs, and the intracellular signal cascade induced in the responding cells.

Heat Shock Proteins and Inflammation, edited by Willem van Eden

HSPs belong to the growing family of pathogen-associated molecular patterns

HSPs are divided into different classes according to their molecular weight. Among the various subgroups of HSPs we will focus on HSP60, HSP70 and gp96 in our discussion of the immunostimulatory role of HSPs.

HSP60, also called chaperonin based on its essential role in the folding of cellular proteins, is well known to serve as a target for T- and B-cell mediated adaptive immune responses [1, 2]. The presence of antibodies specific for human and bacterial HSP60 were demonstrated in the serum of certain patients [3, 4] indicating that HSP60 might exist in soluble form where it could act as a whole protein to stimulate other cell types. Indeed, a recent study reported that human HSP60 is present in the plasma of healthy humans [5]. Chen et al., reported that human HSP60 stimulates a pro-inflammatory response in human and murine macrophages [6]. These results were confirmed and extended by others reporting that chlamydial and human HSP60 activate macrophages, dendritic cells, endothelial cells and smooth muscle cells [7, 8]. Importantly, polymyxin B, a well known and effective endotoxin-antagonist, was unable to impair HSP60-induced cellular responses [6]. Heat treatment of human and chlamydial HSP60 destroyed their stimulatory capacity, also arguing against a role of contaminating endotoxin which is heat-resistant [8]. A recent publication of our group showed that degradation of the chlamydial HSP60 preparation by proteinases impeded completely the stimulation of murine dendritic cells [9]. Based on this information, it is rather unlikely that the activity of HSP60 to stimulate cells of the innate immune system is due to contaminating endotoxin. In addition, cells of the innate immune system are able to recognize HSP60 *via* TLR2, whereas endotoxin is detected *via* TLR4. However, all arguments discussed above do not rule out the possibility that a contaminating protein, firmly bound to HSP60, represents the stimulatory principle. In this case HSP60 would act as a carrier comparable to HSP70 and gp96 which were shown to transport immunogenic peptides for cross-presentation [10]. With this caveat in mind host-derived endogenous HSP60 as well as exogenous bacterial HSP60s are now considered to be important stimuli of the innate immune system and are therefore added to the growing list of molecular-defined PAMPs.

The same arguments which were raised to define HSP60 as PAMP are applicable for HSP70 and gp96. Both molecules are able to stimulate macrophages and dendritic cells [11–13], and their activity is heat-sensitive [11, 13]. Furthermore, HSP70 and gp96 display a second important function. The pioneering work of Srivastava and co-workers demonstrated that these HSPs are able to bind immunogenic peptides and thereby induce specific T-cell responses after cross-presentation [14, 15]. In other words, HSP70 and gp96 act as adjuvant in that they activate innate immune cells *via* intracellular pathways described below, while they simultaneously chaperone peptides into the MHC Class I presentation pathway in order to provide T-cells with their specific peptide ligand and the optimal activation environment.

Stimulation of cells of the innate immune system by HSPs

Stimulation of dendritic cells by HSP60, HSP70 or gp96 results in the increased expression of costimulatory molecules like CD40, CD80, CD86 as well as MHC Class II molecules [9, 11, 16,17]. Macrophages, but also vascular endothelium and smooth muscle cells, up-regulate adhesion molecules like endothelial-leukocyte adhesion molecule-1 (E-selectin), intercellular adhesion molecule-1 (ICAM-1), and vascular cell adhesion molecule-1 (VCAM-1) [8]. Dendritic cells are also induced to secrete pro-inflammatory cytokines like TNF, IL-12p40, or IL-1β [9, 16] and macrophages express TNF, IL-1, IL-6, GM-CSF and IL-8 [18]. Thus, HSPs appear to function as danger signals like other PAMPs, e.g., endotoxins, peptidoglycan or bacterial DNA. However, HSP60 and HSP70 are unique compared to other PAMPs in that they are expressed in procaryotic and eucaryotic cells. Both, exogenous (bacterial) and endogenous (human) HSP60 or HSP70 share in each case a high degree of structural similarity which may explain that bacterial and eucaryotic HSP60 or HSP70 also share the ability to activate dendritic cells or macrophages [19]. Taken together, pathogen-derived HSP60 and HSP70 could not only signal danger in infectious bacterial diseases, but eucaryotic HSP60 and HSP70 could also signal sterile inflammatory processes in host tissues during autoimmune diseases or malignancies. Of note, a recent publication demonstrated that HSPs are released during necrotic but not apoptotic cell death and that the released HSPs are able to activate and mature dendritic cells [10].

Cell surface receptors specific for HSPs expressed on macrophages and dendritic cells

As discussed above, HSPs efficiently induce activation of macrophages and dendritic cells. Therefore, one or several cellular receptors were postulated that bind the different HSPs to initiate an intracellular signal cascade. Indirect evidence supported this postulate since gp96 and HSP90 showed a specific and saturable binding to CD11b-positive cells [20]. Interestingly, the gp96-related HSP90 competed with gp96 binding, whereas HSP70 (although interacting effectively with CD11b-positive cells) was poor in this respect. The molecular definition of a HSP-receptor was achieved first by Srivastava and colleagues. They reported, that CD91, also known as α2-macroglobulin receptor, is able to bind eucaryotic HSP70, HSP90 and gp96 [21, 22]. The interaction is direct and not mediated *via* other ligands [21]. In contrast to human HSP70, mycobacterial HSP70 appears to use a different receptor. As demonstrated recently, stimulation of the monocytic cell line THP1 with mycobacterial HSP70 to produce the chemokine RANTES is dependent on CD40 since antibodies specific for CD40 suppress activation [23]. Further, only CD40-transfected HEK293 cells are able to produce RANTES upon stimulation with mycobacterial HSP70. To explain the difference to human HSP70 the authors speculated that eucaryotic and procaryotic HSP70 may use different receptors. Still another receptor, unrelated to CD91, appears to be involved in the binding of HSP60. A recent publication reported that the interaction of fluorescence dye-labeled HSP60 with macrophages was saturable and could only be blocked by unlabeled HSP60, but not by HSP70, HSP90, or gp96, indicating specificity for a putative receptor for HSP60 [24]. Interestingly, HSP60 binding seemed to occur in the absence of TLR4 (see below), whereas HSP60-induced cytokine release depended on TLR4 [24, 25]. Since the HSP60-induced intracellular signaling also depends on CD14 [26], the monocyte binding receptor for endotoxin, it is tempting to speculate that CD14 might also bind HSP60, but data in support of this hypothesis are lacking. Interestingly, HSP70-induced activation of NF-κB reporter constructs also appears to be dependent on CD14 [16]. In summary, albeit HSP60, HSP70 and gp96 share the ability to stimulate monocytes and dendritic cells, HSP60 appears to interact with a receptor distinct from CD91, the binding receptor for HSP70 and gp96.

Toll-like receptors are involved in the recognition of HSPs

The discovery of the Toll-like receptors pointed out that innate immune cells possess a sophisticated system to recognize conserved and vital key structures of microorganisms. As mentioned above ligand(s) for most of the TLRs, with the exception of TLR10, are defined [27]. TLR2 is remarkable in that it appears to be involved in the recognition of multiple ligands. This contrasts to TLR9 which until today has only one defined ligand, bacterial DNA [28]. The ability of TLR2 to sense a variety of ligands may at least in part be explained by its ability to form heterodimers with TLR1 or TLR6, thus displaying a new ligand specificity. For example, TLR2 co-transfected with TLR6 recognizes a diacylated lipoprotein from mycoplasma, the mycoplasmal macrophage-activating lipopeptide-2 kD (MALP-2) [29], whereas TLR2 together with TLR1 detects triacylated lipoproteins [30]. The formation of heterodimers seems to be required for TLR2-initiated intracellular signaling. This is inferred from experiments where transfection of the cytoplasmic domain of TLR2 does not result in TNF-production in macrophages; however, co-transfection of the cytoplasmic domains of TLR2 together with TLR1 or TLR6 induces TNF-secretion in macrophages [31]. In contrast to TLR2, the cytoplasmic domain of TLR4 appears to be autonomous for the initiation of signaling [31]. The potential involvement of TLRs in the recognition of HSPs was first described by Ohashi et al. who demonstrated that human HSP60 stimulated murine macrophages to secrete TNF and nitric oxide only in the presence of a functional TLR4 [25]. Also, recombinant HSP60 from *Chlamydia pneumoniae* activates macrophages in a TLR4-dependent fashion [32], implying that the recognition of HSP60 from procaryotic and eucaryotic cells is similar. In addition, HSP60 is also recognized *via* TLR2. Thus, chlamydial or human HSP60 activates an NF-κB-reporter construct in TLR2- or TLR4-transfected HEK293 cells [7] and dendritic cells derived from TLR2- or TLR4-deficient mice are impaired to recognize chlamydial HSP60 [9]. Overall these experiments led to the conclusion that TLR2 as well as TLR4 are involved in the recognition of bacterial and human HSP60. Also, the recognition of human HSP70 appears to rely on TLR2 and TLR4 since only HEK293 cells transfected with either TLR2 or TLR4 activate NF-κB upon stimulation with human HSP70 [13, 16]. Finally, the same TLR-dependency was described for the activation of dendritic cells by gp96 using TLR4- or TLR2/4-double-deficient cells [33].

Several points of HSP-recognition by dendritic cells (or macrophages) are remarkable. First, the cells detect apparently three structurally-unrelated HSPs using identical molecular tools, i.e., TLR2 and TLR4. This raises the question whether the different HSPs share a common motif which is recognized by TLRs. Alternatively, since CD14 is also involved in the detection of at least HSP60 and HSP70, this molecule might be responsible to recruit TLR2 and TLR4 after interaction with one of the three different HSPs. Second, do TLR2 and TLR4 co-operate for the detection of HSPs? As discussed above TLR4 was described to signal independently of other TLRs whereas TLR2 appears to need a partner. Therefore one could envisage that TLR4 could act autonomously and TLR2 may interact with other, yet undefined TLRs which are specific for HSP60, HSP70, or gp96. In an alternative scenario TLR2 and TLR4 may form a heterodimer for the detection of HSPs. Third, the importance of TLR2 or TLR4 to initiate cellular responses after detection of HSPs may vary with the type of cellular response. For instance, IL-12p40 production by dendritic cells after stimulation with HSP60 is equally dependent on TLR2 and TLR4. However, TNF-secretion depends almost completely on TLR2 [9]. This implies that TLRs possess individual signaling qualities, even though they appear to induce a common signal cascade (see below).

The intracellular signal cascade induced by HSPs in macrophages and dendritic cells

The intracellular signal cascade triggered by HSP60, HSP70 and gp96 in monocytic cell lines or dendritic cells appears to be similar. This is not surprising as all three HSPs are sensed by TLR2 and TLR4. Both human and chlamydial HSP60 activate the stress-induced protein kinases p38 and JNK1/2, the mitogen-activated protein kinases ERK1/2, and the IκB kinase (IKK) in the macrophage cell line RAW264.7 [7]. Activation of JNK1/2 and IKK depends on myeloid differentiation factor 88

(MyD88) [7, 32], an important adaptor molecule of the Toll/IL-1R signal pathway which is recruited to the cytoplasmic TIR-domain (Toll/Interleukin 1 receptor) of TLRs upon TLR-ligand interaction (Fig. 1). Activation of both kinases requires in addition TNF-receptor associated factor 6 (TRAF6) [7], another crucial member of the Toll/IL-1R signal pathway downstream of MyD88 (Fig. 1). Similarly, HSP70-induced activation of IL-12 and ELAM-1 promotors in RAW264.7 macrophages depends on MyD88 and TRAF6 [13] and HEK293 cells, that stably express TLR4, activate NF-κB reporter constructs MyD88-dependently upon stimulation with HSP70 [16]. Finally, gp96 also triggers the activation of JNK, p38, and ERK [33], presumably *via* MyD88 and TRAF6, although this is not proven experimentally. Interestingly, endocytosis of HSP60 and gp96 by macrophages and dendritic cells, respectively, appears to be a prerequisite for the induction of the intracellular signal cascade by these two HSPs [7, 33], whereas other microbial stimuli like endotoxin do not require endocytosis for signaling. The subsequent signal steps, however, of bacterial and human HSP60, human HSP70 and human gp96 appear to be identical to each other and to other TLR-ligands.

How do dendritic cells or macrophages get access to HSPs?

HSPs of eucaryotic cells are intracellular molecules, which are not secreted. The primary localization of human HSP60 is the mitochondrium [34], human HSP70 is a cytosolic protein [10], and gp96 resides in the endoplasmic reticulum [35]. Thus, the question arises how do HSPs come into contact with dendritic cells or macrophages to induce their activation. One possible way was reported recently in that cells dying due to necrosis but not apoptosis release HSPs, in particular HSP70, HSP90, and gp96 [10]. In parallel necrotic but not apoptotic cells are able to stimulate dendritic cells [10], implying the possibility that the cells were activated *via* HSPs, although this is not proven. On the other hand a minor fraction of human HSP60 found in CHO cells seems to be localized at the cell membrane as shown by biotin-labeling of the cell surface and subsequent immunoprecipitation [34]. Furthermore, some tumor cells express HSP70 on the cell surface [36]. This was also shown for gp96 [35]. The latter observations raise the possibility that intact cells expressing HSPs at the cell membrane might directly stimulate cells of the innate immune system.

In the case of bacterial HSP60, there is evidence that micro-organisms shed their HSP. For instance, immunoelectron microscopy revealed that a GroEL-like molecule from *Actinobacillus actinomycetemcomitans* can be found in the extracellular space surrounding the bacterium [37]. In contrast, Dna-K like molecules were mostly found in the cytoplasm of *A. actinomycetemcomitans* indicating that this observation was specific for the GroEL-like molecule. Furthermore, HspB, a GroEL homolog from *Helicobacter pylori*, was associated with the outer membrane of the bacterium [38]. In addition, other bacteria like a virulent Legionella strain,

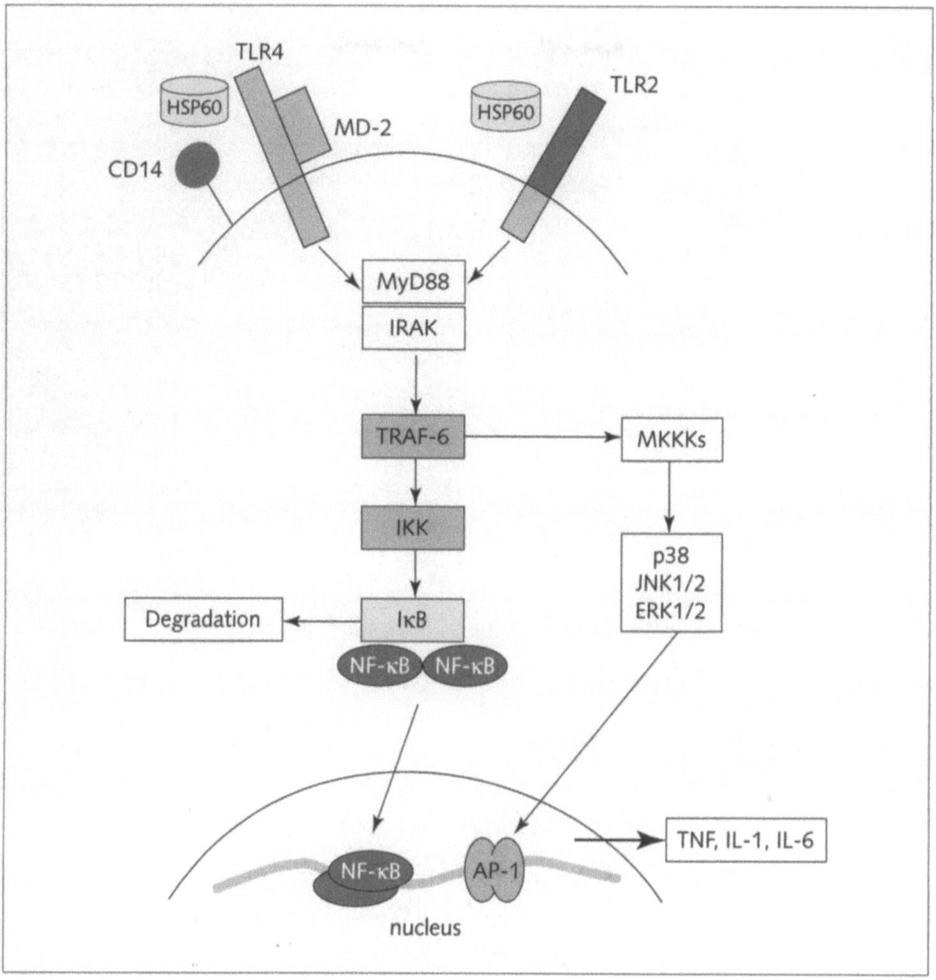

Figure 1
Upon ligand stimulation with HSP60, MyD88 is recruited to the cytoplasmic domain of the TLR (TIR-domain) and autophosphorylation of interleukin-1-receptor associated kinase (IRAK) is triggered through the binding of its death-domain to that of MyD88. Activated IRAK is then released from the receptor-complex, binds and activates TRAF-6 (TNF receptor-associated factor 6) to stimulate the IκB kinase complex (IKK). Subsequently, IκB is phosphorylated and degraded, leading to the translocation of NF-κB into the nucleus. In parallel, TRAF-6 also activates the mitogen-activated protein kinase pathway (MKKKs), namely the stress-activated protein kinases p38 and janus-kinase 1/2 (JNK1/2) and the mitogen-activated kinases ERK1/2 (extracellular signal-regulated kinases) leading to binding of the transcription factors to the AP-1 site. Hereby, expression of pro-inflammatory cytokines is initiated.

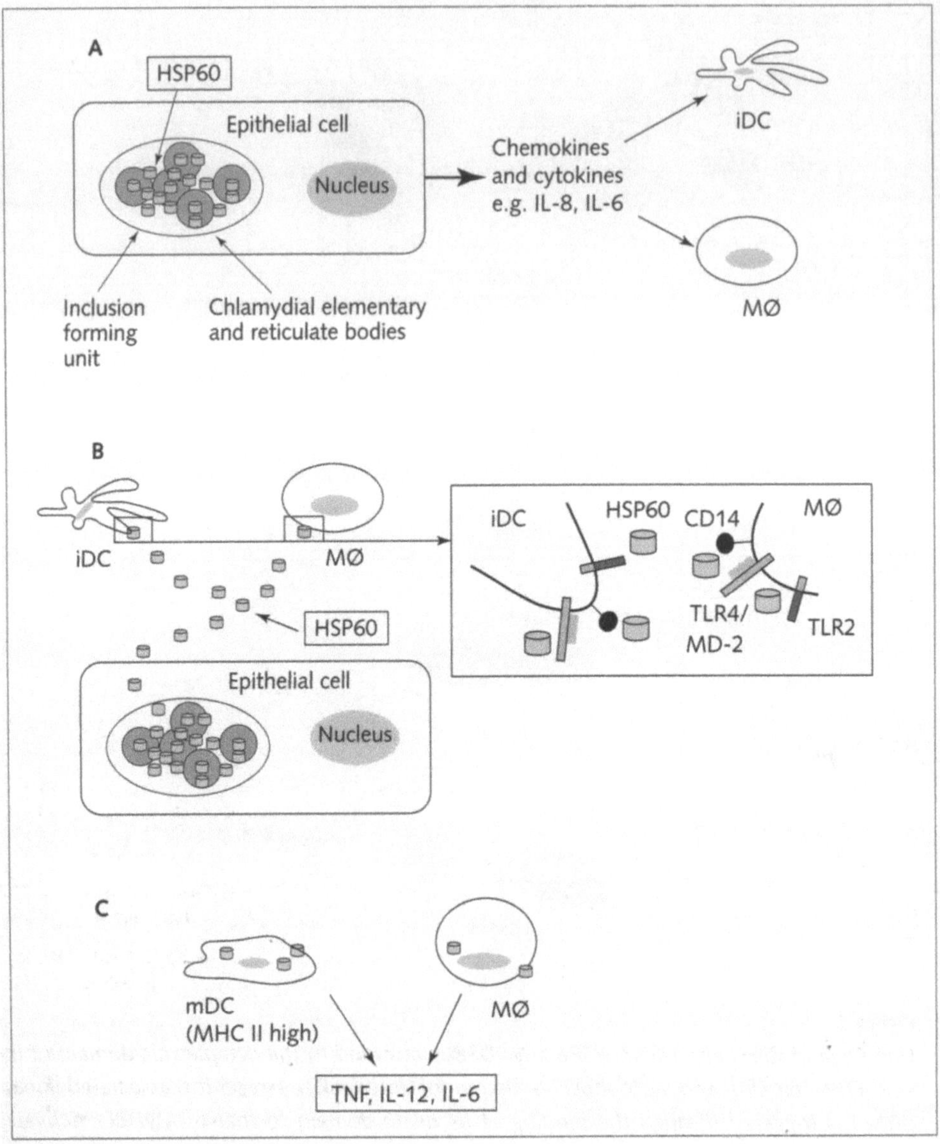

Figure 2

(A) Infected epithelial cell secretes chemokines and cytokines to attract macrophages and dendritic cells (MØ: macrophages, iDC: immature dendritic cells, HSP60: heat shock protein 60); (B) The recruited macrophages and dendritic cells are activated by HSP60 released from the infected cell via *TLR2 and TLR4/MD-2 and CD14 (TLR2/TLR4: toll-like receptor 2/4); (C) Activated dendritic cells and macrophages release pro-inflammatory cytokines e.g., TNF, IL-12, IL-6 (mDC: mature dendritic cells).*

Mycobacterium leprae and *Salmonella typhimurium* were reported to express HSP60-homologs on the cell surface [39–41]. We demonstrated recently that HSP60 of the obligate intracellular bacterium *Chlamydia pneumoniae* is present in the culture supernatant of infected epithelial HEp2 cells and that transfer of the supernatant to dendritic cells stimulates the cells in a TLR2/4 dependent fashion [9]. A proposition of how bacterial HSP60 could activate innate immune cells during an infection, in this case with *C. pneumoniae*, is demonstated in Figure 2. Thus, *C. pneumoniae*-infected HEp2 cells secrete IL-8 and probably other chemokines to attract macrophages and dendritic cells to the site of infection. In parallel, the ongoing chlamydial infection produces HSP60 which through ill-defined mechanisms is present in the extracellular space and triggers immigrating cells *via* TLR2 and TLR4 to secrete pro-inflammatory cytokines. Similar scenarios can be envisaged for other bacterial infections as exemplified above.

Concluding remarks

Recent years have witnessed an amazing progress in our knowledge of the mechanism how HSPs activate the innate immune system, in particular macrophages and dendritic cells. We now know some of the surface receptors which bind HSPs and the receptors involved to sense HSPs in order to initiate the intracellular signal cascades resulting in pro-inflammatory cellular responses. However, our knowledge is still limited as to how *in vivo* endogenous or exogenous (bacterial) HSPs become liberated to be able to act as ligands for TLRs expressed by cells of the innate immune system. Furthermore, scarce information is available whether HSPs purified to homogeneity are still able to trigger inflammatory responses *in vivo*. Until now only subcutaneous injection of gp96 was reported to induce immigration of dendritic cells in the local draining lymph node [17]. Thus, it remains to be seen whether all the pro-inflammatory reactions induced by HSPs *in vitro* will also occur *in vivo*. Only this approach will allow us to analyze the immunopathological potential of HSPs.

Acknowledgement
Thomas C. Miethke is supported by the Deutsche Forschungsgemeinschaft (MI 471/1-1).

References

1 Kaufmann SH, Schoel B, van Embden JD, Koga T, Wand-Wurttenberger A, Munk ME, Steinhoff U (1991) Heat-shock protein 60: Implications for pathogenesis of and protection against bacterial infections. *Immunol Rev* 121: 67–90

2 Ranford JC, Henderson B (2002) Chaperonins in disease: Mechanisms, models, and treatments. *Mol Pathol* 55: 209–213

3 Schett G, Xu Q, Amberger A, Van der ZR, Recheis H, Willeit J, Wick G (1995) Autoantibodies against heat shock protein 60 mediate endothelial cytotoxicity. *J Clin Invest* 96: 2569–2577

4 Mahdi OS, Horne BD, Mullen K, Muhlestein JB, Byrne GI (2002) Serum immunoglobulin G antibodies to chlamydial heat shock protein 60 but not to human and bacterial homologs are associated with coronary artery disease. *Circulation* 106: 1659–1663

5 Lewthwaite J, Owen N, Coates A, Henderson B, Steptoe A (2002) Circulating human heat shock protein 60 in the plasma of British civil servants: Relationship to physiological and psychosocial stress. *Circulation* 106: 196–201

6 Chen W, Syldath U, Bellmann K, Burkart V, Kolb H (1999) Human 60-kDa heat-shock protein: A danger signal to the innate immune system. *J Immunol* 162: 3212–3219

7 Vabulas RM, Ahmad-Nejad P, da Costa C, Miethke T, Kirschning CJ, Hacker H, Wagner H (2001) Endocytosed HSP60s use toll-like receptor 2 (TLR2) and TLR4 to activate the toll/interleukin-1 receptor signaling pathway in innate immune cells. *J Biol Chem* 276: 31332–31339

8 Kol A, Bourcier T, Lichtman AH, Libby P (1999) Chlamydial and human heat shock protein 60s activate human vascular endothelium, smooth muscle cells, and macrophages. *J Clin Invest* 103: 571–577

9 Prazeres da Costa C, Kirschning CJ, Busch D, Dürr S, Jennen L, Heinzmann U, Prebeck S, Wagner H, Miethke T (2002) Role of chlamydial heat shock protein 60 in the stimulation of innate immune cells by *Chlamydia pneumoniae*. *Eur J Immunol* 32: 2460–2470

10 Basu S, Binder RJ, Suto R, Anderson KM, Srivastava PK (2000) Necrotic but not apoptotic cell death releases heat shock proteins, which deliver a partial maturation signal to dendritic cells and activate the NF-kappa B pathway. *Int Immunol* 12: 1539–1546

11 Singh-Jasuja H, Scherer HU, Hilf N, Arnold-Schild D, Rammensee HG, Toes RE, Schild H (2000) The heat shock protein gp96 induces maturation of dendritic cells and down-regulation of its receptor. *Eur J Immunol* 30: 2211–2215

12 Asea A, Kraeft SK, Kurt-Jones EA, Stevenson MA, Chen LB, Finberg RW, Koo GC, Calderwood SK (2000) HSP70 stimulates cytokine production through a CD14-dependant pathway, demonstrating its dual role as a chaperone and cytokine. *Nat Med* 6: 435–442

13 Vabulas RM, Ahmad-Nejad P, Ghose S, Kirschning CJ, Issels RD, Wagner H (2002) HSP70 as endogenous stimulus of the toll/interleukin-1 receptor signal pathway. *J Biol Chem* 277: 15107–15112

14 Udono H, Srivastava PK (1993) Heat shock protein 70-associated peptides elicit specific cancer immunity. *J Exp Med* 178: 1391–1396

15 Suto R, Srivastava PK (1995) A mechanism for the specific immunogenicity of heat shock protein-chaperoned peptides. *Science* 269: 1585–1588

16 Asea A, Rehli M, Kabingu E, Boch JA, Bare O, Auron PE, Stevenson MA, Calderwood

SK (2002) Novel signal transduction pathway utilized by extracellular HSP70: Role of toll-like receptor (TLR) 2 and TLR4. *J Biol Chem* 277: 15028–15034

17 Binder RJ, Anderson KM, Basu S, Srivastava PK (2000) Cutting edge: Heat shock protein gp96 induces maturation and migration of CD11c+ cells *in vivo*. *J Immunol* 165: 6029–6035

18 Retzlaff C, Yamamoto Y, Hoffman PS, Friedman H, Klein TW (1994) Bacterial heat shock proteins directly induce cytokine mRNA and interleukin-1 secretion in macrophage cultures. *Infect Immun* 62: 5689–5693

19 Wallin RP, Lundqvist A, More SH, von Bonin A, Kiessling R, Ljunggren HG (2002) Heat-shock proteins as activators of the innate immune system. *Trends Immunol* 23: 130–135

20 Binder RJ, Harris ML, Menoret A, Srivastava PK (2000) Saturation, competition, and specificity in interaction of heat shock proteins (hsp) gp96, hsp90, and hsp70 with CD11b+ cells. *J Immunol* 165: 2582–2587

21 Binder RJ, Han DK, Srivastava PK (2000) CD91: A receptor for heat shock protein gp96. *Nat Immunol* 1: 151–155

22 Basu S, Binder RJ, Ramalingam T, Srivastava PK (2001) CD91 is a common receptor for heat shock proteins gp96, hsp90, hsp70, and calreticulin. *Immunity* 14: 303–313

23 Wang Y, Kelly CG, Karttunen JT, Whittall T, Lehner PJ, Duncan L, MacAry P, Younson JS, Singh M, Oehlmann W et al (2001) CD40 is a cellular receptor mediating mycobacterial heat shock protein 70 stimulation of CC-chemokines. *Immunity* 15: 971–983

24 Habich C, Baumgart K, Kolb H, Burkart V (2002) The receptor for heat shock protein 60 on macrophages is saturable, specific, and distinct from receptors for other heat shock proteins. *J Immunol* 168: 569–576

25 Ohashi K, Burkart V, Flohe S, Kolb H (2000) Cutting edge: Heat shock protein 60 is a putative endogenous ligand of the toll-like receptor-4 complex. *J Immunol* 164: 558–561

26 Kol A, Lichtman AH, Finberg RW, Libby P, Kurt-Jones EA (2000) Cutting edge: Heat shock protein (HSP) 60 activates the innate immune response: CD14 is an essential receptor for HSP60 activation of mononuclear cells. *J Immunol* 164: 13–17

27 Medzhitov R (2001) Toll-like receptors and innate immunity. *Nat Rev Immunol* 1: 135–145

28 Hemmi H, Takeuchi O, Kawai T, Kaisho T, Sato S, Sanjo H, Matsumoto M, Hoshino K, Wagner H, Takeda K et al (2000) A toll-like receptor recognizes bacterial DNA. *Nature* 408: 740–745

29 Takeuchi O, Kawai T, Muhlradt PF, Morr M, Radolf JD, Zychlinsky A, Takeda K, Akira S (2001) Discrimination of bacterial lipoproteins by toll-like receptor 6. *Int Immunol* 13: 933–940

30 Takeuchi O, Sato S, Horiuchi T, Hoshino K, Takeda K, Dong Z, Modlin RL, Akira S (2002) Cutting edge: Role of toll-like receptor 1 in mediating immune response to microbial lipoproteins. *J Immunol* 169: 10–14

31 Ozinsky A, Underhill DM, Fontenot JD, Hajjar AM, Smith KD, Wilson CB, Schroeder

L, Aderem A (2000) The repertoire for pattern recognition of pathogens by the innate immune system is defined by co-operation between toll-like receptors. *Proc Natl Acad Sci USA* 97: 13766–13771

32 Bulut Y, Faure E, Thomas L, Karahashi H, Michelsen KS, Equils O, Morrison SG, Morrison RP, Arditi M (2002) Chlamydial heat shock protein 60 activates macrophages and endothelial cells through toll-like receptor 4 and MD2 in a MyD88-dependent pathway. *J Immunol* 168: 1435–1440

33 Vabulas RM, Braedel S, Hilf N, Singh-Jasuja H, Herter S, Ahmad-Nejad P, Kirschning CJ, da Costa C, Rammensee HG, Wagner H et al (2002) The endoplasmic reticulum-resident heat shock protein Gp96 activates dendritic cells *via* the toll-like receptor 2/4 pathway. *J Biol Chem* 277: 20847–20853

34 Soltys BJ, Gupta RS (1997) Cell surface localization of the 60 kDa heat shock chaperonin protein (hsp60) in mammalian cells. *Cell Biol Int* 21: 315–320

35 Altmeyer A, Maki RG, Feldweg AM, Heike M, Protopopov VP, Masur SK, Srivastava PK (1996) Tumor-specific cell surface expression of the-KDEL containing, endoplasmic reticular heat shock protein gp96. *Int J Cancer* 69: 340–349

36 Multhoff G, Botzler C, Jennen L, Schmidt J, Ellwart J, Issels R (1997) Heat shock protein 72 on tumor cells: A recognition structure for natural killer cells. *J Immunol* 158: 4341–4350

37 Goulhen F, Hafezi A, Uitto VJ, Hinode D, Nakamura R, Grenier D, Mayrand D (1998) Subcellular localization and cytotoxic activity of the GroEL-like protein isolated from *Actinobacillus actinomycetemcomitans. Infect Immun* 66: 5307–5313

38 Phadnis SH, Parlow MH, Levy M, Ilver D, Caulkins CM, Connors JB, Dunn BE (1996) Surface localization of *Helicobacter pylori* urease and a heat shock protein homolog requires bacterial autolysis. *Infect Immun* 64: 905–912

39 Garduno RA, Faulkner G, Trevors MA, Vats N, Hoffman PS (1998) Immunolocalization of Hsp60 in *Legionella pneumophila. J Bacteriol* 180: 505–513

40 Gillis TP, Miller RA, Young DB, Khanolkar SR, Buchanan TM (1985) Immunochemical characterization of a protein associated with *Mycobacterium leprae* cell wall. *Infect Immun* 49: 371–377

41 Ensgraber M, Loos M (1992) A 66-kilodalton heat shock protein of *Salmonella typhimurium* is responsible for binding of the bacterium to intestinal mucus. *Infect Immun* 60: 3072–3078

Eukaryotic HSP60: A "danger signal" for T- and natural killer cells

Arne von Bonin, Minka Breloer and Solveig H. Moré

Bernhard-Nocht Institute for Tropical Medicine, Bernhard-Nochtstr. 74, 20359 Hamburg, Germany

HSP60 – A molecular link in innate and adaptive immunity

The main functions ascribed to heat shock proteins (HSP) are to act as chaperones of nascent or aberrantly folded proteins [1]. However, it has been shown that HSP purified from tumour and virus infected cells are also capable of eliciting a protective CTL-mediated immunity [2–4]. Preparations of HSP70 and gp96 are able to activate antigen specific T-cells and the uptake of HSP/peptide complexes is most likely mediated by receptors that are expressed on the cell surface of antigen-presenting cells (APC) including macrophages and dendritic cells [5, 6]. Besides the effective delivery of HSP-associated peptides into the major histocompatibility complex (MHC) Class I restricted antigen-presenting pathway, human and murine monocytes have been found to mount pro-inflammatory responses when incubated with HSP molecules [7–9].

The addition of human recombinant HSP60 induced the up-regulation of co-stimulatory molecules on APC and the release of cytokines like IL-1, IL-6 and TNF-α. HSP60 signalling is mediated by CD14 and toll-like receptor (TLR)-4 [9–11]. TLR-4 is a pattern recognition receptor (PRR). This class of receptor recognizes different sets of homologous molecules, pathogen-associated molecular patterns (PAMP) with housekeeping functions in pathogens [12]: PAMPs are essential for pathogen survival and therefore, in contrast to virulence factors, highly conserved among micro-organisms of a given class. It has been shown before that toll-like receptors are involved in the recognition of PAMPs, e.g., TLR-4 mediates lipopolysaccharide (LPS) induced signalling events [13], whereas other TLRs recognize bacterial products like CPG-motifs and flagellin [14] or viral products like double-stranded RNA [15].

In addition to its ability to activate professional APC, human HSP60 also influences the activation of T-cells [16–18]. We have recently shown that the addition of HSP60 to cultures containing peritoneal exudate cells (PEC) and *ex vivo* purified T-cells, expressing transgenic MHC Class I or MHC Class II restricted T-cell receptors (TCR), specifically induce a release of IFN-γ. Here we discuss the impact of HSP60 on the activation of cells from the innate and adaptive immune system.

Heat Shock Proteins and Inflammation, edited by Willem van Eden

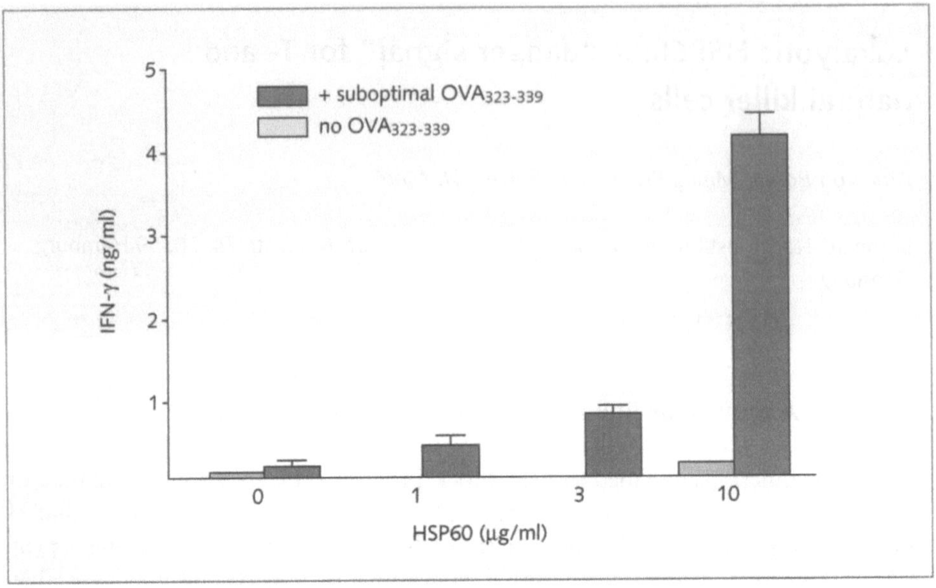

Figure 1
HSP60-induced IFN-γ secretion in DO11.10 spleen cells.
DO11.10 mice-derived spleen cells 2×10^5/well were incubated with (black bars) or without (hatched bars) synthetic $OVA_{323\text{-}339}$ peptide (10 ng/ml) in the presence of indicated amounts of HSP60 for three days. IFN-γ in the supernatant was determined by ELISA. Results are presented as means +/– SD of triplicates.

HSP60 predominantly activates naïve T-cells

Our earlier results revealed that HSP70 and gp96 preparations can activate long-term cultured cytotoxic T-cells irrespective of the source of HSP and irrespective of the HSP associated peptides [19]. Since it has been shown in the literature that HSP60 preparations stimulate professional APC [9], we asked whether a naïve T-cell response is also influenced by HSP60. To this end, we employed the mouse strain DO11.10 [20], which expresses a transgenic T-cell receptor (TCR), as a source of naïve T-cells with defined specificity. As shown in Figure 1, titrated amounts of HSP60 lead to an increased release of IFN-γ by DO11.10 spleen cells that had been stimulated with a suboptimal dose of the antigenic peptide. In contrast to the elevated levels of IFN-γ in the presence of HSP60, proliferation and IL-2 secretion were not changed in cultures containing HSP60.

Naïve and effector T-cells display marked differences in their activation kinetics and in their requirements to pass a given threshold for complete activation [21].

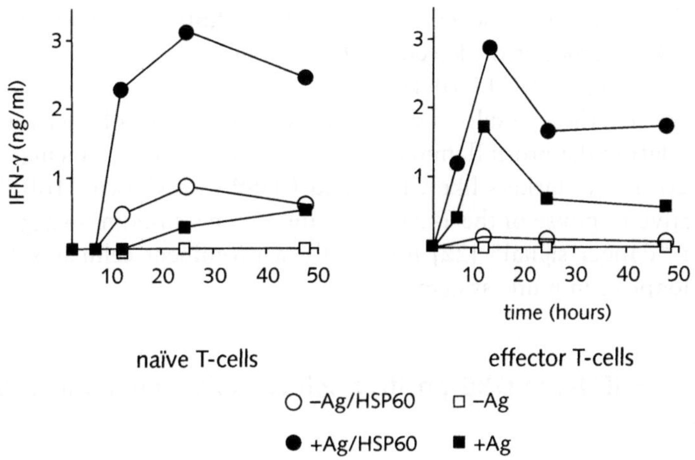

Figure 2
Accelerated and enhanced IFN-γ secretion in naïve CD4-positive T-cells in HSP60 containing cultures.
DO11.10 naïve (5×10⁴/well) or effector (2.5×10⁴/well) T-cells were incubated with 5×10⁴ OVA pulsed (filled symbols) or unpulsed (open symbols) BALB/c-derived (pristane-induced) PEC in the presence (circles) or absence (squares) of HSP60 (10 μg/ml). IFN-γ in the supernatants was determined by ELISA at the indicated time points. Results are presented as means +/– SD of triplicates.

Thus, naïve T-cells in general need a prolonged contact phase with APC, higher amounts of antigen and are strongly dependent on a sufficient expression of co-stimulatory molecules. To determine whether HSP60 as a potential danger signal could lower the activation threshold of naïve T-cells and could accelerate the stimulation kinetics, we used purified naïve and effector DO11.10 transgenic T-cells and incubated these T-cells with defined populations of professional APC in the absence or presence of HSP60. The addition of HSP60 increased the IFN-γ secretion of naïve T-cells to the level produced by effector T-cells and induced IFN-γ at early time points, when the antigenic stimulus alone was not sufficient to trigger significant amounts of IFN-γ. Even in the absence of antigen, the addition of HSP60 induced clearly detectable amounts of IFN-γ in the cultures (Fig. 2). Most of the "antigen independent" IFN-γ in the supernatant was produced by the co-culture of T-cells and PEC, since PEC cultured alone secreted only minor amounts of IFN-γ and T-cells alone no IFN-γ at all. In contrast to naïve T-cells, DO11.10 effector T-cells, which had been re-stimulated once *in vitro*, secreted large amounts of IFN-γ when cultured with PEC in the presence of antigen. The addition of HSP60 to effector T-cells increased the amount of detectable IFN-γ, but to a significantly lower extent than compared to naïve DO11.10 T-cells.

Taken together these results indicated that HSP60 partially activates naïve T-cells, up-regulating the pro-inflammatory cytokine IFN-γ and inducing cell surface-expressed activation markers like CD25 and CD69 (not shown, [16]), but leaving the proliferative response of the T-cells unchanged. These findings suggest a role for HSP60 as a "danger signal" [22] for T-cells, a prominent cellular subpopulation within the adaptive immune system.

HSP60 induces IFN-γ in DX5-positive spleen cells – predominantly natural killer cells

Although the involvement of the evolutionary ancient TLR in HSP60 mediated signal transduction has been clearly demonstrated [10], it was not known whether innate effector cells of the immune system – like natural killer (NK) cells [23] – are activated as a consequence of HSP60 mediated activation of APC, or alternatively directly by HSP60. Since the employed spleen cells of normal or TCR-transgenic mice contained up to eight percent DX5-positive NK or possibly NKT cells [24], we asked whether these cells might contribute to the HSP60 induced production of IFN-γ. To this end we performed a series of double-staining experiments, shown in Figure 3, employing the surface molecule DX5 to identify NK and NKT cells and the TCR clonotypic mAb KJI-26 ("DO11.10 TCR") to stain the antigen-specific T-cells. Analysis of the stained cells revealed that the addition of HSP60 does not affect the secretion of IL-2, nor do DX5-positive cells express IL-2 at all. Although the amounts of IFN-γ detectable in the supernatants in the presence of HSP60 are more than doubled in most of the experiments (compare Fig. 1), the absolute number of KJI-26 or DX5-positive cells producing IFN-γ is low. As shown in Figure 3, only a minority of KJI-26 or DX5-positive cells express IFN-γ intracellularly. In order to determine whether DX5-positive cells are the main responsive population to HSP60, we analyzed the response in mice lacking NKT cells and on the other hand we made use of $Rag^{-/-}$ mice which lack conventional lymphocytes, but do contain enriched numbers of NK cells.

In T-cells expressing a transgenic TCR, the DX5-positive cells may represent NK or activated T-cells [25]. To address if NKT cells, which mainly express a conserved Vα14 and Jα281 containing T-cell receptor α-chain [26], can contribute to HSP60-dependent production of IFN-γ, we made use of Jα281 gene targeted mice which lack conventional NKT cells [27]. As shown in Figure 4A, spleen cells which were activated with titrated amounts of a CD3-specific mAb responded with a compara-

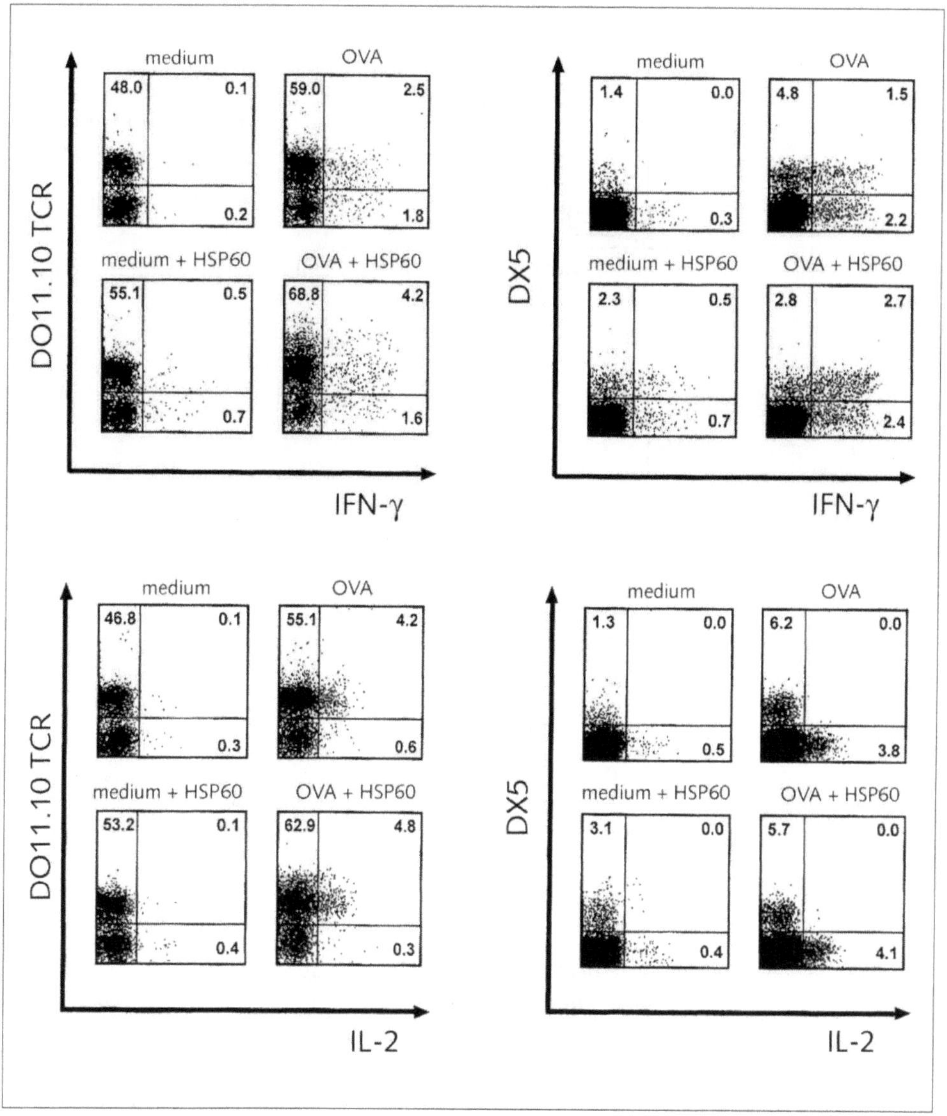

Figure 3

Activation of DX5-positive cells by HSP60.

Nylon wool enriched DO11.10 T-cells (2×10⁶/well) were incubated with 1×10⁶ OVA pulsed ("OVA") or unpulsed ("medium") PEC in the absence or presence of 10 µg/ml HSP60 ("+HSP60"). Brefeldin-A was added for the last five hours of cultures and non-adherent cells were collected and IL-2 and IFN-γ were detected by intracellular FACS-stainings. To stain DO11.10 transgenic T-cells the TCR-clonotypic mAb KJI-26 was used; to detect NKT and NK-cells we employed a DX5-specific mAb (Pharmingen, Heidelberg, Germany).

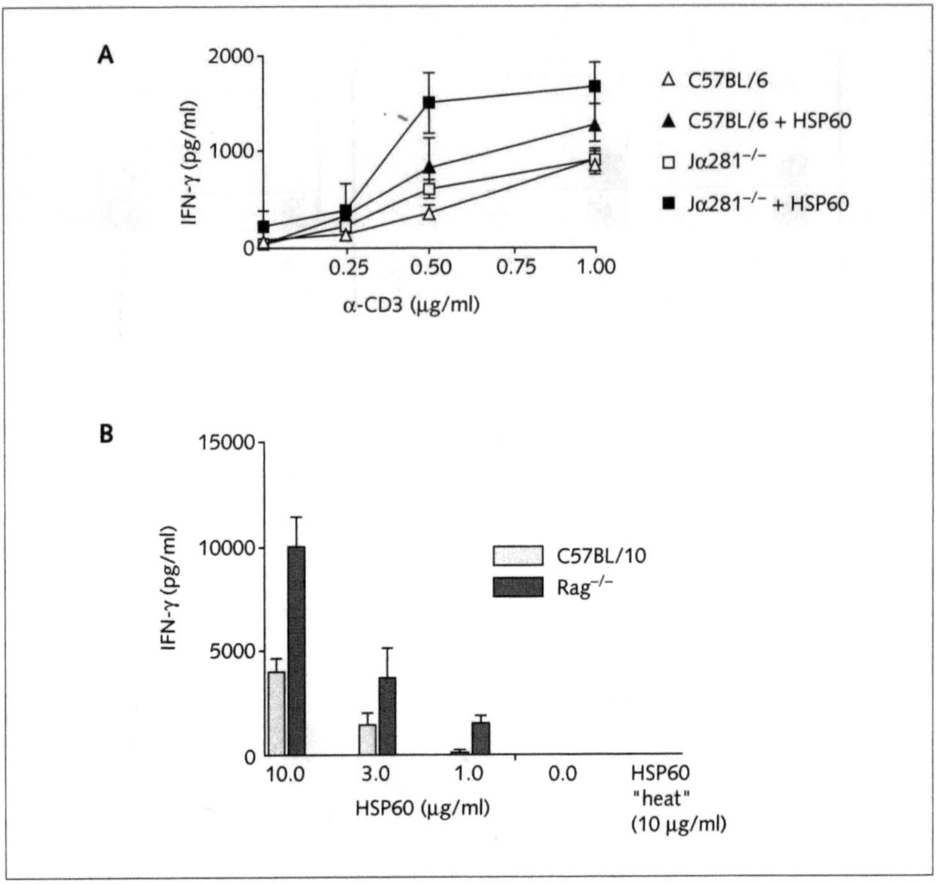

Figure 4
Jα281$^{-/-}$ spleen cells produce high amounts of IFN-γ in response to HSP60.
(A) Spleen cells (2$\times$10^5/well) from C57BL/6 and Jα281$^{-/-}$ mice were incubated together with titrated amounts of a CD3-specific mAb (145-2C11) in the absence or presence of HSP60 (10 μg/ml).
(B) C57BL/10 (grey bars) and Rag-1$^{-/-}$ (dark grey bars) spleen cells (2$\times$10^5/well) were incubated in the absence or presence of titrated, native or heat inactivated HSP60 (HSP60"heat"; 10 μg/ml).
The levels of IFN-γ in the supernatants were analyzed with cytokine specific ELISA.

ble increase of IFN-γ secretion in the presence of HSP60. Thus, probably NK cells – and not primarily NKT cells – are responsible for the observed secretion of IFN-γ in the presence of HSP60. To further investigate whether NK cells respond to HSP60, we employed spleen cells from Rag$^{-/-}$ mice as a pure T- and NKT cell-free source for

NK cells. Titrated amounts of HSP60 led to an increased release of IFN-γ into the culture supernatants of Rag-1$^{-/-}$ spleen cells in the absence of any further stimulation. In spleen cells from C57BL/10 control mice, significantly less IFN-γ could be detected when HSP60 was added to the cells (Fig. 4B). These results sustain the previous experiments, showing that HSP60 can stimulate NK cells in the absence of T-cells to secrete IFN-γ.

CD14 – a crucial receptor for HSP60-mediated signalling events

CD14 interacts with microbial products like LPS and can mediate activation of macrophages and dendritic cells promoting an innate immune response [28]. Recent studies have shown that HSP60-induced activation of monocytes is also related to the expression of the CD14 molecule [11]. To analyze whether HSP60 also binds to CD14 directly, we used CHO cells transfected with a cDNA encoding for the human CD14 [29] molecule and Alexa488-labelled human HSP60 (HSP60-ALEXA). The binding of HSP60 to CHO-CD14 cells is saturable (Fig. 5A), indicating a specific receptor/ligand interaction, especially since at high HSP60 concentrations no unspecific binding to control CHO-DHFR cells was observed. The binding of labelled HSP60 to CHO-CD14 cells does not depend on LPS-binding protein (LBP, [30]) or soluble CD14 molecules which are components of the serum, since washed cells in the absence of FCS also exhibited a comparable binding to CHO-CD14 cells, whereas binding of FITC-labelled LPS was completely abolished. These results reveal that CD14 not only plays an important role for HSP60-mediated signalling events in cells from the innate immune system, but also represents a specific binding receptor for HSP60 [31]. Binding of HSP60 to CD14 is independent of serum components like LBP and thus can be differentiated from the mode of LPS interactions with the CD14 receptor.

To analyze whether CD14 is the only receptor for HSP60 or whether multiple HSP60 receptors exist, we made use of CD14 gene-targeted (CD14$^{-/-}$) mice which have been shown to be defective in response to LPS. As shown in Figure 5B, surface-staining of macrophages from CD14-positive mice (CD14$^{+/+}$) with HSP60-ALEXA can be competed with an excess of unlabelled HSP60, but only weakly with HSP70 and not with BSA as a control. Bone marrow macrophages from homozygous CD14$^{-/-}$ mice could also be stained with HSP60-ALEXA. Again, HSP60-ALEXA binding could be competed with an excess of unmodified HSP60 but not with HSP70. These results indicate that, independently of CD14, a second receptor (-complex) on the surface of macrophages exists which can specifically interact with HSP60. The biological activity of HSP60, thus, is not exclusively dependent on the presence of CD14, although there is a quantitative correlation between CD14 expression and the activation of APC. Hence, the results obtained with the help of CD14$^{-/-}$ mice allowed the identification of a second CD14 receptor, the molecular

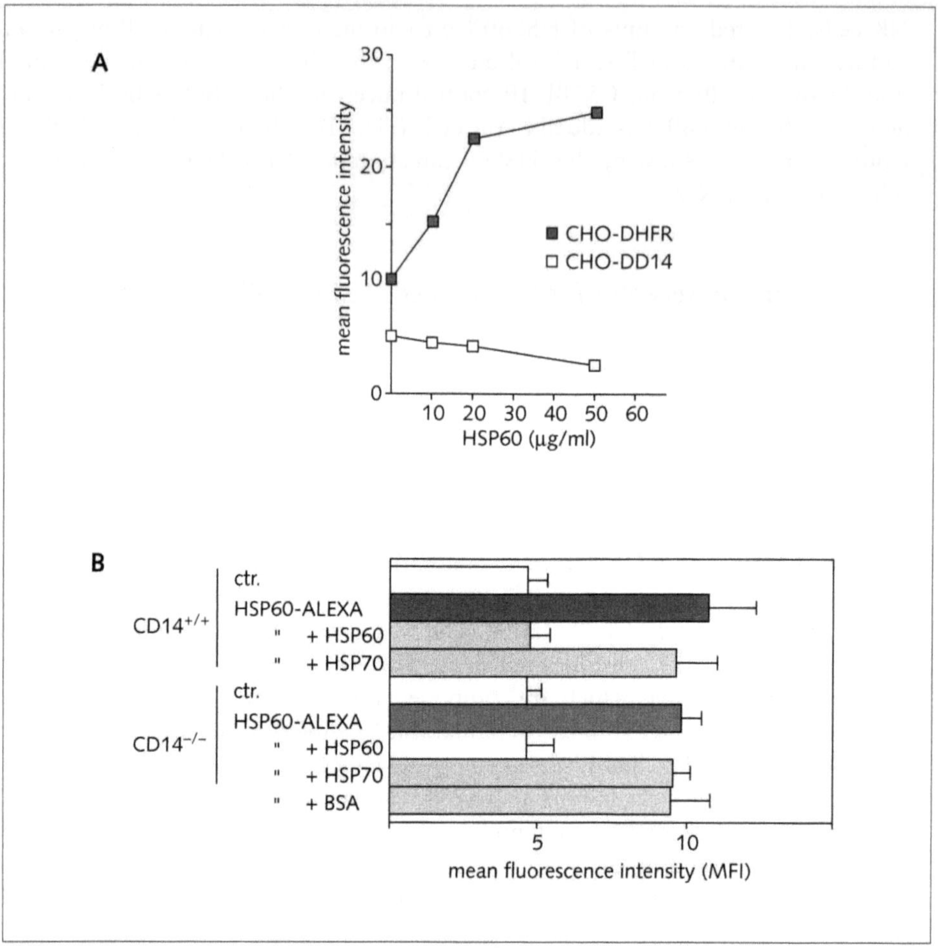

Figure 5
HSP60 binds human CD14.
(A) Titrated amounts of biotinylated HSP60 were added to CHO-CD14 and control CHO-DHFR cells. Shown is the mean fluorescence intensity of the analyzed cells.
(B) 2 µg HSP60-ALEXA was incubated together with 1×10^5 bmMØ from $CD14^{+/-}$ or $CD14^{-/-}$ mice. FACS-analysis ("mean fluorescence intensity") of CD14 bmMØ in the absence of competitor or in the presence of unlabelled HSP60 (100 µg), HSP70 (100 µg) or BSA (100 µg) is shown.

structure of which is awaiting further characterisation. In this context, it is interesting to note that, for LPS, an alternative receptor complex has recently been identified [32].

Concluding remarks

HSP60 plays a dual role in the course of T-cell activation: HSP60 stimulates cytokine secretion in antigen-presenting cells and furthermore, it enhances and accelerates the activation of predominantly naïve T- and NK-cells, mainly by inducing IL-12 in macrophages. Our results – that effector T-cells are influenced by HSP60 to a much lower extent than naïve T-cells – are in accordance with the model of a danger signal that is necessary for enhancing primary T-cell responses [33].

The activation of professional APC by HSP60 is mediated by evolutionary-conserved proteins, like CD14 and toll-like receptors, and has immediate consequences for the cellular immune response. The release of chemo-attractant chemokines from the APC recruits T-cells and the infiltrating T-cells are then completely activated by recognizing processed peptide fragments of bacterial or viral pathogens in a pro-inflammatory environment. HSP60, thus, would act as an initiator of an adaptive T-cell response. We take our results as an indication that HSP60 may act as a molecular link between the innate immune system, employing toll-like receptors on APC and the adaptive immune system, and enhancing and accelerating the stimulation of naïve T-cells (Fig. 6).

While interactions of cells with HSP60, either directly with professional APC, or indirectly with T-cells *via* cytokines (e.g., IL-12, [34]) released from APC, have been established in previous reports, HSP-mediated effects on NK and NKT-cells have not been addressed. NK cells are not only stimulated to secrete IFN-γ in the presence of HSP60 and APC but represent the major source for the HSP60-induced IFN-γ. The HSP60-mediated activation occurs *via* an IL-12 dependent mechanism, since IL-12 can reconstitute the effects of HSP60 in the presence of professional APC, whereas APC from IL-$12^{-/-}$ mice cannot induce elevated levels of IFN-γ [16] in the presence of HSP60. Although IL-12 is of central importance for HSP60 induced IFN-γ production, other factors are also required, since recombinant IL-12 in the absence of APC could not reconstitute the observed HSP60 effects on T- and NK cells.

The use of recombinant proteins in cellular assay systems requires a careful analysis of potential bacterial contaminants. To this end we performed several controls to minimize the contributions of bacterial products in the analysis of our results: First, heat denaturing of the HSP60 preparations abrogated the induction of cytokines, whereas LPS-induced effects were still detectable. Second, the LPS inhibitor Polymyxin-B had no effect when added to HSP60-containing cultures, whereas LPS-mediated effects were completely blocked. Third, the comparison of HSP60 preparations containing different amounts of LPS and "low endotoxin" HSP60, that was virtually LPS-free, revealed that the biologic activity of the various HSP60 preparations did not correlate with the degree of LPS contamination [35]. Fourth, incubation of PEC in serum-free medium completely abrogated the release of pro-inflammatory cytokines induced by LPS, since the LPS-binding protein is

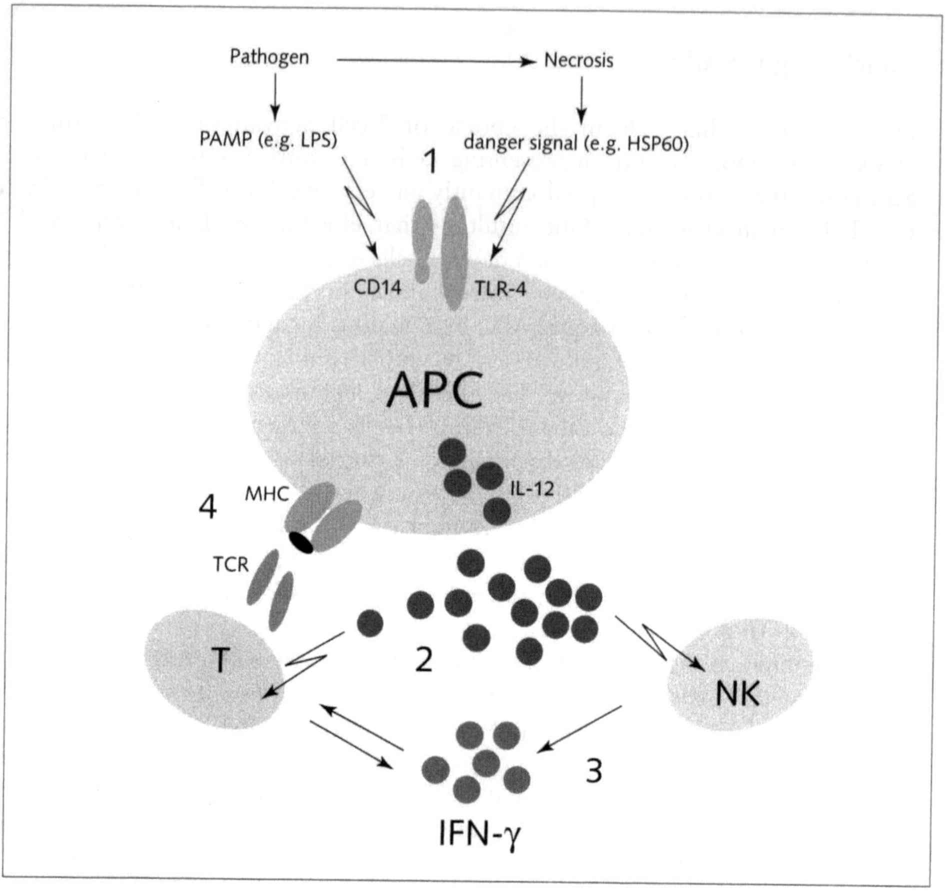

Figure 6
Scheme of HSP60 mediated T-cell activation.
HSP60 released from necrotic cells (or LPS as a pathogen associated molecular pattern, "PAMP") binds and trigger pattern recognition receptors (PRR) CD14 and TLR-4 ("1"). Activation of the APC leads to IL-12 secretion ("2", and to the secretion of other cytokines) which in turn stimulates T- and NK cells to release IFN-γ. NK cell-derived IFN-γ ("3") may further contribute to the activation of T-cells and accelerate T-cell triggering. As a direct consequence of released HSP60, naïve T-cells can be fully stimulated following recognition of MHC/peptide complexes by the antigen-specific T-cell receptor ("4").

missing under these conditions. In contrast, HSP60 still activated PEC in serum-free cultures, albeit at slightly reduced levels. This result again indicates that HSP60 – and not contaminating LPS – is responsible for the observed increased levels of cytokines. Moreover, at least concerning the HSP60 binding assays, minute contaminations of bacterial products in the HSP60 preparations can be neglected for the interpretation of the data, since at least 200 ng of fluorescein isothiocyanate (Fitc)-LPS was needed to show direct binding to CD14-transfected CHO cells and even larger amounts of unlabelled LPS were necessary to compete biotinylated or ALEXA488-labelled HSP60. Although we cannot rule out that other bacterial contaminants are contributing to the interaction with CD14, the high purity of the employed HSP60 batches as well as the "non-binding" of recombinant HSP70 and purified BSA control proteins argue against contaminations within the employed HSP60 batches being responsible for the observed effects.

HSP60 obviously acts on a variety of different immune cells, and thus might alert immunological responses. A central player in the HSP60-mediated IFN-γ secretion is IL-12, which is known to have important functions in inducing protective TH1 responses in mice. It will be interesting to learn how the expression of a secreted as well as a membrane-anchored eukaryotic form of HSP60 induce IL-12 (and other cytokines) in a defined microenvironment *in vivo* to investigate its stimulatory capacity in tumour and infectious disease models, too. Future findings employing different eukaryotic HSP60 expression-constructs may provide further insight into the molecular mechanisms by which extracellular HSP in general activate the immune system.

Acknowledgements
We thank C. Steeg, and S. Ehrlich for perfect technical assistance and B. Fleischer for continuous support. We are grateful to H.-G. Ljunggren, R. Wallin and other members of the MTC, Karolinska Institutet, Stockholm, Sweden for many helpful and inspiring discussions.

References

1 Hartl FU (1996) Molecular chaperones in cellular protein folding. *Nature* 381: 571–580
2 Blachere NE, Udono H, Janetzki S, Li Z, Heike M, Srivastava PK (1993) Heat shock protein vaccines against cancer. *J Immunother* 14: 352–356
3 Li Z, Srivastava PK (1994) A critical contemplation on the role of heat shock proteins in transfer of antigenic peptides during antigen presentation. *Behring Inst Mitt* 94: 37–47
4 Srivastava PK, Menoret A, Basu S, Binder RJ, McQuade KL (1998) Heat shock proteins come of age: Primitive functions acquire new roles in an adaptive world. *Immunity* 8: 657–665
5 Binder RJ, Han DK, Srivastava PK (2000) CD91: A receptor for heat shock protein gp96. *Nature Immunol* 1: 151–155

6 Basu S, Binder RJ, Ramalingam T, Srivastava PK (2001) CD91 is a common receptor for heat shock proteins gp96, hsp90, hsp70, and calreticulin. *Immunity* 14: 303–313
7 Singh-Jasuja H, Scherer HU, Hilf N, Arnold-Schild D, Rammensee HG, Toes RE, Schild H (2000) The heat shock protein gp96 induces maturation of dendritic cells and down-regulation of its receptor. *Eur J Immunol* 30: 2211–2215
8 Asea A, Kraeft SK, Kurt-Jones EA, Stevenson MA, Chen LB, Finberg RW, Koo GC, Calderwood SK (2000) HSP70 stimulates cytokine production through a CD14-dependant pathway, demonstrating its dual role as a chaperone and cytokine. *Nat Med* 6: 435–442
9 Chen W, Syldath U, Bellmann K, Burkart V, Kolb H (1999) Human 60-kDa heat-shock protein: A danger signal to the innate immune system. *J Immunol* 162: 3212–3219
10 Kol A, Lichtman AH, Finberg RW, Libby P, Kurt-Jones EA (2000) Heat shock protein (HSP) 60 activates the innate immune response: CD14 is an essential receptor for HSP60 activation of mononuclear cells. *J Immunol* 164: 13–17
11 Ohashi K, Burkart V, Flohe S, Kolb H (2000) Heat shock protein 60 is a putative endogenous ligand of the toll-like receptor-4 complex. *J Immunol* 164: 558–561
12 Medzhitov R, Janeway CA Jr (1997) Innate immunity: The virtues of a non-clonal system of recognition. *Cell* 91: 295–298
13 Poltorak A, He X, Smirnova I, Liu MY, Huffel CV, Du X, Birdwell D, Alejos E, Silva M, Galanos C et al (1998) Defective LPS signaling in C3H/HeJ and C57BL/10ScCr mice: Mutations in Tlr4 gene. *Science* 282: 2085–2088
14 Hemmi H, Takeuchi O, Kawai T, Kaisho T, Sato S, Sanjo H, Matsumoto M, Hoshino K, Wagner H, Takeda K et al (2000) A Toll-like receptor recognizes bacterial DNA. *Nature* 408: 740–745
15 Hayashi F, Smith KD, Ozinsky A, Hawn TR, Yi EC, Goodlett DR, Eng JK, Akira S, Underhill DM, Aderem A (2001) The innate immune response to bacterial flagellin is mediated by toll-like receptor 5. *Nature* 410: 1099–1103
16 Breloer M, Dorner B, More SH, Roderian T, Fleischer B, von Bonin A (2001) Heat shock proteins as 'danger signals': Eukaryotic Hsp60 enhances and accelerates antigen-specific IFN-gamma production in T cells. *Eur J Immunol* 31: 2051–2059
17 More SH, Breloer M, von Bonin A (2001) Eukaryotic heat shock proteins as molecular links in innate and adaptive immune responses: Hsp60-mediated activation of cytotoxic T cells. *Int Immunol* 13: 1121–1127
18 Breloer M, More SH, Osterloh A, Stelter F, Jack RS, Bonin Av A (2002) Macrophages as main inducers of IFN-gamma in T-cells following administration of human and mouse heat shock protein 60. *Int Immunol* 14: 1247–1253
19 Breloer M, Fleischer B, von Bonin A (1999) *In vivo* and *in vitro* activation of T-cells after administration of Ag- negative heat shock proteins. *J Immunol* 162: 3141–3147
20 Murphy KM, Heimberger AB, Loh DY (1990) Induction by antigen of intrathymic apoptosis of CD4+CD8+TCRlo thymocytes *in vivo*. *Science* 250: 1720–1723
21 Iezzi G, Karjalainen K, Lanzavecchia A (1998) The duration of antigenic stimulation determines the fate of naïve and effector T-cells. *Immunity* 8: 89–95

22 Matzinger P (1994) Tolerance, danger and the extended family. *Annu Rev Immunol* 12: 991–1045
23 Karre K (2002) NK cells, MHC class I molecules and the missing self. *Scand J Immunol* 55: 221–228
24 Iwabuchi C, Iwabuchi K, Nakagawa K, Takayanagi T, Nishihori H, Tone S, Ogasawara K, Good RA, Onoe K (1998) Intrathymic selection of NK1.1(+)alpha/beta T-cell antigen receptor (TCR)+ cells in transgenic mice bearing TCR specific for chicken ovalbumin and restricted to I-Ad. *Proc Natl Acad Sci USA* 95: 8199–8204
25 Kambayashi T, Assarsson E, Chambers BJ, Ljunggren HG (2001) Expression of the DX5 antigen on CD8+ T-cells is associated with activation and subsequent cell death or memory during influenza virus infection. *Eur J Immunol* 31: 1523–1530
26 Bendelac A (1995) Mouse NK1$^+$ T-cells. *Curr Opin Immunol* 7: 367–374
27 Kawano T, Cui J, Koezuka Y, Toura I, Kaneko Y, Motoki K, Ueno H, Nakagawa R, Sato H, Kondo E et al (1997) CD1d-restricted and TCR-mediated activation of v alpha14 NKT cells by glycosylceramides. *Science* 278: 1626–1629
28 Hoshino K, Takeuchi O, Kawai T, Sanjo H, Ogawa T, Takeda Y, Takeda K, Akira S (1999) Cutting edge: Toll-like receptor 4 (TLR4)-deficient mice are hypo-responsive to lipopolysaccharide: Evidence for TLR4 as the Lps gene product. *J Immunol* 162: 3749–3752
29 Stelter F, Bernheiden M, Menzel R, Jack RS, Witt S, Fan X, Pfister M, Schutt C (1997) Mutation of amino acids 39-44 of human CD14 abrogates binding of lipopolysaccharide and *Escherichia coli*. *Eur J Biochem* 243: 100–109
30 Jack RS, Fan X, Bernheiden M, Rune G, Ehlers M, Weber A, Kirsch G, Mentel R, Furll B, Freudenberg M et al (1997) Lipopolysaccharide-binding protein is required to combat a murine gram-negative bacterial infection. *Nature* 389: 742–745
31 Habich C, Baumgart K, Kolb H, Burkart V (2002) The receptor for heat shock protein 60 on macrophages is saturable, specific, and distinct from receptors for other heat shock proteins. *J Immunol* 168: 569–576
32 Triantafilou K, Triantafilou M, Dedrick RL (2001) A CD14-independent LPS receptor cluster. *Nat Immunol* 2: 338–345
33 Matzinger P (1998) An innate sense of danger. *Semin Immunol* 10: 399–415
34 Azzoni L, Kanakaraj P, Zatsepina O, Perussia B (1996) IL-12-induced activation of NK and T-cells occurs in the absence of immediate-early activation gene expression. *J Immunol* 157: 3235–3241
35 Breloer M, Moré SH, Dorner B, Fleischer B, von Bonin A (2001) Hitzeschock-Proteine als Gefahrensignale: Hsp60 verstärkt die antigenspezifische Aktivierung primärer T-Zellen. *Modern Aspects of Immunol* 2: 64–66

22. Matzinger P (1994) Tolerance, danger and the extended family. *Annu Rev Immunol* 12: 991–1045
23. Kärre K (2002) NK cells, MHC class I molecules and the missing self. *Scand J Immunol* 55: 221–228
24. Iwabuchi C, Iwabuchi K, Nakagawa K, Takayanagi T, Nishihori H, Tone S, Ogasawara K, Good RA, Onoe K (1998) Intrathymic selection of NK1.1(+)alpha/beta T cell antigen receptor (TCR)+ cells in transgenic mice bearing TCR specific for chicken ovalbumin and restricted to I-Ad. *Proc Natl Acad Sci USA* 95: 8199–8204
25. Kimbauer I, Asgeirsson E, Chambers BJ, Ljunggren HG (2001) Expression of the CD

Heat shock proteins and experimental arthritis

Rebecca J. Brownlie and Stephen J. Thompson

Department of Life Sciences, King's College London, Franklin-Wilkins Building, 150 Stamford Street, London SE1 9NN, UK

Introduction

Rheumatoid arthritis (RA) is a multi-factorial, human autoimmune disease, characterised by inflammatory and proliferative synovitis that exhibits invasive and destructive features [1]. As yet, the underlying pathogenic mechanisms of RA remain poorly understood. In order to help gain a better understanding of the disease, several animal models of RA have been developed and have been widely studied. Although these models can be very instructive, no one model reflects all aspects of the human disease illustrating the complexity and heterogeneous nature of RA (Tab. 1).

Initial evidence for the involvement of HSP in arthritis

Heat shock proteins (HSP) are ubiquitous, housekeeping proteins essential for many cellular processes in prokaryotic and eukaryotic organisms [17]. Based on their molecular weight, HSP are divided into four main families of structurally related proteins: HSP90, HSP70, HSP60 and small HSP. The expression of HSP is elevated under a variety of stressful conditions such as a rise or fall in temperature, oxidative stress, irradiation, infection and exposure to toxic chemicals [18]. A prominent feature of these proteins is the extraordinary high conservation of amino acid sequence throughout the prokaryotic and eukaryotic kingdoms. For example, mycobacterial HSP65 is 48% homologous with human HSP60 [19]. However, despite this sequence conservation, HSP have become immunodominant target antigens for the mammalian immune system [20].

Evidence for a role of heat shock proteins (HSP) in autoimmunity first came to light in the mycobacteria-induced model of adjuvant arthritis (AA). Seminal work by Cohen and colleagues identified a T-cell line raised against *Mycobacterium tuberculosis*, namely A2, which could evoke disease when transferred to sub-lethally irra-

Heat Shock Proteins and Inflammation, edited by Willem van Eden

Table 1 - Widely studied experimental models of arthritis

Non-specific induction
Spontaneous onset MRL-lpr/lpr mouse [2] Ageing male DBA-1 mice [3]
Injection of mineral oil or non-immunogenic adjuvants Mice injected with the paraffin oil pristane (pristane induced arthritis: PIA) [4] Rats injected with avridine (lipidal amine CP20961) in oil (avridine induced arthritis: AIA) [5]
Other oil arthritides: Rats injected with Incomplete Freund's adjuvant (IFA) (oil induced arthritis: OIA) [6] Rats injected with pristane (rat pristane induced arthritis: PIA) [6, 7] Rats injected with squalene in oil (squalene induced arthritis: SIA) [8]
Specific induction
Characterised inducing immunogen Mice [9], rats [10] and primates [11] injected with heterologous Type II collagen (collagen induced arthritis: CIA) Mice injected with heterologous cartilage proteoglycan [12]
Microbial agent Rats injected with heat killed *Mycobaterium tuberculosis* [13] or the mycobacterial cell wall component muramyl dipeptide (MDP) [14] Rats injected with streptococcal cell walls (streptococcal cell wall-induced arthritis: SCWA) [15] Rabbits injected with *E. coli* [16]

diated, naïve recipient rats. Paradoxically, when this same T-cell line was administered to non-irradiated recipients, the animals became resistant to the induction of disease [21]. After dissection of the parental cell line into individual clones, it was found that one clone, designated A2b, was arthritogenic and another, A2c, conferred protection from arthritis [22]. Further important studies by van Eden and colleagues identified that both clones responded to the same amino acid sequence 180–188 of mycobacterial HSP65 [23].

These findings not only suggested an essential role for antigen-specific T-cells in disease induction, but also proposed a pathogenic and more importantly, a protec-

Table 2 - Models of arthritis protected from disease development by HSP administration

HSP	Model	Reference
Mycobacterial HSP65	Adjuvant arthritis	[23]
Mycobacterial HSP65	SCWA	[26]
Mycobacterial HSP65	Pristane induced arthritis	[27]
Mycobacterial HSP65	Collagen induced arthritis	[24]
Mycobacterial HSP65	Avridine arthritis	[24]
Mycobacterial HSP65 DNA	Adjuvant arthritis	[28]
Mycobacterial HSP65 delivered by *Vaccinia*	Adjuvant arthritis	[29]
Human HSP60 DNA	Adjuvant arthritis	[30]
Mycobacterial HSP10	Adjuvant arthritis	[31]
Mycobacterial HSP70	Adjuvant arthritis	[32]
Escherichia coli HSP70	Adjuvant arthritis	[33]
BiP (HSP70 family)	Collagen induced arthritis	[34]
BiP (HSP70 family)	Adjuvant arthritis	[34]

tive role for HSP65 in AA. Further evidence for a role of HSP in experimental arthritis came from the finding that rats mounted significant anti-HSP65 T-cell responses during AA. However, many attempts to induce arthritis by administering HSP65 to susceptible strains of rats were not successful [24]. Instead, protection from subsequent disease induction was established. Furthermore, these effects were proven to be T-cell-mediated after it was demonstrated that protection from arthritis, afforded by transfer of cells from immunised animals to naïve recipients, was dependent on the presence of HSP65-specific T-cells [25].

The discovery that HSP65 plays an important role in AA led to extensive investigations into the protective role of HSP in various other models of arthritis (Tab. 2).

As shown in Table 2, the protective effect of mycobacterial HSP65 is not just limited to AA, a model in which mycobacteria are administered for the induction of disease; rather protection is also seen in models where non-microbial agents are used to induce disease. Indeed, protective roles for HSP65 have been reported for pristane-induced arthritis (PIA) in mice [27] and to a more variable degree in collagen-induced arthritis (CIA) [24] and avridine arthritis [24]. Furthermore, HSP65 has a protective effect in other experimental autoimmune diseases, such as diabetes in the non-obese diabetic mouse [35] and experimental autoimmune encephalomyelitis [36]. In addition, more recent studies have shown that protection from arthritis can be afforded by pre-immunisation with additional members of the HSP gene families, including HSP70, HSP10 and BiP [31, 32, 34]. This illustrates the apparent general protective role of HSP in experimental inflammatory diseases. This protective capacity appears to be specific to the HSP family as other highly evolutionary-conserved proteins such as bovine myosin, bacterial superoxide dismutase, glyceraldehyde-3-phosphate dehydrogenase (GAPDH) and aldolase fail to prevent disease in AA [37]. Furthermore, extensive studies of HSP-reactive cells identified immunodominant T-cell epitopes within the parent molecules. In certain cases, the administration of these epitopes as synthetic peptides could induce antigen-specific protection against several of the disease models [38–41]. For example, subcutaneous administration of the HSP65 peptides, amino acids 180–188 or 256–265, prevented the development of AA in Lewis rats [23, 38]. In addition, the delivery of peptide 180–188 from HSP65 and peptide 111–125 from HSP70 *via* the tolerogenic intra-nasal route ameliorated adjuvant arthritis [39, 41]. As far as PIA in CBA mice is concerned, the immunodominant epitope (corresponding to amino acids 261–271) from HSP65 was found to be protective when administered either intraperitoneally, subcutaneously or intramuscularly either before or after the induction of arthritis [40 and unpublished observations]. Paradoxically, frequent intranasal administration of this same peptide exacerbated the development of PIA [42]. In these experiments, administration of this non-conserved, bacterial-specific epitope led to priming of pathogenic Th1 cells rather than the induction of T-cell unresponsiveness or tolerance. These data have important implications for the design of antigen or T-cell specific therapy for inflammatory diseases. Thus it is important to identify antigens and dosing regimes that counteract but do not activate adverse immune responses.

Environmental trigger of immune cell reactivity to HSP65 during disease

Adaptive immune cell responses to HSP65 during experimental arthritis have been extensively investigated in order to gain a better understanding of the precise role this molecule plays in disease. Although in some models in which HSP65 has been shown to protect, such as CIA and avridine arthritis, T-cell responses to HSP65 are not significantly increased in diseased animals. By contrast, extensive studies have shown that lymph node and splenic T-cells taken from rats with AA [25] and streptococcal cell wall-induced arthritis (SCWA) [26] have been shown to respond to *in vitro* stimulation with HSP65. One could argue that this is not surprising since in the case of AA these data could be explained by experiments showing that HSP65 is a potent immunogen, and is a component of the mycobacterial preparation that is injected into rats to induce disease. Similarly, it has been suggested that streptococcal cell wall antigens share epitopes with HSP65 [43], and further studies have shown that there is antigenic similarity between streptococcal M proteins and HSP65 [44].

Adaptive immune responses generated against HSP65 during the course of AA and SCWA can potentially be explained by antigenic cross-reactivity between endogenous HSP and the preparations used to induce disease. However, the finding that the development of PIA (where no antigenic components are used to induce disease) is also associated with increased levels of specific antibody and T-cell responses to HSP65 is less intuitive and suggests a more general role for these proteins [27, 45]. In addition, data indicate that the anti-HSP65 antibody response of non-manipulated naïve mice increases with age, directly correlating with the age-related susceptibility to PIA [46].

Pristane induced arthritis is elicited by the intraperitoneal injection of the paraffin oil pristane (2,6,10,14-tetramethylpentadecane), which itself is non-immunogenic [4, 47]. A number of theories have been proposed to explain how oil injected in to the peritoneum can lead to increased T-cell responses to HSP65 and furthermore, induce arthritis. Pristane is an irritant and therefore may, due to intestinal inflammation, expose the immune system to gut microbes thus initiating a cross-reactive response to HSP65. Alternatively, inflammation caused by pristane injection could lead to the increased expression of endogenous HSP60 resulting in cross-reactive immune responses. The former supposition is given credence by the finding that specific-pathogen free mice maintained under sterile conditions in an isolator are resistant to the development of PIA. However, when these animals are returned to the normal "dirty" environment, disease susceptibility is restored [45]. Furthermore, mice maintained under sterile conditions show reduced levels of anti-HSP65 antibodies and T-cell reactivity [unpublished observations]. These studies demonstrate the importance of environmental antigens in priming for disease susceptibility.

It is apparent that the role of the environment in disease induction differs depending on which experimental model is utilised. For example, studies of Fischer rats (a strain closely related to the AA-susceptible Lewis rat) have shown opposite findings to those reported in PIA. Fischer rats raised in a barrier facility are susceptible to the induction of AA, whereas animals raised in conventional conditions show considerably reduced disease [48]. Moudgil and colleagues have proposed an explanation for these findings. These investigators have reported that certain peptides encoded in the C-terminus of HSP65 can down-modulate AA when administered to naïve Lewis rats [49]. Further studies by this group have indicated that conventionally housed naïve Fischer rats, but not rats raised in a barrier facility, mount a T-cell response against these C-terminus derived peptides. They speculate that exposure to microbial agents in the environment primes T-cells that cross-react with these disease-modulating peptides [50]. In contrast, other studies comparing the role of bowel flora in three different models of arthritis induced in the rat (AA, CIA and OIA) report that disease induction is independent of gut flora [51]. Collectively, these reports highlight the complex relationship between environmental antigens, the immune responses to HSPs and disease development.

Mechanisms of protection in experimental arthritis

Immunological cross-recognition of bacterial and endogenous HSP

The expression of endogenous HSP60 in arthritic joints initially led many investigators to the hypothesis that this protein was an important target auto-antigen under immunological attack during disease progression. However, more recent findings have not supported this supposition. Indeed the induction of cross-reactive immunity to endogenous HSP has now been proposed to underlie the disease-modulating effects of HSP65. Thus a substantial body of evidence that strongly favours this theory has accumulated, and suggests that it is the nature of the cross-reactive response that determines the protective outcome. It is known that inflammation leads to local up-regulation of endogenous HSP60 and HSP70 in the joints of rheumatoid patients [52, 53] and similarly, HSP60 expression is up-regulated in the joints of arthritic mice [46] and rats [54]. Furthermore, it has been shown by immunohistochemical analysis that HSP60 is expressed in murine thymic medullary epithelium [55], suggesting that the self HSP60-responsive T-cells detectable in healthy individuals are low affinity, positively selected, thymic-derived T-cells. These data support the idea that exposure to bacterial HSP65, either by immunisation, infection or from the gut flora, causes cross-stimulation of naturally, pre-existing, circulating self-HSP reactive T-cells. This is thought to lead to the expansion of T-cells with an anti-inflammatory or regulatory phenotype that cross react with up-regulated self HSP in inflamed joints and, subsequently, mediate suppression of inflammation (discussed in full below).

Alternatively, an injection of incomplete Freund's adjuvant (IFA) alone into the footpads of BALB/c mice resulted in acute inflammation and furthermore draining lymph nodes from these mice contained T-cells that responded to HSP65 [56]. The generation of T-cell clones from these lymph nodes enabled further analysis of the cell specificity. It was found that over 80% of the T-cell clones also responded to human HSP60. Therefore, self-HSP60 specific T-cells can be activated as part of a normal inflammatory response without exposure to bacterial HSP65. However, injecting mice with IFA activates HSP65-responsive T-cells that respond to different epitopes than T-cells derived from mice previously challenged with HSP65 [56]. Thus the precise nature of the two HSP-reactive T-cell populations is likely to be different. Studies performed on human peripheral blood mononuclear cells (PBMC) support the idea that inflammation activates self-HSP60 reactive T-cells. PBMC obtained from patients with established RA were shown to proliferate upon *in vitro* stimulation with HSP60 [57]. Furthermore, this study also reported HSP60-driven proliferation in cultures of PBMC from normal individuals but this response was of a lower magnitude and peaked later than the responses seen in PBMC from RA patients. Since the peak proliferative response of human peripheral blood T-cells to recall antigens is earlier than that to non-recall antigens [58], then it is reasonable to suggest that PBMC from RA patients are primed to HSP60. These results are further supported by the observation that T-cells from normal individuals responsive to HSP60 are confined to the CD45RA$^+$ (naïve) subset of T-cells which have delayed kinetics as compared with the response of the CD45RO$^+$ (memory) T-cell subset to recall antigens [59]. Taken together, these findings suggest that priming to HSP60 *in vivo* occurred during the disease process where increased expression of HSP60 in the inflamed joint leads to activation and expansion of these pre-existing HSP-responsive T-cells.

Immune deviation

The importance of the quality of the immune response directed against HSP65 and the outcome on disease progression is highlighted in studies carried out in PIA. Previous work from our laboratories reported that mice with PIA and those protected from disease by pre-immunisation with HSP65 both possess raised T-cell responses to HSP65. This subsequently led to investigations into the nature of these responses, resulting in the demonstration that T-cells derived from the two groups of mice produced different cytokine profiles upon stimulation with HSP65 *in vitro* [60]. Splenic cells from all mice produced Th1-type cytokines, IL-2 and IFN-γ, whereas only splenic cells from mice that had been pre-immunised with HSP65 produced the Th2-type cytokines IL-4, IL-5 and IL-10. Furthermore, there was a dramatic increase in the IgG1 to IgG2a ratio of anti-HSP65 antibodies from arthritic to protected mice. These findings suggested that disease protection was mediated by the secretion of anti-inflammatory cytokines from HSP65-specific T-cells. This was confirmed *in vivo* by administering recombinant IL-12 prior to the induction of PIA in order to polarise the HSP65-specific Th2 biased immune response towards a Th1 response. The incidence of PIA was recorded 200 days after the first injection of pristane and was found to be significantly higher in the animals treated with IL-12, HSP65 and pristane as compared to the protected group (HSP65 and pristane treated). Alongside these studies, analysis of T-cell specificity illustrated that cells from HSP65-protected mice only respond to a limited number of HSP65 peptides (predominantly to the immunodominant epitope, corresponding to amino acids 261–271). Hence the T-cell repertoire to HSP65 from protected animals is limited or controlled (repertoire limitation). By contrast, T-cells from arthritic mice respond to many peptides (epitope spreading). We suggested that epitope spreading and Th1 responses are related, and are the means by which quiescent autoreactive T-cells are stimulated to potentiate disease. Conversely, repertoire limitation is mediated by regulatory or Th2 cytokines (for full discussion see [61]). Indeed, a number of articles attribute these changes in T-cell specificity to enhanced or altered processing and presentation of antigens under inflammatory conditions [49, 62–64]. Finally, a recent study examining the protective role of HSP-reactive antibodies presents evidence for B-cell determinant spreading. Antibodies from Lewis rats that had been injected with CFA to induce AA (a Th1-mediated disease) recognised more epitopes from HSP65 than those from naïve Lewis rats [65].

Induction of regulatory T-cells by HSP

As discussed above, bacterial and endogenous HSP are highly homologous and have been shown to activate cross-reactive T-cells. Recent attention has focussed on the regulatory nature of these cross-reactive T-cells and their role has been implicated in controlling disease. For example, studies carried out in AA have shown that T-cells that recognise the highly conserved epitope 256–270 of HSP65, which can confer protection from AA, cross-react with endogenous rat HSP60 [66], although, in this report, the precise rat HSP60 epitope was not fully characterised. However, further studies showed that the cross-reactive T-cells were highly proliferative when stimulated with syngeneic heat-shocked APC and that they secreted IFN-γ, IL-4 and IL-10 and also up-regulated expression of the co-stimulatory molecule CD86 [67]. Another report from this group indicated that the 256–265-specific T-cells do in fact cross react with the rat homologue of this peptide, which differs in sequence at three amino acid residues. These findings led to the suggestion that the self 256–265 peptide acts as an altered peptide ligand (APL) resulting in the induction of a regulatory phenotype in the HSP65 256–265-specific T-cells. Such cells are rendered non-proliferative, produce no IFN-γ or IL-10, low levels of IL-4 and but express increased levels CD86 which could preferentially interact with CTLA-4 on pro-inflammatory T-cells resulting in disease amelioration [68]. Lending credence to the possibility that APL could induce regulatory T-cells is a recent report by Prakken and colleagues. These investigators performed mutational analyses of the arthritogenic HSP65 epitope 180–188, substituting a single alanine for a lysine residue at position 183. Importantly, nasal administration of rats with the altered peptide resulted in the prevention of AA and this was shown to be more efficacious than the native peptide. Furthermore, this prevention of arthritis was predominantly mediated by IL-10, although IL-4 and TGF-β were also produced [69]. These data indicate that minor alterations in the peptide sequence of HSP may have profound effects on the modulation of arthritis.

Comparable experiments have been performed using mycobacterial HSP70 in an attempt to suppress adjuvant arthritis. These studies have been successful, showing that either subcutaneous administration of whole HSP70 [32] or HSP70 peptide (comprising residues 234–252) [70] could protect rats from AA. The later study presents data suggesting that protection from disease is mediated by the production of IL-10, due to the fact that arthritis susceptibility was restored in rats upon treatment with a blocking IL-10 antibody. Furthermore, it has recently been shown that by using an alternative peptide from HSP70 (residues 276–290), that AA could be prevented by intranasal inhalation of the peptide, yet when this peptide was administered subcutaneously only minimal protection was observed [41]. The resulting HSP70-specific T-cells were shown to secrete considerable levels of IL-10 upon *in vitro* stimulation with not only the bacterial 234–252 peptide, but also the mammalian HSP70-peptide homologue [41].

Finally, unpublished work from our laboratory has suggested that self-HSP60 in the joints of mice with PIA could act as a target for regulatory T-cells. As previously stated, immunisation of mice with HSP65 or the immunodominant T-cell epitope (261–71) protects mice from the development of PIA and this protection is mediated by Th2-type cytokines [27, 40, 60]. Since the expression of self-HSP60 is up-regulated in the joints of mice with PIA we examined responses of T-cells derived from HSP65 and peptide (261–71) immunised mice to recombinant bacterial HSP65 or self-HSP60. From these studies it was clear that 261–71 primed T-cells produced high levels of regulatory cytokines, in particular IL-10, when stimulated with self-HSP60. These cells also showed an increased expression of CTLA-4, a marker of regulatory T-cells.

In conclusion, understanding the exquisite specificity and nature of the HSP-reactive cells and the factors controlling their evolution, along with their intricate role in immunological homeostasis, bodes well for the future design and development of novel vaccines and immunomodulators for the treatment of inflammatory diseases.

References

1 Janossy G, Panayi GS, Duke O, Bofill M, Poulter LW, Goldstein G (1981) Rheumatoid arthritis. A disease of T-lymphocyte/macrophage immunoregulation. *Lancet* 2: 839–842

2 Hang L, Theofilopoulus AN, Dixon FJ (1982) A spontaneous rheumatoid arthritis like disease in MRL/1 mice. *J Exp Med* 155: 1690

3 Nordling C, Karlsson-Parra A, Jansson L, Holmdalh R, Klareskog L (1992) Characterisation of a spontaneously occurring arthritis in male DBA/1 mice. *Arthritis Rheum* 35: 717–722

4 Potter M, Wax JS (1981) Genetics of susceptibility to pristane-induced plasmacytomas in BALB/cAn: Reduced susceptibility in BALB/cj with a brief description of pristane-induced arthritis. *J Immunol* 127: 1591–1595

5 Chang YH, Pearson CM, Abe C (1980) Adjuvant polyarthritis. IV. Induction by a synthetic adjuvant: Immunologic, histopathologic, and other studies. *Arthritis Rheum* 23: 62–71

6 Kleinau S, Erlandsson H, Holmdahl R, Klareskog L (1991) Adjuvant oils induce arthritis in the DA rat. I. Characterization of the disease and evidence for an immunological involvement. *J Autoimmunity* 4: 871–880

7 Holmdahl R, Goldschmidt TJ, Kleinau S, Kvick C, Jonsson R (1992) Arthritis induced

in rats with adjuvant oil is a genetically restricted, alpha beta T-cell dependant autoimmune disease. *Immunology* 76: 197
8 Carlson BC, Jansson AM, Larsson A, Bucht A, Lorentzen JC (2000) The endogenous adjuvant squalene can induce a chronic T-cell-mediated arthritis in rats. *Am J Pathol* 156: 2057–2065
9 Courtenay JS, Dallman MJ, Dayan AD, Martin A, Mosedale B (1980) Immunisation against heterologous type II collagen induces arthritis in mice. *Nature* 283: 666–8
10 Trentham DE, Townes AS, Kang AH (1977) Autoimmunity to collagen: an experimental model of arthritis. *J Exp Med* 146: 857
11 Cathcart ES, Hayes KC, Gonnerman WA, Lazzari AA, Franzblau C (1986) Experimental arthritis in a nonhuman primate. I. Induction of bovine type II collagen. *Lab Invest* 54: 26–31
12 Mikecez K, Glant TT, Poole AR (1987) Immunity to cartilage proteoglycans in BALB/c mice with progressive polyarthritis and ankylosing spondylytis induced by injection of human cartilage proteoglycan. *Arthritis Rheum* 30: 306
13 Pearson CM (1956) Development of arthritis, periarthritis and periostitis in rats given adjuvant. *Proc Soc Exp Biol Med* 91: 95–101
14 Kohashi O, Kutwata J, Umehara K, Uemura F, Takahashi T, Ozawa A (1979) Susceptibility to adjuvant-induced arthritis among germfree, specific-pathogen free, and conventional rats. *Infect Immun* 26: 791
15 Cromartie WJ, Craddock JG, Schwab JH, Anderle SK, Yang CH (1977) Arthritis in rats after systemic injection of streptococcal cells or cell walls. *J Exp Med* 146: 1585–1602
16 Aoki S, Ikuta K, Aoyma G (1972) Induction of chronic polyarthritis in rabbits. *Nature* 237: 168
17 Young RA (1990) Stress proteins and immunology. *Annu Rev Immunol* 8: 401
18 Lindquist S, Craig EA (1988) The heat shock proteins. *Annu Rev Genet* 22: 631–677
19 Jindal S, Dudani AK, Singh B, Harley CB, Gupta RS (1989) Primary structure of a human mitochondrial protein homologous to the bacterial and plant chaperonins and to the 65-kilodalton mycobacterial antigen. *Mol Cell Biol* 9: 2279
20 Kaufmann SH (1990) Heat shock proteins and the immune response. *Immunol Today* 11: 129–136
21 Holoshitz J, Naparstek Y, Ben-Nun A, Cohen IR (1983) Lines of T lymphocytes induce or vaccinate against autoimmune arthritis. *Science* 219: 56
22 Holoshitz J, Matitiau A, Cohen IR (1984) Arthritis induced in rats by clones of T lymphocytes responsive to mycobacteria but not to collagen type II. *J Clin Invest* 73: 211
23 Van Eden W, Thole J, van der Zee R, Zoordzij A, van Embden J, Hensen E, Cohen IR (1988) Cloning of the mycobacterial epitope recognized by T lymphocytes in adjuvant arthritis. *Nature* 331: 171–173
24 Billingham MEJ, Carney S, Butler R, Colston MJ (1990) A mycobacterial 65-kDa heat shock protein induces antigen-specific suppression of adjuvant arthritis, but is not itself arthritogenic. *J Exp Med* 171: 339–344
25 Hogervorst EJM, Wagenaar JPA, Boog CJP, van der Zee R, van Embden JDA, van Eden

W (1992) Adjuvant arthritis and immunity to the mycobacterial 65kDa heat shock protein. *Int Immunol* 4: 719

26 Van den Broek M, Hogervorst EJ, van Bruggen MC, van Eden W, van der Zee R, van den Berg WB (1989) Protection against streptococcal cell wall-induced arthritis by prevention with the 65-kD mycobacterial heat shock protein. *J Exp Med* 170: 449–466

27 Thompson SJ, Rook GAW, Brealey RJ, van der Zee R, Elson CJ (1990) Autoimmune reactions to heat-shock proteins in pristine-induced arthritis. *Eur J Immunol* 20: 2479

28 Ragno S, Colston M, Lowrie D, Winrow V, Blake D, Tascon R (1997) Protection of rats from adjuvant arthritis by immunisation with naked DNA encoding for mycobacterial heat shock protein 65. *Arthritis Rheum* 40: 277

29 Hogervorst EJM, Schouls L, Wagenaar JP, Boog CJ, Spaan WJ, van Embden JD, van Eden W (1991) Modulation of experimental autoimmunity treatment of adjuvant arthritis by immunisation with a recombinant vaccinia virus. *Infect Immunol* 59: 2029–2035

30 Quintana FJ, Carmi P, Mor F, Cohen IR (2002) Inhibition of adjuvant arthritis by a DNA vaccine encoding human heat shock protein 60. *J Immunol* 169: 3422–3428

31 Ragno S, Winrow VR, Mascagna P, Lucietto P, Di Pierro F, Morris CJ, Blake DR (1996) A synthetic 10-kD heat shock protein (hsp10) from *Mycobacterium tuberculosis* modulates adjuvant arthritis. *Clin Exp Immunol* 103:384–390

32 Kingston AE, Hicks CA, Colston MJ, Billingham ME (1996) A 71-kD heat shock protein (hsp) from *Mycobacterium tuberculosis* has modulatory effects on experimental rat arthritis. *Clin Exp Immunol* 103: 77–82

33 Bloemendal A, van der Zee R, Rutten VP, van Kooten PJ, Farine JC, van Eden W (1997) Experimental immunization with anti-rheumatic bacterial extract OM-89 induced T cell responses to heat shock protein (hsp)60 and hsp70; modulation of peripheral immunological tolerance as its possible mode of action in the treatment of rheumatoid arthritis (RA). *Clin Exp Immunol* 110: 72–78

34 Corrigall VM, Bodman-Smith MD, Fife MS, Canas B, Myers LK, Wooley P, Soh C, Staines NA, Pappin DJ, Berlo SE et al (2001) The human endoplasmic reticulum molecular chaperone BiP is an auto-antigen for rheumatoid arthritis and prevents the induction of experimental arthritis. *J Immunol* 166: 1492–1498

35 Elias D, Markovitis D, Rshef T, van der Zee, Cohen IR (1990) Induction and therapy of autoimmune diabetes in the non-obese diabetic (NOD/1t) mouse by a 65-kDa heat shock protein. *Proc Natl Acad Sci USA* 87: 1576

36 Birnbaum G, Kotilinek L, Schlievert P (1996) Heat shock proteins and experimental autoimmune encephalomyelitis (EAE). I. Immunization with a peptide of the myelin 2',3' cyclic nucleotide 3'phosphodiesterase that is cross-reactive with a heat shock protein alters the course of EAE. *J Neuroscience Res* 44: 381

37 Prakken BJ, Wendling U, van der Zee R, Rutten VP, Kuis W, van Eden W (2001) Induction of IL-10 and inhibition of experimental arthritis are specific features of microbial heat shock proteins that are absent for other evolutionarily conserved immunodominant proteins. *J Immunol* 167: 4147–4153

38 Anderton SM, van der Zee R, Noordzij A, van Eden W (1994) Differential mycobacterial 65 kDa heat shock protein T cell epitope recognition after adjuvant-inducing or protective immunization protocols. *J Immunol* 152: 3656
39 Prakken BJ, van der Zee R, Anderton SM, van Kooten PJ, Kuis W, van Eden W (1997) Peptide induced nasal tolerance for a mycobacterial heat shock protein 60 T cell epitope in rats suppresses both adjuvant arthritis and non-microbially induced experimental arthritis. *Proc Natl Acad Sci USA* 94: 3284–3289
40 Thompson SJ, Francis JN, Khai Siew L, Webb GR, Jenner PJ, Colston MJ, Elson CJ (1998) An immunodominant epitope from mycobacterial 65-kDa heat shock protein protects against pristane-induced arthritis. *J Immunol* 160: 4628–4634
41 Wendling U, Paul L, van der Zee R, Prakken B, Singh M, van Eden W (2000) A conserved mycobacterial heat shock protein (hsp) 70 sequence prevents adjuvant arthritis upon nasal administration and induces IL-10-producing T cells that cross-react with the mammalian self-hsp 70 homologue. *J Immunol* 164: 2711–2717
42 Francis JN, Lamont AG, Thompson SJ (2000) The route of adminstration of an immunodominant peptide derived from heat-shock protein 65 dramatically affects disease outcome in pristine-induced arthritis. *Immunol* 99: 338–344
43 Dejoy SQ, Ferguson KM, Sapp TM, Zabriskie JB, Oronsky AL, Kerwar SS (1989) Streptococcal cell wall arthritis. Passive transfer of disease with a T cell line and cross-reactivity of streptococcal cell wall antigens with *Mycobacterium tuberculosis*. *J Exp Med* 170: 369–382
44 Quinn A, Shinnick TM, Cunningham MW (1996) Anti-Hsp65 antibodies recognise M proteins of group A streptococci. *Infect Immunity* 64: 818–824
45 Thompson SJ, Elson CJ (1993) Susceptibility to pristane-induced arthritis is altered with changes in bowel flora. *Immunology Lett* 36: 227
46 Barker RN, Wells AD, Ghoraishian M, Easterfield AJ, Hitsumoto Y, Elson CJ, Thompson SJ (1996) Expression of mammalian 60-kD heat shock protein in the joints of mice with pristine-induced arthritis. *Clin Exp Immunol* 103: 83–88
47 Bedwell AE, Elson CJ, Hinton CE (1987) Immunological involvement in the pathogenisis of pristane-induced arthritis. *Scand J Immunol* 25: 393–398
48 Kohashi O, Kohashi Y, Takahashi T, Ozawa A, Shigematsu N (1985) Reverse effect of gram-positive bacteria vs. gram-negative arthritis in germfree rats. *Microbiol Immunol* 29: 487–497
49 Moudgil KD, Chang TT, Eradat H, Chen AM, Gupta RS, Brahn E, Sercarz EE (1997) Diversification of T cell responses to carboxy-terminal determinants within the 65–kD heat shock protein is involved in regulation of autoimmune arthritis. *J Exp Med* 185: 1307
50 Moudgil KD, Kim E, Yun OJ, Chi HH, Brahn E, Sercarz EE (2001) Environmental modulation of autoimmune arthritis involves the spontaneous microbial induction of T cell responses to regulatory determinants within heat shock protein 65. *J Immunol* 166: 4237–4243
51 Bjork J, Kleinau S, Midtvedt T, Klareskog L, Smedegard G (1994) Role of bowel flora

for development of immunity to hsp65 and arthritis in three experimental models. *Scand J Immunol* 40: 648–652
52 Karlson-Parra A, Soderstrom K, Ferm M, Ivanyi J, Kiessling R, Klareskog L (1990) Presence of human 65kD heat shock protein (hsp) in inflamed joints and subcutaneous nodules of RA patients: A clue to pathogenicity of anti-65kD hsp immunity. *Scand J Immunol* 31: 283
53 Schett G, Redlich K, Xu P, Bizan P, Goeger M, Tohidast-Akrad M, Kiener H, Smolen J, Steiner G (1998) Enhanced expression of heat shock protein 70 (hsp70) and heat shock factor 1 (HSF1) activation in rheumatoid arthritis synovial tissue. *J Clin Invest* 102: 302
54 Kleinau S, Soderstrom K, Kiessling R, Klareskog L (1991) A monoclonal antibody to the mycobacterial 65 kDa heat shock protein (ML 30) binds to cells in normal and arthritic joints of rats. *Scand J Immunol* 33: 195–202
55 Birk OS, Douek DC, Elias D, Takacs K, Dewchand H, Gur SL, Walker MD, van der Zee R, Cohen IR, Altman DM (1996) A role of Hsp60 in autoimmune diabetes: Analysis in a transgenic model. *Proc Natl Acad Sci USA* 93: 1032–1037
56 Anderton SM, van der Zee R, Goodacre JA (1993) Inflammation activates self hsp60-specific T cells. *Eur J Immunol* 23: 33–38
57 Matcht LM, Elson CJ, Kirwan JR, Gaston JSH, Lamont AG, Thompson JM, Thompson SJ (2000) Relationship between disease severity and responses by blood mononuclear cells from patients with rheumatoid arthritis to human heat-shock protein 60. *Immunol* 99: 208–214
58 Plebanski M, Saunders M, Burtles SS, Crowe S, Hooper DC (1992) Primary and secondary human *in vitro* T cell responses to soluble antigens are mediated by subsets bearing different CD45 isoforms. *Immunol* 75: 86
59 Ramage JM, Young JL, Goodall JC, Gaston JSH (1999) T cell responses to heat- shock protein 60. Differential responses by $CD4^+$ T cell subsets according to their expression of CD45 isotypes. *J Immunol* 162: 704
60 Beech JT, Khai Siew L, Ghoraishian M, Stasiuk LM, Elson CJ, Thompson SJ (1994) $CD4^+$ Th2 cells specific for mycobacterial 65-kilodalton heat shock protein protect against pristane induced arthritis. *J Immunol* 159: 3692–3697
61 Elson CJ, Barker RN, Thompson SJ, Williams NA (1995) Immunologically ignorant autoreactive T cells, epitope spreading and repertoire limitation. *Immunol Today* 16: 71–75
62 Drakesmith H, O'neil D, Schneider SC, Binks M, Medd P, Sercarz E, Beverley P, Chain B (1998) *In vivo* priming of T cells against cryptic determinants by dendritic cells exposed to interleukin 6 and native antigens. *Proc Natl Acad Sci USA* 25: 14903–14908
63 Opdennakker G, van Damme J (1994) Cytokine-regulated proteases in autoimmune diseases. *Immunol Today* 15: 103
64 Siew LK, Beech JT, Thompson SJ, Elson CJ (1998) Effect of T-helper cytokine environment on specificity of T-cell responses to mycobacterial 65,000 MW heat-shock protein. *Immunology* 93: 493–497
65 Ulmansky R, Cohen C, Szafer F, Moallem E, Fridlender Z, Kashi Y, Naparstek Y (2002)

Resistance to adjuvant arthritis is due to protective antibodies against heat shock protein surface epitopes and the induction of IL-10 secretion. *J Immunol* 168: 6463–6469
66 Anderton SM, van der Zee R, Prakken B, Noordzij A, Van Eden W (1995) Activation of T cells recognizing self 60-kD heat shock protein can protect against experimental arthritis. *J Exp Med* 181: 943
67 Paul AG, van Kooten PJ, van Eden W, van der Zee R (2000) Highly autoproliferative T cells specific for 60-kDa heat shock protein produce IL-4/IL-10 and IFN-γ and are protective in adjuvant arthritis. *J Immunol* 165: 7270–7277
68 Paul AG, van der Zee R, Taam LS, van Eden W (2000) A self-hsp60 peptide acts as a partial agonist inducing expression of B7-2 on mycobacterial hsp60-specific T cells: A possible mechanism for inhibitory T cell regulation of adjuvant arthritis? *Int Immunol* 12: 1041–1050
69 Prakken BJ, Roord S, van Kooten PJ, Wagenaar JP, van Eden W, Albani S, Wauben MH (2002) Inhibition of adjuvant-induced arthritis by interleukin-10-driven regulatory cells induced *via* nasal administration of a peptide analog of an arthritis-related heat-shock protein 60 T cell epitope. *Arthritis Rheum* 46: 1937–1946
70 Tanaka S, Kimura Y, Mitani A, Yamamoto G, Nishimura H, Spallek R, Singh M, Noguchi T, Yoshikai Y (1999) Activation of T cells recognizing an epitope of heat-shock protein 70 can protect against rat adjuvant arthritis. *J Immunol* 163: 5560–5565

Heat shock proteins and reactive arthritis

J.S. Hill Gaston, Richard C. Duggleby, Jane C. Goodall, Roberto Raggiaschi, and Mark S. Lillicrap

Department of Medicine, University of Cambridge School of Clinical Medicine, Addenbrooke's Hospital, Hills Road, Cambridge CB2 2QQ, UK

Introduction

Reactive arthritis (ReA) is unusual amongst inflammatory joint diseases because the cause of arthritis has been identified, namely preceding infection with certain bacteria, whereas the aetiology of rheumatoid arthritis or ankylosing spondylitis, or at least the triggering event, remains unknown. Most subjects infected by *Chlamydia*, *Salmonella*, *Campylobacter*, *Yersinia* or *Shigella* develop symptoms due to the effects of the infection in the genito-urinary (*Chlamydia*) or gastro-intestinal tracts, but resolve the infection without any long-term sequelae. A minority develop inflamed joints, and also inflammation at the sites of insertion of tendons and ligaments – a process termed enthesitis [1] – but the factors which confer susceptibility to arthritis have not been completely identified. HLA alleles play a part with HLA-B27$^+$ individuals being more likely to develop reactive arthritis, and to have more severe and long-lasting disease – indeed a proportion develop chronic arthritis, and these are nearly all HLA-B27$^+$ [2, 3]. Other HLA alleles including HLA-DRB1*0103 have also been implicated [4]. The associations between disease incidence and/or severity with HLA imply that T-cell mediated immune responses are likely to be central to the pathogenesis of ReA, and that immune responses in subjects who develop arthritis will differ in some way from those who have uncomplicated infection. It is in this context that immune responses to heat shock proteins (HSP) have been studied in ReA.

Immune responses to bacteria which can cause ReA

There is evidence that T-cell-mediated immunity is essential for satisfactory resolution of infection by ReA-associated bacteria. Much of this is from animal studies, but there are data from human subjects as well. For infection with *Chlamydia trachomatis* (Ct) murine studies have shown that the most important immune response is by CD4$^+$ T-cells which make interferon (IFN)-γ (i.e., the TH1 subset of helper T-cells). This has been established using gene targeted mice deficient in CD4, or IFN-γ or the IFN-γ receptor [5–7]. Protection can be obtained by transfer of organism-

Heat Shock Proteins and Inflammation, edited by Willem van Eden

specific CD8+ T-cells when these make IFN-γ [8], but there is little evidence for a primary role for CD8+ cytolytic T-cells in the murine response to infection. Indeed β_2 microglobulin deficient mice, which lack CD8+ T-cells, clear infection satisfactorily [9]. Likewise, in murine *Salmonella* and *Yersinia* infections, it is CD4+ T-cells which provide protection [10–14]. Interestingly, amongst the T-cells shown to be protective are some which recognize *Yersinia* HSP60 [15]. In humans the evidence is more indirect but individuals who are unable to generate IFN-γ-producing TH1 T-cells through a deficiency in IL-12 or its receptor are particularly susceptible to disseminated *Salmonella* infection [16, 17]. In *Chlamydia* infection, Ct-specific T-cells can be isolated from inflamed upper genital tract tissue, and have been found to make IL-5 or IL-10, suggesting that an inadequate TH1 response to Ct may have been responsible for ascending infection and tubal damage leading to infertility [18, 19]. In summary, infection with ReA-associated organisms normally elicits a vigorous CD4+ TH1 T-cell response, part of which can be directed at HSP, and this is required for successful control of infection.

A current view of the pathogenesis of reactive arthritis

Three kinds of observation underlie current views of the pathogenesis of ReA. Firstly, the organisms seem to persist irrespective of treatment with antibiotics; the first evidence for this was the observation of persistent high titres of antibodies to the organisms in ReA, including organism-specific secretory IgA which turns over rapidly [20, 21]. Additional evidence for persistence stems from the second observation, namely that the organism can be detected in affected joints. The fact that bacteria-specific antigens or nucleic acids can be shown many months or even years after the triggering infection implies persistence, although the precise site where this occurs is not known and is probably not simply the joint. Bacterial antigens – proteins, LPS, cell walls – have been demonstrated by immunofluorescence or immunoblotting with organism-specific mono- or polyclonal antisera [22–24], whilst PCR and RT-PCR have been used to demonstrate nucleic acids; the presence of short-lived RNA has been taken to imply active transcription by the organism whilst in the joint. *Chlamydia* DNA/RNA has been reported in ReA joints by many investigators [25–28] whereas searches for DNA/RNA from enteric pathogens have rarely been positive [29, 30]. Recently however rRNA was clearly identified in synovial fluid from a patient with ReA secondary to *Yersinia pseudotuberculosis* some 19 months after the initial infection [31].

The third observation is that the joint contains organism-specific T-cells, with many reports of CD4+ T-cells, but also some of CD8+ T-cells. The T-cell population in the affected joint is activated in terms of phenotype, and produces pro-inflammatory cytokines spontaneously and even more so following *in vitro* challenge with the organism responsible for triggering the infection [32–34]. Again, among the antigens which are recognized by organism-specific T-cells in the joint, HSP60 is particularly prominent – indeed a surprisingly high proportion of all the studies of immune responses to HSP60 in humans have been carried out in the context of arthritis. T-cell clones specific for *Chlamydia* HSP60 ([35] and for *Yersinia* HSP60 [36, 37] have been characterized in ReA patients, and the epitopes which they recognize have been mapped – with implications which are discussed below. In addition, much of the animal work on T-cell responses to HSP60 and HSP70 has been done in adjuvant arthritis, as discussed elsewhere in this volume. It is reasonable to regard adjuvant arthritis as a model of ReA – challenge with bacterial antigens results in an arthritis which is generally self-limiting. In humans an ReA-like illness has been described following intra-vesical challenge with *M. bovis* BCG as part of the therapy for bladder tumours, and mycobacterial HSP60-specific T-cells were readily detectable in the synovial fluid in such a case [38].

In summary, the joint affected by reactive arthritis contains antigens from disease-triggering pathogens and, on occasion, transcriptionally-active bacteria which may synthesize antigens locally, along with organism-specific T lymphocytes amongst which those directed at bacterial heat shock proteins, especially HSP60, are particularly prominent. It follows therefore that joint inflammation could be driven by T-cell recognition of bacterial antigens in the affected joint. In this case persistence of arthritis would be related to the availability of bacterial antigens, so that successful eradication of infection and/or clearance of bacterial antigens would limit its duration. This model might well account for mild cases of ReA with limited joint involvement which lasts less than one month – such cases are in fact rather common in community surveys, in which it has been possible to follow up complete cohorts exposed to infection by *Salmonella* or *Campylobacter* through contaminated food [39]. It is less successful in accounting for longer lasting ReA, particularly those cases which fail to resolve, and where the influence of HLA-B27 is most evident. In these circumstances it is tempting to speculate that additional factors are at work in pathogenesis. These could include:

- Priming of responses to the ReA-associated bacterium by previous encounter with other pathogens or commensal bacteria;
- Failure to eliminate the organism due to an inadequate immune response;
- Sustained activation of bacteria-specific immune responses by antigens derived from commensal bacteria;
- Transformation of an immune response initially directed against a bacterial antigen into one directed against self antigens in the joint or enthesis;
- Failure to regulate the anti-bacterial immune response.

Immune responses to heat shock proteins may be relevant to all of these outcomes which might predispose to the development of arthritis, and a detailed study of responses (mainly those to HSP60) has produced provocative data.

Table 1 - Epitopes recognized by HSP60-specific T-cell clones from: (a) a patient with Chlamydia trachomatis *(ct)-induced ReA (SF derived clones); (b) a patient with* Yersinia enterocolitica *(yer)-induced ReA (SF-derived clones); (c) a normal individual (PB-derived clones). In each case the sequence of the mapped epitope is given and compared with the equivalent sequence in related organisms (Other abbreviations: cp,* C. pneumoniae; *ec,* E. coli, *salm,* Salmonella typhi*).*

Epitope	Species
(a) Ct HSP60-sepcific clones	
epitope 1	
G R H V V I D K S F G S P Q V T	ct
* * * * * * * * * * * * * * * *	cp
* * N * * L * * * * * A * T I *	ec and salm
* * T * I * E Q * W * * * K * *	human
epitope 2	
K V V V D Q I R K I S K P V Q H	ct
* * * * * E L K * * * * * * * *	cp
T A A * E E L K A L * V * C S D	ec (and salm A*****....)
D A * I A E L K * Q * * * * T T	human
(b) Yersinia HSP60-specific clones	
epitope 1	
R V V I N K D T T I I I	yer
* * * * * * * * * T * *	ec and salm
P H D L G * V G E V * V	human
epitope 2	
S P L R Q I V V N A G E E A S V I A N N V K A G	yer
A * * * * * * L * C * * * P * * V * * T * * G *	ec and salm
I * A M T * A K * * * V * G * L * V E K I M Q S	human
epitope 3	
I K V G A A T E V E M K E K K A R V E D A L H A	yer
* *	ec and salm
* R * * * * * * I * * * * * * D * * D * * Q * *	ct
L * * * G T S D * * V N * * * D * * T * * * N *	human

Table 1 - continued

Epitope	Species
(c) E. coli *HSP60 (GroEL)-specific clones*	
D A R V K M L R G V N V L A D A	ec and salm
* * * I * * * * * * * * I * * * *	yer
E * * K * I Q K * * K T * * E *	ct
* * * *A L* * * *Q* * * *D L* * * * *	*human*

Properties of HSP-reactive T-cells in ReA

The majority of this work concerns recognition of ReA-associated bacterial antigens by $CD4^+$ T-cells, but there is additional evidence for organism-specific $CD8^+$ T-cells. In both cases HSP60 has been implicated as a target antigen. Detailed characterization of HSP60-specific T-cell clones from ReA patients has been carried out for both *Yersinia* and *Chlamydia* HSP60, but necessarily these studies have involved very small numbers of patients though polyclonal responses to HSP60 have been clearly demonstrated in much larger numbers of patients. Two conclusions can be drawn from this work.

1. Epitopes recognized by organism-specific T-cells are often conserved in other bacteria.

Firstly, the epitopes which have been mapped in HSP60 in both *Yersinia* and *Chlamydia trachomatis* are conserved in other pathogens likely to have been encountered by ReA patients. For example, as shown in Table 1 (a), the DR4-restricted epitope (epitope 1) in CtHSP60 is identical in HSP60 from *Chlamydia pneumoniae* (Cpn), an organism which infects a significant proportion of the population, establishes a persistent infection, and has been implicated in atherosclerosis [40]. We have recently mapped a second epitope (epitope 2), restricted by HLA-DP, and again this is almost completely conserved in *C. pneumoniae* and the T-cell clone cross-reacts strongly. Whilst HSP60's sequence conservation makes this kind of cross-reactivity likely, it is not confined to HSP60. We have recently mapped an epitope in the *Chlamydia* 60 kD membrane protein OMP2 which is conserved in Ct and Cpn, and also a conserved epitope in Ct enolase. When the sequences of the two HSP60 epitopes were compared with the equivalent regions in *E. coli* HSP60 (GroEL), epitope 2 showed no sequence conservation but epitope 1 was highly conserved. We were able to show that, even when the three residues in the core of the epitope which differ in Ct and *E. coli* (H/N, I/L and S/A) were changed to those in

E. coli, this peptide could still be recognized, albeit with much lower affinity. In addition experiments with alanine substituted peptides showed that the four critical residues for T-cell recognition (KSFG) were wholly conserved in *E. coli*. Thus, even encounter with *E. coli* might conceivably prime a response to this Ct HSP60 epitope. Conversely, *C. pneumoniae* infection might prime responses to *Salmonella* HSP60 since the sequence of the epitope in *Salmonella* is identical to that found in *E. coli*.

Cross-reactivity with *E. coli* is even more relevant when considering T-cell responses to HSP60 from the enteric pathogens *Yersinia* and *Salmonella*. Interestingly, for each of three epitopes in *Yersinia enterocolitica* HSP60 defined by Mertz and co-workers, there was complete or very strong conservation of sequence in *E. coli* HSP60 (Tab. 1b). In one case there was also substantial sequence conservation of the epitope in Ct HSP60, and indeed the clone cross-reacted with Ct HSP60, i.e., a mirror image of the situation described above for Ct HSP60 epitope 1.

E. coli is a gut commensal, and whilst it can cause clinical infection particularly urinary tract infection and even ReA [41], we have obtained evidence that there is T-cell reactivity to *E. coli* HSP60 even in normal individuals with no history of infection. The epitope, shown in Table 1 (c), was mapped to amino acids 11–26, again a sequence highly conserved in *Yersinia* and *Salmonella*.

For B27-restricted $CD8^+$ T-cells in ReA, an epitope has been mapped in *Yersinia* HSP60; its sequence is completely conserved in *E. coli* but not in Ct or human HSP60 [42]. In addition an HLA-A2 restricted epitope has recently been mapped in mycobacterial HSP60 and is highly conserved in *E. coli*, *Yersinia* and *Chlamydia* HSP60 [43]. Although priming a Class I MHC-restricted response by prior infection would usually require the infection to be intracellular, the same principle applies i.e., that prior experience with pathogens or commensals could influence responses to ReA-associated bacteria.

2. T-cells which recognize epitopes in bacterial HSP do not generally cross-react with self.

There is clear support from experimental models for the idea that HSP-reactive T-cells, including those which recognize self HSP60, can produce inflammatory pathology, and both $CD4^+$ and $CD8^+$ T-cells have been implicated [44, 45]. However, despite the attractive hypothesis that in ReA bacterial HSP60-specific T-cells cross-react with self HSP60 and produce arthritis, this idea has not been borne out experimentally. In none of the cases shown in Table 1 has cross-reactivity with self HSP60 been demonstrated – certainly not cross-reactivity defined as inducing proliferative responses (see below for discussion of other possible forms of cross-reactivity). In the case of the GroEL-specific epitope, the homology with the human sequence is quite striking, and we have investigated this in some detail (Lillicrap et al., manuscript in preparation). The frequent low level contamination of recombinant preparations of full-length human HSP60 protein with GroEL combined with the exquisite sensitivity of the T-cell clones made it difficult to exclude cross-reactivity. However, human HSP60-derived peptides were not convincingly stimulatory, and having mapped the critical residues in the GroEL peptide by means of alanine replacements, it is now clear that three of the five amino acid differences between the *E. coli* and human sequences involve residues which are critical for T-cell responses. This explains the lack of a proliferative response to the human HSP60-derived peptide. There are examples in the literature of bacterial HSP60-specific clones which do appear to cross-react with self HSP60 [36, 46], but they may be the exception, and their relevance to disease has not been established. In the report from Quayle and colleagues, a peptide was mapped in mycobacterial HSP60, but the sequence of the epitope is identical in GroEL, and highly conserved in Ct and human HSP60. Whether the clone could respond to intact human HSP60 was not tested. In the report by Hermann et al., the evidence for responses to self HSP60 was based on the response to heat-shocked antigen presenting cells and the precise epitope was not mapped.

The significance of cross-reactivity of HSP-specific T-cells in ReA

Priming responses to ReA-triggering infection

The cross-reactive nature of the response to ReA-associated bacteria, which has been clearly established, has implications for our understanding of the pathogenesis of ReA. Thus many patients infected for the first time with, for example, *Chlamydia* or *Yersinia*, may already have T-cells which are primed to recognize epitopes in HSP60 or other antigens of the new pathogen, because of previous infection. This might have a profound effect on the nature of the immune response which is generated, and the cytokines which result. For instance, T-cells primed by encounter with *C. pneumoniae* or *E. coli* might rapidly generate large quantities of pro-inflammatory cytokines and drive inflammation following infection with *Chlamydia* or *Yersinia*. However, there is an alternative possibility; since *C. pneumoniae* establishes persistent infection, and *E. coli* is normally encountered in the gut as a commensal rather that an invasive pathogen, the responses to cross-reactive epitopes may be down-regulated or modulated towards the production of immunoregulatory (e.g., IL-10, TGF-β) rather than pro-inflammatory (IFN-γ, TNF-α) cytokines. This could result in a failure to mount a sufficiently robust inflammatory response to the pathogen, particularly adequate levels of IFN-γ, and could contribute to pathogen persistence. In *Chlamydia* infection an inadequate response to HSP60 has been correlated with pelvic inflammatory disease, i.e., persistence and dissemination of the organism [47], but similar findings in ReA have not been reported. Some studies of synovial T-cells have reported on a higher proportion of IL-4 producing cells in ReA as compared to RA, suggesting a failure to generate an adequate TH1 response to infection [48]. *Yersinia* HSP60-specific T-cell clones which make IL-10 and IL-4 have also been described, but the predominant cytokines found in these studies were IFN-γ and TNF-α [37, 49].

Amplification of immune responses to ReA-associated organisms by traffic of commensal bacteria to the joint

Bacteria can be detected with great sensitivity by PCR or RT-PCR using "universal primers" which amplify 16S rRNA from all bacteria. This exploits the conservation of sequence seen in all bacterial 16S rRNA genes; the identity of the bacteria whose rRNA has been amplified can be obtained by sequencing the product to reveal sequences within the 16SrRNA product which are species-specific. This technique has been applied to synovium from inflamed joints, including ReA. The general point which has emerged from several studies is that many commensal bacteria from skin and gut can be detected within the joint both by PCR and RT-PCR [50, 51]. These most likely traffic into synovium within macrophages or polymorphs which are continuously recruited to the joint. This means that organism-specific T cells which recognize conserved epitopes may be stimulated by antigens from commensal bacteria such as *E. coli* – commonly detected in inflamed joints. This would amplify and maintain an immune response initially directed against the triggering infection, with potential for a vicious circle: – immune response to triggering organism – inflammation – recruitment of bacteria laden macrophages – cross-reactive secondary immune response – further inflammation.

The ability of conserved peptides to act as altered peptide ligands

In considering cross-reactivity, we have considered examples where peptides from one bacterium stimulate responses which are qualitatively similar to those from another – T-cell proliferation or cytokine production. However, for conserved epitopes, there is another possibility because peptides which differ by a few amino acids can stimulate qualitatively different responses from a T-cell. The peptide which elicits a different or partial response from that of the cognate peptide is termed an "altered peptide ligand" and can modulate immune responses [52]. The altered response reflects changes in the amino acids which contact the T-cell receptor and therefore affect the degree of activation through the T-cell receptor. In some cases APL can antagonize or anergise T-cell responses to cognate peptide. When considering conserved proteins, there is the possibility that the amino acids which differ in proteins from different bacterial species, or in the human homologue, may act as altered peptide ligands. This has not been investigated thoroughly in ReA and warrants further attention.

Table 2 - Comparison of the epitopes in HSP60 recognized by the regulatory cells described in [53] – induced by immunization with the mycobacterial peptide (myco) and cross-reactive with rat HSP60. The sequence conservation of the epitope in HSP60 from ReA-associated bacteria is also shown. Abbreviations are as in the legend to Table 1.

Epitope	Species
A L S T L V VN K I	myco
* * * * * * L * R L	human/rat
* * A * * * * * T M	ec/salm/yers
* * A * * * * * R *	ct/cp

There is evidence for this kind of mechanism in studies of adjuvant arthritis, reviewed in more detail elsewhere. In summary, priming rats with a mycobacterial HSP60 peptide from a conserved region resulted in two populations of T-cells [53]. One, which recognized the N-terminal 10 amino acids, responded to autologous antigen presenting cells in the absence of antigen – a response enhanced by heat shock and interpreted as presentation of endogenous (rat) HSP60. This T-cell line showed some ability to ameliorate adjuvant arthritis, and produced both IL-4 and IL-10, in addition to some IFN-γ. Oral feeding with mycobacterial HSP60 also results in IL-10 production by T-cells and protection from arthritis [54]. These results are relevant to our present discussion since the sequence of the epitope recognized in rat HSP60 is identical in human HSP60, whilst the epitope is also highly conserved in *Mycobacteria*, *E. coli* and *Chlamydia* HSP60 (Tab. 2). Thus, whilst experimentally the protective T-cells were induced by immunization with a mycobacterial HSP60 peptide, the same process can readily be imagined as part of the response *in vivo* to infection with other bacteria or commensals. An example of commensal bacteria priming a protective HSP60-specific immune response is seen in the Fisher rat which fails to develop adjuvant arthritis unless maintained under germ-free conditions i.e., protection is effected by commensal bacteria, and there is evidence in this system that a HSP60-reactive T-cell population confers the protection [55].

Therefore the outcome of infection with ReA-associated organisms will depend to some extent, as far as inflammation is concerned, on the efficiency with which regulatory HSP60-reactive cells are generated. This in turn will depend on the dominant epitopes, and whether these are in conserved regions, since the regulatory T-cell population responds to the self protein. The precise epitopes identified in rat are not likely to be relevant to humans since they require rat Class II MHC molecules for their presentation, but there are many peptides in the HSP60 sequence recognized by human T-cells which are equivalent to those described in rat, e.g., the peptide noted above in Table 1(c). The influence of Class II HLA alleles on ReA which has been observed [4] might reflect epitope selection and the generation of pro-inflammatory or protective responses.

Although the evidence for HSP60-specifc regulatory T-cells has been reviewed, similar mechanisms have been demonstrated for HSP70 [56], and could in principle apply to other conserved antigens such as enolase, shown to be a target antigen in *Chlamydia*-induced ReA [57], although experiments in adjuvant arthritis make it clear that not all conserved antigens are associated with generation of regulatory T-cells [58].

Conclusions

Immune responses to heat shock proteins are an important component of the way in which the body deals with infection. HSP are amongst the most prominent target antigens in the response, and their properties of conservation of sequence, immunogenicity, and ability to generate immunoregulatory T-cell population, mean that there are several ways in which immune responses to HSP can influence the occurrence of reactive arthritis, which is essentially a form of aberrant response to infection. Many of the possibilities discussed above are still speculative or rest on insufficient experimental data, but the role of immune responses to HSP in ReA will be clarified by continuing clinical and experimental investigations.

References

1 McGonagle D, Gibbon W, Emery P (1998) Classification of inflammatory arthritis by enthesitis. *Lancet* 352: 1137–1140

2 Leirisalo M, Skylv G, Kousa M, Voipio-Pulkki LM, Suoranta H, Nissila M, Hvidman L, Nielsen ED, Svejgaard A, Tilikainen A et al (1982) Followup study on patients with Reiter's disease and reactive arthritis, with special reference to HLA-B27. *Arthritis Rheum* 25: 249–259

3 Ekman P, Kirveskari J, Granfors K (2000) Modification of disease outcome in Salmonella-infected patients by HLA-B27. *Arthritis Rheum* 43: 1527–1534

4 Orchard TR, Thiyagaraja S, Welsh KI, Wordsworth BP, Gaston JSH, Jewell DP (2000) Clinical phenotype is related to HLA genotype in the peripheral arthropathies of inflammatory bowel disease. *Gastroenterology* 118: 274–278

5 Su H, Caldwell HD (1995) CD4(+) T cell play a significant role in adoptive immunity to *Chlamydia trachomatis* infection of the mouse genital tract. *Infect Immun* 63: 3302–3308

6 Cotter TW, Ramsey KH, Miranpuri GS, Poulsen CE, Byrne GI (1997) Dissemination of

Chlamydia trachomatis chronic genital tract infection in gamma interferon gene knockout mice. *Infect Immun* 65: 2145–2152

7 Johansson M, Schon K, Ward M, Lycke N (1997) Genital tract infection with *Chlamydia trachomatis* fails to induce protective immunity in gamma interferon receptor- deficient mice despite a strong local immunoglobulin A response. *Infect Immun* 65: 1032–1044

8 Lampe MF, Wilson CB, Bevan MJ, Starnbach MN (1998) Gamma interferon production by cytotoxic T lymphocytes is required for resolution of *Chlamydia trachomatis* infection. *Infect Immun* 66: 5457–5461

9 Magee DM, Williams DM, Smith JG, Bleicker CA, Grubbs BG, Schachter J, Rank RG (1995) Role of CD8 T cells in primary *Chlamydia* infection. *Infect Immun* 63: 516–521

10 Kempf VA, Bohn E, Noll A, Bielfeldt C, Autenrieth IB (1998) *In vivo* tracking and protective properties of *Yersinia*-specific intestinal T cells. *Clin Exp Immunol* 113: 429–437

11 Autenrieth I, Tingle A, Reske KA, Heesemann J (1992) T lymphocytes mediate protection against *Yersinia enterocolitica* in mice: Characterization of murine T cell clones specific for *Y. enterocolitica*. *Infect Immun* 60: 1140–1149

12 Noll A, Roggenkamp A, Heesemann J, Autenrieth IB (1994) Protective role for heat shock protein-reactive alpha beta T cells in murine yersiniosis. *Infect Immun* 62: 2784–2791

13 Hess J, Ladel C, Miko D, Kaufmann SHE (1996) Salmonella typhimurium aroA(–) infection in gene-targeted immunodeficient mice – Major role of CD4(+) TCR-alpha beta cells and IFN-gamma in bacterial clearance independent of intracellular location. *J Immunol* 156: 3321–3326

14 McSorley SJ, Asch S, Costalonga M, Reinhardt RL, Jenkins MK (2002) Tracking salmonella-specific CD4 T cells *in vivo* reveals a local mucosal response to a disseminated infection. *Immunity* 16: 365–377

15 Noll A, Autenrieth IB (1996) Yersinia-hsp60-reactive T cells are efficiently stimulated by peptides of 12 and 13 amino acid residues in a MHC class II (I-A(b))-restricted manner. *Clin Exp Immunol* 105: 231–237

16 Dejong R, Altare F, Haagen IA, Elferink DG, Deboer T, Vriesman P, Kabel PJ, Draaisma JMT, Vandissel JT, Kroon FP et al (1998) Severe mycobacterial and *Salmonella* infections in interleukin-12 receptor-deficient patients. *Science* 280: 1435–1438

17 Picard C, Fieschi C, Altare F, Al-Jumaah S, Al-Hajjar S, Feinberg J, Dupuis S, Soudais C, Al-Mohsen IZ, Genin E et al (2002) Inherited interleukin-12 deficiency: IL-12B genotype and clinical phenotype of 13 patients from six kindreds. *Am J Hum Genet* 70: 336–348

18 Kinnunen A, Molander P, Laurila A, Rantala I, Morrison R, Lehtinen M, Karttunen R, Tiitinen A, Paavonen J, Surcel HM (2000) *Chlamydia trachomatis* reactive T lymphocytes from upper genital tract tissue specimens. *Hum Reprod* 15: 1484–1489

19 Kinnunen A, Molander P, Morrison R, Lehtinen M, Karttunen R, Tiitinen A, Paavonen J, Surcel HM (2002) Chlamydial heat shock protein 60-specific T cells in inflamed salpingeal tissue. *Fert Steril* 77: 162–166

20 Granfors K, Toivanen A (1986) IgA-anti-*Yersinia* antibodies in *Yersinia*-triggered reactive arthritis. *Ann Rheum Dis* 45: 561–565
21 Granfors K, Viljanen M, Tiilikainen A, Toivanen A (1980) Persistence of IgM, IgG, and IgA class *Yersinia* antibodies in *Yersinia* arthritis. *J Infect Dis* 141: 424–429
22 Granfors K, Jalkanen S, von Essen R, Lahesmaa-Rantala R, Isomaki O, Pekkola-Heino K, Merilahti-Palo R, Saario R, Isomaki H, Toivanen A (1989) *Yersinia* antigens in synovial fluid cells from patients with reactive arthritis. *N Engl J Med* 320: 216–221
23 Granfors K, Jalkanen S, Lindberg AA, Maki-Ikola O, von Essen R, Lahesmaa-Rantala R, Isomaki H, Saario R, Arnold WJ, Toivanen A (1990) *Salmonella* lipopolysaccharide in synovial cells from patients with reactive arthritis. *Lancet* 335: 685–688
24 Granfors K, Jalkanen S, Toivanen P, Lindberg A (1992) Bacterial lipopolysaccharide in synovial fluid cells in Shigella triggered reactive arthritis. *J Rheumatol* 19: 500
25 Gerard HC, Branigan PJ, Schumacher HR, Hudson AP (1998) Synovial *Chlamydia trachomatis* in patients with reactive arthritis/ Reiter's syndrome are viable but show aberrant gene expression. *J Rheumatol* 25: 734–742
26 Freise J, Gerard HC, Bunke T, WhittumHudson JA, Zeidler H, Kohler L, Hudson AP, Kuipers JG (2001) Optimised sample DNA preparation for detection of *Chlamydia trachomatis* in synovial tissue by polymerase chain reaction and ligase chain reaction. *Ann Rheum Dis* 60: 140–145
27 Schnarr S, Putschky N, Jendro MC, Zeidler H, Hammer M, Kuipers JG, Wollenhaupt J (2001) *Chlamydia* and *Borrelia* DNA in synovial fluid of patients with early undifferentiated oligoarthritis – Results of a prospective study. *Arthritis Rheum* 44: 2679–2685
28 Keat A, Thomas B, Dixey J, Osborn M, Sonnex C, Taylor-Robinson D (1987) *Chlamydia trachomatis* and reactive arthritis: the missing link. *Lancet* 1: 72–74
29 Nikkari S, Merilahti-Palo R, Saario R, Soderstrom K-O, Granfors K, Skurnik M, Toivanen P (1992) *Yersinia* triggered reactive arthritis. Use of polymerase chain reaction and immunocytochemical staining in the detection of bacterial components from synovial specimens. *Arthritis Rheum* 35: 682–687
30 Nikkari S, Rantakokko K, Ekman P, Mottonen T, Leirisalo-Repo M, Virtala M, Lehtonen L, Jalava J, Kotilainen P, Granfors K et al (1999) *Salmonella*-triggered reactive arthritis: Use of polymerase chain reaction, immunocytochemical staining, and gas chromatography-mass spectrometry in the detection of bacterial components from synovial fluid. *Arthritis Rheum* 42: 84–89
31 Gaston JSH, Cox C, Granfors K (1999) Clinical and experimental evidence for persistent *Yersinia* infection in reactive arthritis. *Arthritis Rheum* 42: 2239–2242
32 Gaston JSH, Life PF, MerilahtiPalo R, Bailey L, Consalvey S, Toivanen A, Bacon PA (1989) Synovial T lymphocyte recognition of organisms that trigger reactive arthritis. *Clin Exp Immunol* 76: 348–353
33 Sieper J, Braun J, Wu P, Kingsley G (1993) T-cells are responsible for the enhanced synovial cellular immune response to triggering antigen in reactive arthritis. *Clin Exp Immunol* 91: 96–102
34 Hermann E, Yu DTY, Meyer zum Buschenfelde KH, Fleischer B (1993) HLA-B27-

restricted CD8 T-cells derived from synovial fluids of patients with reactive arthritis and ankylosing spondylitis. *Lancet* 342: 646–650

35 Deane K, Jecock R, Pearce J, Gaston J (1997) Identification and characterization of a DR4-restricted T cell epitope within *chlamydia* hsp60. *Clin Exp Immunol* 109: 439–445

36 Hermann E, Lohse AW, vanderZee R, vanEden W, Mayet WJ, Probst P, Poralla T, MeyerzumBuschenfelde KH, Fleischer B (1991) Synovial fluid derived *Yersinia* reactive T cells responding to human 65 kDa heat shock protein and heat stressed antigen presenting cells. *Eur J Immunol* 21: 2139–2143

37 Mertz AK, Wu P, Sturniolo T, Stoll D, Rudwaleit M, Lauster R, Braun J, Sieper J (2000) Multispecific $CD4^+$ T cell response to a single 12-mer epitope of the immunodominant heat-shock protein 60 of *Yersinia enterocolitica* in *Yersinia*-triggered reactive arthritis: Overlap with the B27-restricted CD8 epitope, functional properties, and epitope presentation by multiple DR alleles. *J Immunol* 164: 1529–1537

38 Gaston J, Young J, Ramsey S, Scott D (1996) Specificity of synovial fluid T cells in iatrogenic human adjuvant arthritis. *Arthritis Rheum* 39: S149

39 Hannu T, Mattila L, Siitonen A, LeirisaloRepo M (2002) Reactive arthritis following an outbreak of *Salmonella typhimurium* phage type 193 infection. *Ann Rheum Dis* 61: 264–266

40 Saikku P (2000) Epidemiologic association of *Chlamydia pneumoniae* and atherosclerosis: The initial serologic observation and more. *J Infec Dis* 181: S411–S413

41 Laasila K, Leirisalorepo M (1999) Recurrent reactive arthritis associated with urinary tract infection by *Escherichia coli*. *J Rheumatol* 26: 2277–2279

42 Ugrinovic S, Mertz A, Wu PH, Braun J, Sieper J (1997) A single nonamer from the *Yersinia* 60-kDa heat shock protein is the target of HLA-B27-restricted CTL response in *Yersinia*-induced reactive arthritis. *J Immunol* 159: 5715–5723

43 Charo J, Geluk A, Sundback M, Mirzai B, Diehl AD, Malmberg KJ, Achour A, Huriguchi S, van Meijgaarden KE, Drijfhout JW et al (2001) The identification of a common pathogen-specific HLA class I A*0201-restricted cytotoxic T cell epitope encoded within the heat shock protein 65. *Eur J Immunol* 31: 3602–3611

44 vanEden W, Thole J, vander Zee R, Noordzij A, vanEmbden J, Hensen E, Cohen I (1988) Cloning of the mycobacterial epitope recognized by T lymphocytes in adjuvant arthritis. *Nature* 331: 171–173

45 Steinhoff U, Brinkmann V, Klemm U, Aichele P, Seiler P, Brandt U, Bland PW, Prinz I, Zugel U, Kaufmann SHE (1999) Autoimmune intestinal pathology induced by hsp60-specific CD8 T cells. *Immunity* 11: 349–358

46 Quayle AJ, Black WK, Li SG, Kjeldson-Kragh J, Oftung F, Sioud M, Forre O, Capra D, Natvig JB (1992) Peptide recognition, T cell receptor usage and HLA restriction of human heat-shock protein 60 and mycobacterial 65-kDa hsp-reactive T cell clones from rheumatoid synovial fluid. *Eur J Immunol* 22: 1315–1322

47 Debattista J, Timms P, Allan J (2002) Reduced levels of gamma-interferon secretion in response to chlamydial 60 kDa heat shock protein amongst women with pelvic inflam-

matory disease and a history of repeated *Chlamydia trachomatis* infections. *Immunol Lett* 81: 205–210

48 Simon AK, Seipelt E, Sieper J (1994) Divergent T-cell cytokine patterns in inflammatory arthritis. *Proc Natl Acad Sci USA* 91: 8562–8566

49 Lahesmaa R, Yssel H, Batsford S, Luukkainen R, Mottonen T, Steinman L, G P (1992) *Yersinia enterocolitica* activates a T helper type 1 like T cell subset in reactive arthritis. *J Immunol* 148: 3079–3085

50 Kempsell KE, Cox CJ, Hurle M, Wong A, Wilkie S, Zanders ED, Gaston JSH, Crowe JS (2000) Reverse transcriptase-PCR analysis of bacterial rRNA for detection and characterization of bacterial species in arthritis synovial tissue. *Infec Immunity* 68: 6012–6026

51 Wilbrink B, Vanderheijden IM, Schouls LM, Vanembden JDA, Hazes JMW, Breedveld FC, Tak PP (1998) Detection of bacterial DNA in joint samples from patients with undifferentiated arthritis and reactive arthritis, using polymerase chain reaction with universal 16S ribosomal RNA primers. *Arthritis Rheum* 41: 535–543

52 Sloan Lancaster J, Allen P (1996) Altered peptide ligand-induced partial T cell activation. *Ann Rev Immunol* 14: 1–28

53 Paul AGA, van Kooten PJS, van Eden W, van der Zee R (2000) Highly auto-proliferative T cells specific for 60-kDa heat shock protein produce IL-4/IL-10 and IFN-gamma and are protective in adjuvant arthritis. *J Immunol* 165: 7270–7277

54 Cobelens PM, Heijnen CJ, Nieuwenkuis EES, Kramer PPG, van der Zee R, van Eden W, Kavelaars A (2000) Treatment of adjuvant-induced arthritis by oral, administration of mycobacterial Hsp65 during disease. *Arthritis Rheum* 43: 2694–2702

55 Moudgil KD, Kim E, Yun OJ, Chi HH, Brahn E, Sercarz EE (2001) Environmental modulation of autoimmune arthritis involves the spontaneous microbial induction of T cell responses to regulatory determinants within heat shock protein 65. *J Immunol* 166: 4237–4243

56 Wendling U, Paul L, Vanderzee R, Prakken B, Singh M, Vaneden W (2000) A conserved mycobacterial heat shock protein (HSP) 70 sequence prevents adjuvant arthritis upon nasal administration and induces IL-10-producing T cells that cross-react with the mammalian self-HSP70 homologue. *J Immunol* 164: 2711–2717

57 Goodall JC, Yeo G, Huang M, Raggiaschi R, Gaston JSH (2001) Identification of *Chlamydia trachomatis* antigens recognized by human $CD4^+$ T lymphocytes by screening an expression library. *Eur J Immunol* 31: 1513–1522

58 Prakken BJ, Wendling U, van der Zee R, Rutten V, Kuis W, van Eden W (2001) Induction of IL-10 and inhibition of experimental arthritis are specific features of microbial heat shock proteins that are absent for other evolutionarily conserved immunodominant proteins. *J Immunol* 167: 4147–4153

The development of immune therapy with HSP60 for juvenile idiopathic arthritis

Ismé M. de Kleer[1], Berent P. Prakken[1], Salvatore Albani[2] and Wietse Kuis[1]

[1]University Medical Center Utrecht, Wilhelmina Children's Hospital, Department of Pediatric Immunology, P.O. Box 85090, 3508 AB Utrecht, The Netherlands; [2]University of California, Departments of Medicine and Pediatrics, 9500 Gilmore Drive, La Jolla, CA 92093, USA

Introduction

Immune regulation is an important mechanism in controlling the severity and duration of an immune response. Ongoing and destructive inflammation suggests a dysfunction in the immune regulation, while a spontaneous relapse or remission of disease suggests a successful immune-regulatory process. Understanding the regulatory mechanisms that control the balance between immunity and tolerance can provide potential new targets for therapeutic intervention.

Juvenile idiopathic arthritis

Juvenile idiopathic arthritis (JIA) is the most common rheumatic disease of childhood [1]. Local inflammation in the joints results in joint destruction and overall, an estimated 49% of affected children end up with severe functional limitation because of JIA [2]. JIA is not a homogenous disease but consists of various subtypes with three principle types of onset: (a) oligoarthritis; (b) polyarthritis; (c) systemic-onset disease [3]. Though the histopathological abnormalities found in these three subtypes are similar, there is a striking difference in both severity and outcome of the three subtypes. In oligoarticular JIA the disease has a relative benign course; the disease is restricted to one to four joints and is often self remitting. In polyarticular and systemic JIA five or more joints are affected and in contrast to oligoarticular JIA, polyarticular and systemic JIA are usually non-remitting and crippling diseases requiring aggressive immunosuppressive therapy. This difference in clinical course between the various subsets of JIA is an intriguing phenomenon and provides us with ideal study groups to investigate the immune-regulatory processes that determine the varying clinical prognoses.

Heat Shock Proteins and Inflammation, edited by Willem van Eden

Pathogenesis of JIA

Although the pathogenesis of JIA is still unknown, studies suggest excessive immune reactivity of several types of cells in predisposed children, suspected but not proved to be in reaction to exposure to certain infectious pathogens [4, 5]. Studies of T-cell receptor expression confirm recruitment of T-cells specific for (unknown) antigens present in joint synovium [6, 7]. The recruitment of these T-cells is made possible by certain HLA-types found with increased frequency in affected children. HLA-DR4 is associated with polyarticular JIA; oligoarticular JIA has been associated with HLA alleles at the DR8 and DR5 loci [8, 9]. In JIA, T-cell activation results in a cascade of events leading to tissue damage in joints and other affected tissues, including B-cell activation, complement consumption, and, in particular, release of interleukin (IL)-6 [10], tumour necrosis factor (TNF)-α, and other pro-inflammatory cytokines [11], possibly under the control of specific genetic alleles. Thus, the synovial lymphocyte infiltrate in JIA exists predominantly of Th1 type cells [6] and like rheumatoid arthritis (RA) is therefore believed to be the result of a polarization towards a persistent pro-inflammatory Th1 response.

Th2 counter-regulation in remitting JIA

Despite this strong Type 1 phenotype in JIA, variable immune components that favour down-regulation of inflammation can be present, especially in early disease. In particular, the presence of a Th2 component, in terms of Type 2 cytokines and expression of chemokine receptors [12], early in the disease process of JIA is suggested to function in an anti-inflammatory capacity and to play a role in determining disease phenotype [12, 13]. Several studies demonstrated that IL-4 and IL-10 are more readily detectable in synovium of patients with limited joint destruction. Murray et al., showed that synovial fluid from patients with oligoarticular JIA significantly over expresses IL-4 messenger RNA relative to synovial fluid from patients with polyarticular JIA [13]. Crawley et al., demonstrated a hereditary predisposition to low IL-10 production in children with a non-remitting form of JIA [14].

Regulatory T-cells in remitting JIA

In addition to regulatory cytokines, multiple populations of T-cells with specialized regulatory capacity have been identified to play a role in JIA. Thompson et al., demonstrated equally increased levels of CCR5 and CCR4 bearing synovial fluid lymphocytes, the latter producing greater IL-4 than IFN-γ [12]. Increased expression of CCR5 has been associated with a Type 1 cytokine profile and is dependent on recent activation events. CCR4 is preferentially expressed on Th2 like cells and is therefore suggested to be a chemokine receptor for regulatory cells. Recently the existence of professional, natural occurring regulatory cells have been identified, which are enriched in the $CD4^+CD25^+$ lymphocyte population [15]. $CD4^+CD25^+$ regulatory T-cells seem to have a pivotal role in the maintenance of peripheral tolerance and evidence is accumulating that they also play a suppressive role in human autoimmune diseases [16]. Recent data show that $CD4^+CD25^+$ regulatory T-cells, expressing high CTLA-4, GITR and CCR4, are abundantly present in the joints of patients with remitting oligoarticular JIA (de Kleer et al., submitted). This strongly suggests that $CD4^+CD25^+$ regulatory T-cells play a role in down-regulating inflammation in these patients and therefore could contribute to remission of disease. Though much progress has been made, many key aspects of regulatory T-cells in human diseases still remain to be resolved, particularly concerning their antigen specificity, mechanisms of action and interrelationship.

Immunoregulation by HSP60: Adjuvant arthritis

Further in this chapter another population of T-cells that is associated with disease remission in oligoarticular JIA will be discussed; namely T-cells responding to self-heat shock protein 60 (HSP60). Heat shock proteins (HSPs) are ubiquitous and abundant proteins, essential for cellular viability [17] and present in cells of almost all living organisms. They are induced when a cell undergoes various types of environmental stresses like heat, cold and oxygen deprivation. HSPs are highly conserved during evolution which has resulted in extensive amino acid sequence identities between mammalian and microbial HSPs. This phylogenetic similarity between prokaryotic and eukaryotic heat shock proteins, together with the capacity of heat shock proteins to induce pro-inflammatory responses has led to the proposition that these proteins provide a link between infection and autoimmune disease. Indeed, both elevated levels of antibodies to heat shock proteins and an enhanced immune reactivity to heat shock proteins have been noted in a variety of pathogenic disease states.

The same observations regarding HSPs make the apparent tolerance of healthy individuals to their own (self) HSPs all the more remarkable; most people don't develop dangerous autoimmune responses to self HSPs. In fact, healthy people do possess T-cells which recognize these self HSPs, but suffer no ill effects [18, 19]. The presence of such self-reactive cells in healthy people therefore suggests that these self HSP specific cells are highly regulated. Even more, evidence is accumulating that T-cells with specificity for self HSPs have an immunoregulatory function themselves and play a role in the down-regulation of pathogenic, inflammatory processes. Most evidence for a protective role for HSP60-specific T-cells came from the experimental animal model adjuvant arthritis (AA). In AA T-cell responses to HSPs play an important role in both the induction of AA and the protection from AA. AA is inducible in susceptible rats by immunization with heat-killed *Mycobacterium tuberculosis* (Mt). Mt 65-kDa heat shock protein (HSP65) is a target of pathogenic T-cells in AA; a T-cell clone (A2b) specific for the epitope contained between amino acids 180–188 of HSP65 could adoptively transfer AA [20]. The epitope was also found to react to an epitope of cartilage proteoglycan, suggesting that targeting of inflammation to the joints might be due to cross-reactivity between P180–188 and a self component in the cartilage [21]. On the other hand, pre-immunization with mycobacterial HSP65 or some of its T-cell epitopes protects against subsequent induction of arthritis. Even more, pre-immunization with mycobacterial HSP65 has also been found to protect in other experimental arthritis models as collagen Type 2 induced arthritis and avridine arthritis. Inhibition of AA by treatment with HSP65 is thought to be mediated by regulatory T-cells cross-reactive with the self-60 kDa heat shock protein (HSP60) [22, 23]. Hence, mycobacterial HSP65 appears to provide epitopes with different immune functions in AA: the cross-reaction of P180–188 with cartilage may be involved in the pathogenic effector mechanism, and a cross-reactivity between HSP65 and self-HSP60 might be involved in regulation of disease.

Adjuvant arthritis as an experimental model for RA and JIA

Adjuvant arthritis is the most extensively used experimental model for rheumatoid arthritis and indeed, the histopathological characteristics are very similar. However, a much more striking resemblance is seen between AA and oligoarticular JIA, because of the following reasons;
(1) there is a similar histopathological resemblance;
(2) AA as well as oligoarticular JIA are both self-remitting diseases;
(3) JIA is thought to be induced by a microbial agent (mycobacteria) as well;
(4) both diseases are complicated by uveitis, and;
(5) as will be discussed in the following section; T-cell responses to HSPs seem to play a crucial role in the disease regulation of JIA as well.

Reactivity to self HSP60 correlates with a good prognosis in JIA

As described above, JIA shares many features with the experimental animal model AA. Besides a clinical and histopathological resemblance evidence is growing that like in AA HSPs play a central role in the immune regulation of JIA as well. Synovial lining cells of patients with JIA show an increased expression of endogenously produced HSP60 [24]. Thus, at the site of inflammation self HSP60 is available as a target for the immune system. Indeed, IgG antibodies to human HSP60 can be detected in both serum and synovial fluid from patients with JIA [25]. Furthermore, both peripheral blood mononuclear cells and synovial fluid derived mononuclear cells from patients with JIA show T-cell reactivity to human as well as mycobacterial HSP60 [26]. Also reactivity to other HSPs, in particular to *E. coli* derived HSP65 (GroEL) [27] and to DnaJ [28] are seen. [29]. Most of the T-cell responses to human HSP60 are found in HLA B27 negative oligoarticular JIA, the subtype of JIA with the best prognosis. These responses to human HSP60 are correlated with responses to mycobacterial HSP60 and HSP70 suggesting that the responses are directed to more conserved parts of the HSP60 molecule.

A prospective longitudinal study in newly diagnosed patients with JIA showed the role of the reactivity to self-HSP60 in the early course of the disease [30]. In oligoarticular JIA patients' significant proliferative responses to human HSP60 are found after the first four weeks and within the first three months after the onset of disease. During phases of remission of disease the majority of the oligoarticular patients lose their previous positive responses to human HSP60, while during an exacerbation of the disease, again positive T-cell responses to human HSP60 are found. A similar pattern of T-cell responses to human HSP60 can not be found in patients with polyarticular and systemic disease. Furthermore, *in vitro* priming of the non-responder cells during remission restores the responsiveness in oligoarticular patients and not in polyarticular and systemic patients. In this prospective study, all patients that showed positive T-cell responses to human HSP60 during an exacerbation of the disease eventually reached a disease remission, while non-responders at the onset of the disease were prone to develop a more severe, non-remitting polyarticular arthritis. This association between early T-cell reactivity to human HSP60 and disease remission suggests that self-HSP60 specific T-cells contribute to regulatory mechanisms that down-regulate inflammation.

Though polyarticular and systemic JIA patients lack T-cell reactivity to human HSP60 at the onset of disease, responses towards self-HSP60 can be found at later stages of disease. It is conceivable that these responses inevitably develop as a result of epitope spreading and bystander activation during ongoing inflammation. Indeed, phenotypic analysis revealed differences between the human HSP60 specific T-cell population of oligoarticular patients, hypothesized to be part of an immune regulatory mechanism, and the human HSP60 specific T-cell population of polyarticular patients, which is presumably the result of epitope spreading [31].

Phenotype of self HSP60 specific T-cells in JIA

Evidence is growing that self-HSP60 specific T-cells in oligoarticular JIA patients are indeed regulatory cells. *In vitro* activation of peripheral blood derived mononuclear cells and synovial fluid derived mononuclear cells of oligoarticular JIA patients with HSP60 induces a high expression of CD30 on $CD4^+$, activated (HLA-DR-positive), memory (CD45RO-positive) T-cells [31]. In contrast, *in vitro* activation with HSP60 fails to induce a similar increased expression of CD30 on PB derived lymphocytes of patients with polyarticular JIA. Although CD30 expression can also be found in Th1 and Th0 clones CD30 surface expression has mainly been described as an *in vitro* feature of Th2-type cells and *in vivo* evidence exists that CD30 has a role in both function and development of Th2 like human $CD4^+$ cells [32, 33]. As the response in terms of CD30 expression to human HSP60 is much higher than to mycobacterial HSP60 it is unlikely that CD30 merely functions as an activation marker. Besides a high expression of CD30, *in vitro* activation of PBMC with mycobacterial or human HSP60 results in a high production of IL-10 and a low production of IFN-γ in oligoarticular JIA patients with active disease, while in polyarticular JIA patients with active disease this ratio is reversed in favour of IFN-γ. Thus, the self-HSP60 specific T-cells in oligoarticular JIA patients clearly show a regulatory phenotype.

A self-HSP60 driven regulatory mechanism controlling inflammation in remitting JIA?

The following self-HSP60 driven mechanism could play a role in oligoarticular JIA (see Fig. 1). Self-HSP60 specific T-cells escape negative selection in the thymus and are kept in the immune repertoire through contacts with microbial heat shock proteins *via* tolerizing gut mucosa, the site known to induce so-called "oral tolerance". While continuously perceiving the presence of their epitopes in the gut associated lymphoid tissue they may develop their regulatory phenotype. During periods of active arthritis, inflammation in the joint leads to local cellular stress and therefore the up-regulation of self HSPs in the synovial tissue. Confronted with their homologues the HSP60 specific T-cells are activated in the inflamed joint, express CD30 and are triggered to high IL-10 and low IFN-γ production. These autoreactive T-cells may then down modulate the pathogenic T-cells either directly through recognition of self HSP60 on the pathogenic T-cells or indirectly through bystander suppression in a suppressive, IL-10 containing environment. In polyarticular- and systemic JIA patients, T-cell responses against human HSP60 are not completely absent but are qualitatively different; they appear later in the course of the disease, probably through bystander activation, and show no regulatory features. In this view, the ongoing and destructive inflammation in polyarticular JIA might be the result of a deficiency in the natural regulatory response induced by self HSPs.

Heat shock proteins as targets for immune therapy?

The current approach to treatment of JIA tends to focus on managing inflammation and pain. These methods have little effect on the autoimmune processes that are

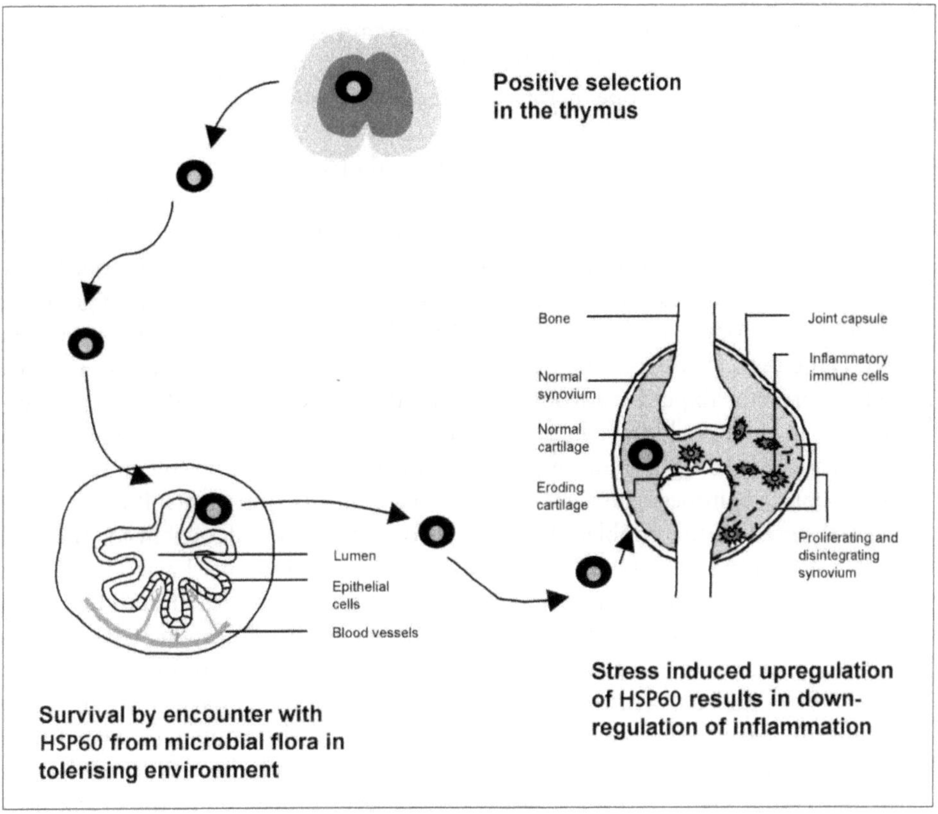

Figure 1
The role of self-HSP60 specific T-cells in oligoarticular JIA.
Self-HSP60 specific T-cells escape negative selection in the thymus, and are kept in the immune repertoire through mucosal contacts with microbial heat shock proteins present in the gastro-intestinal tract. During arthritis self-HSP60 specific T-cells home towards self-HSP60 expressed in the inflamed synovial tissue.

thought to underlie the disease and are of limited use in altering disease progression. Therefore developing novel treatments that target the autoimmune component of the disease is highly desirable. Restoring immunoregulation through manipulation of peripheral tolerance using HSPs may provide such therapies in the future. In several experimental autoimmune models, such as experimental autoimmune encephalomyelitis (EAE) [34], collagen induced arthritis [35], NOD mouse diabetes mellitus [36] and adjuvant arthritis [37] antigen-specific peripheral tolerance leading to disease resistance can be induced by manipulation of the mucosal immune system through oral or nasal administration of a relevant (auto)-antigen. In adjuvant arthritis, for instance, it is possible to obtain specific T-cell tolerance for the AA-associated T-cell epitope (A2B); nasal administration of a peptide containing the arthritogenic T-cell epitope before the induction of AA delayed the onset and reduced the severity of arthritis.

Since patients with polyarticular juvenile idiopathic arthritis seem to lack the regulatory T-cell response to self HSP60 mucosal administration of or immunization with heat shock proteins might be a promising way to re-activate these self-HSP reactive T-cells and restore the balance between immunity and regulation. However, trying to extrapolate the success of mucosal immune therapy in experimental autoimmune models to immunotherapy for JIA one of the major problems encountered is the need to induce mucosal tolerance with peptide epitopes rather than with a whole protein. Experience from animal models showed that different epitopes of the same protein could have opposite effects in immune therapy. However, the heterogeneous HLA background in patients with JIA makes it difficult to predict universal peptide epitopes that are recognized in a majority of patients. It has been possible to overcome this problem by using a novel computer algorithm for the analysis of potential pan-DR binding epitopes (A. Sette, Epimmune, La Jolla, CA). This computer algorithm has identified eight different T-cell epitopes of both human and mycobacterial HSP60. All eight peptides induced clear T-cell responses, as measured by T-cell proliferation and antigen specific cytokine production, in a vast majority (60–90%) of patients with JIA, both in peripheral blood and synovial fluid. Most, but not all, T-cell responses were found in patients with the oligoarticular form of JIA. Most interestingly, one of the peptides identified with the computer algorithm has only minor changes to the peptide that induces protection in the experimental model of AA. The T-cell response to this peptide correlated strongly with the T-cell response to its human homologue peptide, which strongly suggests the identification of a population of cross-reactive T-cells. This peptide could therefore be the primary candidate for use in the human system.

Combining immunotherapy with heat shock proteins and conventional therapy?

Another important area that needs to be addressed is whether self-HSP60 specific regulatory T-cells can effectively down-regulate ongoing immunopathological reactions in JIA. In the adjuvant arthritis model it is possible to treat ongoing arthritis effectively with HSP-derived peptides [38]. However, it is conceivable that, in the more complex human situation, self-HSP60 specific T-cells will be insufficient in down-regulating ongoing disease, especially once massive inflammation and joint destruction has taken place.

In recent years two new advances in the treatment of JIA have been made: anti-tumour necrosis factor therapy, such as etanercept, and autologous stem cell transplantation (ASCT). Though both treatments show great promise in the treatment of children with JIA there are many disadvantages. Etanercept is aimed towards a general suppression of pro-inflammatory pathways and does not restore peripheral tolerance. In addition, the treatment is costly, expected to have serious long-term side effects and most importantly, offers only temporary benefit: once the treatment is discontinued, the disease reoccurs in its original severity.

Interestingly, when in an animal model of experimental arthritis, a single, suboptimal dose of etanercept (soluble TNF-α receptor) was combined with oral tolerance induction with a heat shock protein peptide; this led to a sustained remission of arthritis (Roord et al., submitted).

ASCT is a treatment that is aimed to restore peripheral tolerance by resetting the immune system. Because the procedure carries a significant mortality risk only children with very severe, refractory JIA are eligible for this treatment. ASCT is successful in approximately 55% of the children (de Kleer et al., submitted). It is likely that following ASCT the newly developing T-cell repertoire following encounter with auto-antigens at the site of tissue inflammation is tolerised towards such antigens. In 36% of the children a partial or complete relapse of the disease is seen. It will be interesting to investigate whether mucosal treatment with HSPs following ASCT will induce regulatory T-cells efficient enough in preventing a relapse.

Future research therefore need to be focused on clarifying whether the combination of a temporary resetting of the immune system by etanercept or an autologous bone marrow transplantation, followed by antigen specific immune therapy is the key towards sustained remission of disease in JIA.

Summary

The development of novel treatments that target the autoimmune component of JIA in an immune-modulatory rather than immune-suppressive way is highly desirable. Restoring immunoregulation through manipulation of peripheral tolerance using HSPs may provide such therapies in the future. With the identification of self HSP60 epitopes that are recognized in the majority of JIA patients a Phase I clinical trial has come near.

References

1 Woo P, Wedderburn LR (1998) Juvenile chronic arthritis. *Lancet* 351: 969–973

2 Wallace CA, Levinson JE (1991) Juvenile rheumatoid arthritis: outcome and treatment for the 1990s. *Rheum Dis Clin North Am* 17: 891–905

3 Petty RE, Southwood TR, Baum J, Bhettay E, Glass DN, Manners P, Maldonado-Cocco J, Suarez-Almazor M, Orozco-Alcala J, Prieur AM (1998) Revision of the proposed clas-

sification criteria for juvenile idiopathic arthritis: Durban, 1997. *J Rheumatol* 25: 1991–1994
4 Pugh MT, Southwood TR, Gaston JS (1993) The role of infection in juvenile chronic arthritis. *Br J Rheumatol* 32: 838–844
5 Benoist C, Mathis D (2001) Autoimmunity provoked by infection: how good is the case for T cell epitope mimicry? *Nat Immunol* 2: 797–801
6 Wedderburn LR, Robinson N, Patel A, Varsani H, Woo P (2000) Selective recruitment of polarized T cells expressing CCR5 and CXCR3 to the inflamed joints of children with juvenile idiopathic arthritis. *Arthritis Rheum* 43: 765–774
7 Grom AA, Hirsch R (2000) T-cell and T-cell receptor abnormalities in the immunopathogenesis of juvenile rheumatoid arthritis. *Curr Opin Rheumatol* 12: 420–424
8 Glass DN, Giannini EH (1999) Juvenile rheumatoid arthritis as a complex genetic trait. *Arthritis Rheum* 42: 2261–2268
9 Thomson W, Barrett JH, Donn R, Pepper L, Kennedy LJ, Ollier WE, Silman AJ, Woo P, Southwood T (2002) Juvenile idiopathic arthritis classified by the ILAR criteria: HLA associations in UK patients. *Rheumatology (Oxford)* 41: 1183–1189
10 De Benedetti F, Robbioni P, Massa M, Viola S, Albani S, Martini A (1992) Serum interleukin-6 levels and joint involvement in polyarticular and pauciarticular juvenile chronic arthritis. *Clin Exp Rheumatol* 10: 493–498
11 Lepore L, Pennesi M, Saletta S, Perticarari S, Presani G, Prodan M (1994) Study of IL-2, IL-6, TNF alpha, IFN gamma and beta in the serum and synovial fluid of patients with juvenile chronic arthritis. *Clin Exp Rheumatol* 12: 561–565
12 Thompson SD, Luyrink LK, Graham TB, Tsoras M, Ryan M, Passo MH, Glass DN (2001) Chemokine receptor CCR4 on $CD4^+$ T cells in juvenile rheumatoid arthritis synovial fluid defines a subset of cells with increased IL-4: IFN-gamma mRNA ratios. *J Immunol* 166: 6899–6906
13 Murray KJ, Grom AA, Thompson SD, Lieuwen D, Passo MH, Glass DN (1998) Contrasting cytokine profiles in the synovium of different forms of juvenile rheumatoid arthritis and juvenile spondyloarthropathy: prominence of interleukin 4 in restricted disease. *J Rheumatol* 25: 1388–1398
14 Crawley E, Kon S, Woo P (2001) Hereditary predisposition to low interleukin-10 production in children with extended oligoarticular juvenile idiopathic arthritis. *Rheumatology (Oxford)* 40: 574–578
15 Shevach EM (2000) Regulatory T cells in autoimmunity. *Annu Rev Immunol* 18: 423–449
16 Taams LS, Smith J, Rustin MH, Salmon M, Poulter LW, Akbar AN (2001) Human anergic/suppressive CD4(+)CD25(+) T cells: a highly differentiated and apoptosis-prone population. *Eur J Immunol* 31: 1122–1131
17 Ang D, Liberek K, Skowyra D, Zylicz M, Georgopoulos C (1991) Biological role and regulation of the universally conserved heat shock proteins. *J Biol Chem* 266: 24233-24236

18 Rees A, Scoging A, Mehlert A, Young DB, Ivanyi J (1988) Specificity of proliferative response of human CD8 clones to mycobacterial antigens. *Eur J Immunol* 18: 1881–1887

19 Munk ME, Schoel B, Modrow S, Karr RW, Young RA, Kaufmann SH (1989) T lymphocytes from healthy individuals with specificity to self-epitopes shared by the mycobacterial and human 65-kilodalton heat shock protein. *J Immunol* 143: 2844–2849

20 Holoshitz J, Matitiau A, Cohen IR (1984) Arthritis induced in rats by cloned T lymphocytes responsive to mycobacteria but not to collagen type II. *J Clin Invest* 73: 211–215

21 van Eden W, Holoshitz J, Nevo Z, Frenkel A, Klajman A, Cohen IR (1985) Arthritis induced by a T-lymphocyte clone that responds to *Mycobacterium tuberculosis* and to cartilage proteoglycans. *Proc Natl Acad Sci USA* 82: 5117–5120

22 van Eden W, van der ZR, Paul AG, Prakken BJ, Wendling U, Anderton SM, Wauben MH (1998) Do heat shock proteins control the balance of T-cell regulation in inflammatory diseases? *Immunol Today* 19: 303–307

23 van der ZR, Anderton SM, Prakken AB, Liesbeth Paul AG, van Eden W (1998) T cell responses to conserved bacterial heat-shock-protein epitopes induce resistance in experimental autoimmunity. *Semin Immunol* 10: 35–41

24 Boog CJ, Graeff-Meeder ER, Lucassen MA, van der ZR, Voorhorst-Ogink MM, van Kooten PJ, Geuze HJ, van Eden W (1992) Two monoclonal antibodies generated against human hsp60 show reactivity with synovial membranes of patients with juvenile chronic arthritis. *J Exp Med* 175: 1805–1810

25 Graeff-Meeder ER, Rijkers GT, Voorhorst-Ogink MM, Kuis W, van der ZR, van Eden W, Zegers BJ (1993) Antibodies to human HSP60 in patients with juvenile chronic arthritis, diabetes mellitus, and cystic fibrosis. *Pediatr Res* 34: 424–428

26 Graeff-Meeder ER, van Eden W, Rijkers GT, Prakken BJ, Kuis W, Voorhorst-Ogink MM, van der ZR, Schuurman HJ, Helders PJ, Zegers BJ (1995) Juvenile chronic arthritis: T cell reactivity to human HSP60 in patients with a favorable course of arthritis. *J Clin Invest* 95: 934–940

27 Life P, Hassell A, Williams K, Young S, Bacon P, Southwood T, Gaston JS (1993) Responses to gram negative enteric bacterial antigens by synovial T cells from patients with juvenile chronic arthritis: recognition of heat shock protein HSP60. *J Rheumatol* 20: 1388–1396

28 Albani S, Ravelli A, Massa M, De Benedetti F, Andree G, Roudier J, Martini A, Carson DA (1994) Immune responses to the Escherichia coli dnaJ heat shock protein in juvenile rheumatoid arthritis and their correlation with disease activity. *J Pediatr* 124: 561–565

29 Graeff-Meeder ER, van der ZR, Rijkers GT, Schuurman HJ, Kuis W, Bijlsma JW, Zegers BJ, van Eden W (1991) Recognition of human 60 kD heat shock protein by mononuclear cells from patients with juvenile chronic arthritis. *Lancet* 337: 1368–1372

30 Prakken AB, van Eden W, Rijkers GT, Kuis W, Toebes EA, Graeff-Meeder ER, van der ZR, Zegers BJ (1996) Autoreactivity to human heat-shock protein 60 predicts disease

remission in oligoarticular juvenile rheumatoid arthritis. *Arthritis Rheum* 39: 1826–1832

31 de Kleer IM, Kamphuis S, Rijkers GT, Scholtens L, Gordon G, de Jager W, Häfner R, van de Zee R, van Eden W, Kuis W et al (2003) The spontaneous remission of juvenile idiopathic arthritis is characterized by CD30$^+$ T cells capable of producing the regulatory cytokine IL-10. *Arthritis Rheum* 48: 2001–2010

32 Chilosi M, Facchetti F, Notarangelo LD, Romagnani S, Del Prete G, Almerigogna F, De Carli M, Pizzolo G (1996) CD30 cell expression and abnormal soluble CD30 serum accumulation in Omenn's syndrome: evidence for a T helper 2-mediated condition. *Eur J Immunol* 26: 329–334

33 Manetti R, Annunziato F, Biagiotti R, Giudizi MG, Piccinni MP, Giannarini L, Sampognaro S, Parronchi P, Vinante F, Pizzolo G (1994) CD30 expression by CD8$^+$ T cells producing type 2 helper cytokines. Evidence for large numbers of CD8$^+$CD30$^+$ T cell clones in human immunodeficiency virus infection. *J Exp Med* 180: 2407–2411

34 Metzler B, Wraith DC (1996) Mucosal tolerance in a murine model of experimental autoimmune encephalomyelitis. *Ann NY Acad Sci* 778: 228–242

35 Nagler-Anderson C, Bober LA, Robinson ME, Siskind GW, Thorbecke GJ (1986) Suppression of type II collagen-induced arthritis by intragastric administration of soluble type II collagen. *Proc Natl Acad Sci USA* 83: 7443–7446

36 Aspord C, Thivolet C (2002) Nasal administration of CTB-insulin induces active tolerance against autoimmune diabetes in non-obese diabetic (NOD) mice. *Clin Exp Immunol* 130: 204–211

37 Prakken BJ, van der ZR, Anderton SM, van Kooten PJ, Kuis W, van Eden W (1997) Peptide-induced nasal tolerance for a mycobacterial heat shock protein 60 T cell epitope in rats suppresses both adjuvant arthritis and non-microbially induced experimental arthritis. *Proc Natl Acad Sci USA* 94: 3284–3289

38 Prakken BJ, Roord S, van Kooten PJ, Wagenaar JP, van Eden W, Albani S, Wauben MH (2002) Inhibition of adjuvant-induced arthritis by interleukin-10-driven regulatory cells induced *via* nasal administration of a peptide analog of an arthritis-related heat-shock protein 60 T cell epitope. *Arthritis Rheum* 46: 1937–1946

Heat shock proteins and rheumatoid arthritis

Gabriel S. Panayi and Valerie M. Corrigall

Department of Rheumatology, GKT School of Medicine, King's College London, London SE1 9RT, UK

Introduction

In this chapter we shall review the possible role of heat shock proteins (HSP) and chaperones in the pathogenesis of rheumatoid arthritis (RA). Although HSP/chaperones are found in all cells, their expression is increased at times of cellular or tissue stress. It is under these conditions that new roles or functions for these proteins will become apparent. Hence, the first section of the review deals with the expression of HSP in the RA synovial membrane (SM). There then follow two sections dealing with the immune response to HSP by patients with RA. The first part considers the antibody responses and the second part the response at the T-cell level. In the third section, we consider the proposal by Roudier and his colleagues that bacterial HSP are involved in the pathogenesis of RA because of molecular mimicry with the shared epitope. This is an innovative concept worthy of separate discussion not least because it is being tested in the clinic. Finally, we review our work on BiP, the endoplasmic reticulum chaperone, which focuses on the concept that HSP/chaperones may have immunomodulatory or anti-inflammatory effects when found extra-cellularly rather than solely being involved in pro-inflammatory events, which has been the prevailing view to date.

The expression of HSP in the RA synovium

The concept that molecular mimicry between bacterial and human HSP could be involved in the pathogenesis of RA would be immensely strengthened if the expression of HSP were to be increased in the RA SM. Such an increase would obviously provide more HSP protein to "drive" the disease or to regulate it, depending on the precise mechanisms operating at the time.

The inflamed RA SM has many of the characteristics necessary for the increased expression of HSP and molecular chaperones. The rheumatoid joint is hypoxic [1], undergoes regular and frequent episodes of reperfusion injury [1], produces large

Heat Shock Proteins and Inflammation, edited by Willem van Eden

amounts of reactive oxygen species [2], and is a potent brew of inflammatory cytokines such as interleukin (IL)-1 and tumour necrosis factor (TNF)-α [3]. Thus it is not surprising that the RA SM has many of the features characteristic of a stressed tissue. This has been elegantly described by Schett et al. [4], who showed by gel mobility shift analysis that heat shock transcription factor 1 (HSF1) was activated, hyper-phosphorylated and translocated to the nucleus where it was instrumental in up-regulating HSP70 transcription in RA but not osteoarthritic SM. Further experiments, using cultured RA SM synovial fibroblasts, showed that the pro-inflammatory cytokines (TNF-α, IL-1α and IL-6) but not interferon (IFN)-γ or transforming growth factor (TGF)-β induced activation of HSF1-DNA binding and HSP70 expression in the cultured cells. Glucocorticoids and cytotoxic drugs had no effect on this response although, interestingly, shear stress could induce it.

A summary of the work investigating expression of HSP in the RA SM is presented in Table 1. Several investigators have shown that a variety of HSPs and chaperones are over-expressed in the RA SM including BiP [5], human homologues of the bacterial DnaJ chaperone [6], human HSP60 [7, 8], and HSP65 [9]. However, this is not a universal finding. An equal expression of human HSP65 was found in RA and non-inflamed control SM [10]; there was equal expression of mitochondrial HSP60 by immunohistology in RA and osteoarthritic SM [11]; and the human homologue of the bacterial GroeL HSP was increased in all inflammatory tissues including RA SM, liver and kidney [12]. The findings of the latter study are not surprising as the inimical inflammatory environment, irrespective of the tissue in which it is taking place, is likely to lead to increased expression of HSP in order to protect cells from stress that could ultimately lead to death by apoptosis [13]. Karlsson-Parra et al. [9] made the interesting observation that human (hu) HSP65 was found abundantly at the cartilage-pannus junction. This is of relevance as the failure to demonstrate HSPs by some investigators could be explicable on the basis that maximum expression takes place at the eroding front rather than within the SM itself. This group also demonstrated huHSP65 in rheumatoid nodules – the only report to our knowledge in which HSP expression has been sought in this tissue. Rheumatoid nodules are, of course, the clinical and histological "signature" of RA. When it is recalled that the centre of a nodule has undergone non-caseating necrosis, one could speculate that nodule formation is the consequence of events that lead to increased expression of HSPs such as hypoxia.

The role of HSP in protecting cells from stress is described elsewhere in this volume. This is the classical "intracellular" role of HSP. HSPs, as antigenic peptides, clearly have an additional function in stimulating T-cells that is beyond their classical intracellular role. This aspect of the contribution of HSP to inflammation in general but in RA especially is described below. A third, newly described, function is the ability of HSPs to bind to specific cell membrane receptors and activate the cells to secrete pro-inflammatory cytokines. We have recently reviewed this aspect of HSP function [14].

Table 1 - Expression of HSP in the rheumatoid synovial membrane

Date	Heat shock protein	Method	Reference	Result
1990	Human HSP65	Immunohistology	[76]	Strong reactivity at cartilage-pannus junction and in rheumatoid nodules
1990	Human homologue of bacterial GroeL HSP	Monoclonal antibody to Myco HSP65 used in immunohistology	[77]	Granular pattern of staining Increased in inflammatory tissues including RA SM
1992	Mitochondrial HSP60	Immunohistology	[78]	No significantly increased expression in RA SM compared to normal or osteoarthritic SM
1995	Human HSP65	Semi-quantitative RT-PCR	[79]	Found in equal amounts in RA and non-inflamed SM
1996	Human HSP60	Monoclonal antibody to *Y. enterocolitica* HSP60; Western blot; flow cytometry	[80]	Expression within PB and SF lymphocytes by precipitation commoner than surface expression
1996	Human HSP60	Monoclonal antibody secreted by immortalised B-lymphocytes from RA SM	[23]	Staining of RA SM for HuHSP60 stronger than for normal SM
1999	Human homologues bacterial DnaJ chaperone	Immunohistology; Western blotting	[81]	Over expressed in synovial lining layer in the synoviocytes
2001	BIP (grp 78)	Immunohistology	[69]	Over expressed in RA SM

It is known that HSP may exist free in the circulation of patients. Thus, Lewthwaite et al. [15] showed a correlation between the concentration of circulating huHSP60 in the plasma of British civil servants with physiological and psychosocial stress. Of great interest in view of this finding is the link between high circulating levels of soluble huHSP60 and the amount of carotid atheroma as assessed by high-resolution duplex scanning [16]. Since there is increased mortality in RA and since this increase is primarily due to increased atheromatous disease (for review see Goodson [17]), it would clearly be important to measure circulating HSPs as possible surrogate markers for severity of atheroma and, hence, consequent mortality in patients.

Summary

There is increased expression of human HSPs in RA SM and at the cartilage-pannus junction.

The increased expression is due to the inimical environment in the rheumatoid joint and this may be mediated by increased expression and/or activation of transcription factors such as HSF1.

Soluble cell-free HSP may contribute to some of the extra-articular features of RA, such as rheumatoid nodules, and may be involved in the increased mortality due to more severe atheromatous disease.

Antibody responses to mycobacterial, bacterial or human heat shock proteins in patients with rheumatoid arthritis

Since HSP are highly conserved and necessary for many essential housekeeping tasks, it is possible that molecular mimicry between the microbial HSP and host HSP could give rise to deleterious immune responses leading to autoimmune disease. This possibility is described elsewhere in this volume. We have seen, in the previous section, that some HSP are up-regulated in the RA SM. This is no doubt the consequence of the inimical environment during the course of chronic inflammation. It is of interest that chondrocytes will, in common with other cells, up-regulate HSP expression [18]. Thus, in the RA joint there is up-regulation of the expression of HSP both in the SM and in articular cartilage. This will set the scenario for two processes.

The first process may result in the stimulation of an immune response since the over-expressed host HSP, by mechanisms involving molecular mimicry, will stimulate T-cells and B-cells primed by and responding to bacterial HSP. Stimulation of T-cells by bacterial and human HSP is considered in the next section. Of course, the stimulation of B-cells will lead to the production of anti-HSP antibodies. These will be able to induce inflammation by activating the complement system and by binding to FcγR on monocytes and neutrophils. This receptor/ligand interaction may result in the release of a number of pro-inflammatory factors from the cells including IL-1 and TNF-α. To our knowledge no formal demonstration of HSP containing immune complexes in the peripheral blood (PB), synovial fluid (SF) or SM of patients with RA has been made. However, since HSP can be expressed under appropriate conditions of cellular stress in any cell of any tissue or organ in the body, how can one explain the joint localisation of RA, although it should be remembered that RA is a systemic disease and not just a disease of the joints. A possible explanation may be found in the experimental work of Diane Mathis and her colleagues [19] that showed that immune complexes may localise in the articular cartilage despite the fact that the contained antigen may be widely found in all the cells of the host. Thus it is entirely plausible that over-expression of HSP in the RA SM and the articular cartilage could lead to the localisation of anti-HSP antibodies to these sites, and other systemic sites in which there is a similar over-expression of HSP. No such demonstration has been made either in the SM or the articular cartilage.

The second process may involve the continuing priming of T-cells responding to HSP *via* antigen presentation by B-cells. Since there may be an expansion of B-cells responding to HSP in patients with RA (as described in [20]) these B-cells will capture HSP *via* their surface immunoglobulin receptors, process them and present HSP peptides in the context of surface MHC molecules, to responding T-cells. Again this is a plausible scenario but with no firm basis in experimental work.

We are thus faced with an extremely tantalising situation. On the one hand there are attractive hypotheses some of them based on firm experimental evidence in animal models of arthritis. On the other hand little or no evidence has been produced that such hypotheses are operating in patients who suffer from RA. Indeed, review of the major papers on anti-mycobacteria (myco), bacterial or huHSP antibody responses shows that far from there being a consensus, the negative reports appear to predominate over the positive (Tab. 2). One should have in mind that investigators and journals do not, as a rule, report or publish negative findings. Thus, the situation may be weighted against those published studies that have shown increased HSP antibody response in patients with RA. Some of the positive findings with respect to anti-mycoHSP65 antibodies may have been due to cross-reactivity with *E. coli* HSP60 [21–24]. Some of these studies used mycoHSP65 made in *E. coli* by recombinant techniques but used inadequate methods of purification. Thus, when human cell lysates were used as a source of HSP, so that there was no possibility of contamination with bacterial HSP, no antibodies were detected against huHSP60, huHSP73 or huHSP90 [25]. Furthermore, most investigations failed to find any cross-reactivity between mycoHSP and huHSP60 [21–24, 26–28]. An elegant study on monozygotic twins discordant for RA failed to show any increase in anti-mycoHSP65 in the twin with RA [29]. The antibody response was the same as in

Table 2 - Antibody responses to mycobacterial, bacterial or human heat shock proteins in patients with rheumatoid arthritis

Date	HSP	Method	Conclusion	Refs.	Comment
1977	Various, including foreign antigens	ELISA	Increased frequency of B-cells in the RA joint reacting to Myco HSP60, human type II collagen, and IgG Fc fragments	[82]	Decreased frequency of B-cells responding to foreign antigens suggests that the auto-reactive B-cells may be contributing to the local inflammatory process
1979	Mycobacterial HSP65	Indirect ELISA	No reactivity detected	[83]	Elevation in IgA anti-HSP65 may be due to hypergamma-globulinaemia
1989	Mycobacterial HSP65 and HSP70, *E. coli* HSP65 and 70, human HSP70	ELISA	Increased antibodies to Myco HSP65 only	[27]	No cross-reactivity to human HSP70
1991	Human HSP60, HSP73, HSP90 in human cell lysates	Western blotting	No reactivity detected	[25]	Significance may lie in the fact that human cell lysates used rather than recombinant proteins
1992	Mycobacterial HSP65 and HSP70	ELISA	No clear pattern	[26]	"It is difficult to conclude that antibodies to autologous HSP that cross-react with mycobacterial HSP play a major role in (RA) disease pathogenesis"
1993	Mycobacterial HSP65	ELISA	No association between antibodies and RA	[29]	Monozygotic twins discordant for RA used in this novel study. Compared to healthy controls.
1994	Mycobacterial HSP65	ELISA	Higher concentrations in the synovial fluid	[84]	At variance with other results in literature showing no difference

Table 2 - continued

Date	HSP	Method	Conclusion	Refs.	Comment
1995	Highly purified human HSP60	ELISA	Anti-human HSP60 antibodies exist independently of bacterial anti-HSP60 response	[24]	Previously described cross-reactivities may have been due to contamination of human HSP60 recombinant protein with *E. coli* GroeL (HSP60)
1995	Mycobacterial HSP65	ELISA	No significant difference between RA and healthy controls	[85]	(1) West Africa subjects. High levels in all groups may relate to high level of background mycobacterial exposure. (2) Note that RA is relatively rare in this population
1995	Mycobacterial HSP65	ELISA	(1) No significant difference between RA and healthy control (2) Lower than in patients with tuberculosis	[86]	(1) Study from Taiwan (2) Unlikely to play major role in disease pathogenesis
1996	Human HSP60, *E. coli* HSP60 (GroeL)	ELISA	*E. coli* HSP antibody response responsible for anti-human HSP60 but not mycobacterial HSP60	[22]	Confirms findings of Hirata et al. [21] except for failure to detect anti-mycobacterial HSP cross-reactivity
1996	*Y. enterocolitica* HSP 60	Immortalised B-cells, ELISA	Antibodies specific for bacterial HSP60 and cross-react with human and Myco HSP60	[23]	See [21, 22]
1997	*E. coli* HSP60, mycobacterial HSP60, human HSP60	ELISA, absorption	Increased titre to human and to *E. coli* HSP60	[21]	Absorption experiments suggest that anti-human and anti-mycobacterial HSP response due to cross-reactivity with *E. coli* HSP60

Table 2 - continued

Date	HSP	Method	Conclusion	Refs.	Comment
1999	HSC70, HSP90	ELISA using bovine antigens	Increased	[87]	Increased with disease duration and severity radiological damage
2000	HSP47 (identical to colligin-2)	Purification and sequencing	Prevalence not given Antibodies found in RA sera	[88]	A heat-inducible collagen binding protein
2000	1. *E. coli* GroEl 2. Mycobacterial HSP65 3. Human mitochondrial HSP60 4. Cytosolic chaperonin CCT containing t-complex polypeptide	ELISA, immuno-blotting, absorption	1. No increase 2. No increase 3. Increased 4. Increased	[89]	(1) Marked cross-reactivity of antibodies to the various HSP (2) CCT-reactive auto-antibodies recognise conformational epitopes conserved between CCT and other HSP60 family members
2001	BIP (grp 78)	ELISA	63% RA *versus* 7 % controls	[69]	A specific B-cell response
2001	N-terminal conserved region DnaJ of *A. actinomycetem-comitans*	Antibodies by ELISA	Marginal increase	[90]	
2001	Human HSP60 Mycobacterial HSP65	Antibodies by ELISA	No increase	[91]	
2002	Mycobacterial HSP65	ELISA in *H. pylori* infected individuals	Increased prevalence anti-Myco HSP65 antibodies	[28]	No increase in antibodies to human HSP60 suggesting no molecular cross-reactivity

the healthy controls. Thus there is no conclusive evidence that an antibody response triggered by mycobacterial or bacterial HSP and cross-reacting with human HSP plays any part in the initiation and pathogenesis of RA.

Summary

There are many attractive hypotheses linking an antibody response to mycobacterial, bacterial or human HSP but no firm experimental evidence for any of them.

Technical problems, such as the source and purity of HSP prepared by recombinant techniques in *E. coli*, vitiates many of the reported studies.

There is little or no evidence for cross-reactivity at the antibody level between bacterial HSP and huHSP60 in the pathogenesis of RA.

HSP and T-cell responses in RA

The hypothesis behind the extensive amount of work that was carried out in this area in the 80s and 90s was that molecular mimicry between bacterial and human HSP would activate T-cells in the joints of patients with RA. Such activation would then contribute to the pathogenesis of the disease. One scenario proposed that THI T-cells would be activated that would directly lead to joint inflammation and damage. We have recently reviewed the role of T-cells in promoting and maintaining joint pathology in RA [30]. A different scenario proposes that exogenous bacterial HSP-responsive T-cells meet endogenous HSP up-regulated in the rheumatoid joint. The human peptides are similar but not identical to the homologous bacterial HSP peptides. When these human HSP peptides stimulate the T-cells recognising the bacterial peptides, they behave as altered peptide ligands. The consequence of this is that the proliferation of the relevant T-cells may be decreased and the cytokines secreted altered from a THI to a TH2 profile. These two possibilities – the stimulation of a pro-arthritic THI response or the stimulation of an anti-inflammatory TH2 response – await experimental verification in the setting of RA. The topic, stimulatory differences between bacterial and mammalian HSP peptides on the basis of altered peptide reactivity, is reviewed in detail elsewhere in this volume.

In Table 3, we have reviewed all the major studies with adequate controls that address the response of T-cells from patients with RA to various bacterial HSP or to the homologous human HSP. There are 24 such studies. Most of them compared PB mononuclear cell (MNC) proliferation between patients with RA, disease controls with other inflammatory joint diseases such as ankylosing spondylitis, and healthy controls. Fewer studies included patients with other inflammatory arthritides. MNC were usually used in these studies for convenience but the proliferation seen may be considered to be almost if not entirely due to proliferation of the T-cells contained

Table 3 - T-cell responses to mycobacterial, bacterial or human heat shock proteins in patients with rheumatoid arthritis

Date	Antigen	Culture system and results	Refs.	Comments
1997	*M. bovis* HSP60	Synovial T-cell lines by limiting dilution from a single patient 24% lines were reactive to various antigens including MbHSP60	[48]	Support for the concept of auto-antigen reactive T-cells in joints of patients with RA
1999	Mycobacterial HSP65	$\gamma\delta$ T-cell proliferation	[40]	
1999	Mycobacterial HSP60 *E. coli* HSP60	T-cell proliferation. Decreased.	[46]	In rat adjuvant arthritis Myco HSP60 responsive T-cells protect from disease. Conversely, lack of response in RA could contribute to disease persistence.
2000	Human HSP60 Mycobacterial HSP65	T-cell proliferation HSP60 response increased with duration of disease	[41]	HSP60 response associated with secretion TH2 cytokines and with less severe disease
2001	BIP	T-cell proliferation. Small but significant increase.	[69]	A specific T-cell response
2002	Human HSP60 *Yersinia* HSP19	T-cell response. No causal role.	[39]	

Table 3 - continued

Date	Antigen	Culture system and results	Refs.	Comments
1988	Myco HSP65; tuberculin PPD; acetone precipitate of *M. tuberculosis*	RA SF MNC and from other inflammatory arthritides but not controls, proliferated to Myco antigens. Proliferation negatively correlated with duration RA.	[32]	Propose that this reactivity, possibly *via* a cross-reactive response, may be responsible for triggering chronic arthritis
1990	(1) Myco HSP65, BCG; rat Type II collagen; tetanus toxoid (2) Phytohaemagglutinin	(1) RA SF MNC proliferated better than RA PB MNC to Myco HSP and BCG (2) Response SF and PB MNC to tetanus toxoid and to PHA similar (3) Responding cells in SF were $CD8^+$ with a high proportion of VδI whilst in the PB the $Ti/\gamma A^+/BB3^+$ population was more prominent	[38]	(1) Relevance to pathogenesis of RA not clear (2) These cells could cause joint damage by cytotoxic mechanisms (see also [51])
1990	Acetone-precipitate *M. tuberculosis*; *E. coli* lysate containing Myco HSP65 *E. coli* lysate	Enhanced reactivity seen in MNC from RA SF and pleural exudates	[31]	Enhanced reactivity to Myco and other bacterial HSP may be a feature of chronic inflammation and is not disease-specific
1991	Tuberculin PPD and Myco HSP65	Limiting dilution analysis of T-cells from PB and SF to PPD and to Myco HSP65 revealed no difference between the two compartments and between RA and healthy controls	[44]	No support for pathogenic role T-cells responding to Myco HSP65

Table 3 - continued

Date	Antigen	Culture system and results	Refs.	Comments
1991	Myco HSP65, HuHSP60 and peptides	(1) $CD4^+$ T-cell clones from a single homozyous HLA–DR4 RA patient recognised highly conserved epitope (2) These clones did not respond either to whole HuHSP60 or the homologous human peptide	[91]	(1) No support for cross-reactivity hypothesis between Myco HSP65 and HuHSP60 (2) Restriction by HLA-DP rather than by HLA-DR suggests that T-cell repertoire for recognising Myco HSP65 in context HLA-DR4 deficient (3) Implications for pathogenesis RA not clear.
1991	Mycobacterial antigen	Single cell explants from RA SM organise into pannus-like tissue if T-cells and mycobacterial antigens added	[52]	This may not be unique to mycobacterial antigens. Support for concept that RA synovitis T-cell driven.
1991	Crude Myco antigens, Myco HSP65 and *E. coli* DnaK	Responses to Myco HSP65 or its degradation products. No response to DnaK.	[37]	See also [36]
1991	Myco antigens and Myco HSP65	(1) No T-cell response to Myco HSP65 by PB MNC from RA or controls (2) Proliferation to Myco HSP65 higher in RA SF MNC but also higher to tetanus toxoid	[45]	Results suggest preferential homing of antigen-reactive T-cells to inflamed joints with relative depletion in PB
1992	Mycobacterial BCG and HSP65	(1) Frequency BCG-reactive T-cells similar in RA and control PB. Few such cells found in SF. (2) Lower frequency in PB for HSP65 with no reactive T-cells in RA SF	[43]	(1) Study from Australia where population is not BCG-vaccinated (2) High BCG T-cell responsiveness in PB may reflect natural environmental exposure to mycobacterial or cross-reactivity with bacteria HSP (3) Finding Myco HSP65 T-cell reactivity in RA joint may depend on the population being studied

1992	Mycobacterial HSP65 and human HSP60	(1) SFMNC from two patients with early RA screened for proliferation to peptides from Myco HSP65 with greatest homology to human HSP60 and cultured T-cell clones from one patient (2) One clone recognised HSP65 241–255 and the equivalent human sequence restricted by HLA–DQ (3) Another clone recognised epitope 251–265 but restricted by HLA–DR	[47]	(1) Indirect support for concept that cross-reactivity between bacterial and self protein could trigger auto immunity (2) However, for previous statement to be true the cytokine profile needs to be investigated. This has been done in [50] and a variable pattern obtained. (3) Culture conditions could mean that the T-cells were activated in a primary *in vitro* stimulation rather than being pre-existent, recall responses
1992	BCG-pulsed autologous macrophages	$CD4^+$ and $CD8^+$ T-cell lines and clones from RA PB, SF, and SM lyse BCG-pulsed macrophages. Two such clones also typed macrophages that had been pulsed with recombinant human HSP65.	[51]	"Joint inflammation and destruction might be partly attributable to a cross-reaction of mycobacteria-induced cytotoxic T-cells with self HSP" See also [47] and [50]
1992	Mycobacterial HSP65 and human HSP60	Most RA SF MNC proliferate to Myco HSP65 but only 20% to HuHSP60 and with a lower response	[36]	(1) No support for concept that molecular mimicry between Myco HSP65 and HuHSP60 at the level of proliferation (2) However, RA SF MNC may be responding to HuHSP60 peptides homologous to Myco HSP65 not by proliferation but by an altered cytokine profile
1993	Mycobacterial HSP65	(1) Fifteen CD4 + αβ + T-cell clones raised (2) T-cell clones mainly THI with high IFN-γ production but some clones also produced IL-10 (3) One clone was THO since it produced equal amounts of IFN-γ, IL-10 and IL-4	[50]	Relevance to pathogenesis RA not clear

Table 3 - continued

Date	Antigen	Culture system and results	Refs.	Comments
1993	Bacterial HSP60	Bacterial HSP60-activated RA SF MNC suppressed cartilage proteoglycan synthesis *in vitro via* the secretion of IL-1 and TNF-α	[35]	See [33]
1994	Mycobacterial HSP65	(1) Higher T-cell proliferation in SF than in PB (2) Stimulation index correlated with percentage HLA-DR + $\gamma\delta$ + T-cells	[34]	Relevance to pathogenesis RA not clear
1995	Bacterial HSP60	Stimulation SF mononuclear cells Production IFN-γ, IL-1, TNF-α and induction cartilage damage	[33]	(1) Supportive evidence for concept that THI activation can cause cartilage damage (2) Results not being adduced for role bacterial HSP60 in RA
1995	(1) "Arthritogenic" epitope HSP65 (2) Articular cartilage link protein (3) Bovine Type II collagen	T-cell proliferation from PB and SF (1) Poor reactivity to link protein in RA (2) No reactivity to Myco HSP65-arthritogenic epitope	[42]	(1) Unlikely that T-cell immunity to cartilage link protein plays a role in RA pathogenesis (2) Both RA and controls reacted to whole Myco HSP65 but none to the 17 amino acid "arthritogenic" peptide
1997	Human and bacterial HSP60	Synovial fluid T-cell proliferation in presence IL-4	[90]	"TH2" T-cells suppressed IFN-γ and TNF-α production
1997	Mycobacterial HSP65	T-cell clones from PB and SF TCR αV and βV sequenced	[49]	(1) Clones, from normals and RA, preferentially recognise 1–170 and 303–540 regions of HSP65 (2) No cross-reaction with human HSP60 (3) No *in vivo* clonal activation and expansion of Myco HSP65-reactive T-cells in RA

therein. Some studies compared the responses of T-cells isolated from the joint – usually the SF rather than the SM – but only one studied the response of T-cells from an inflammatory site other than the joint [31]. In this study there was enhanced proliferation of T-cells from RA SF as well as from pleural exudates of patients without RA to acetone-precipitated *M. tuberculosis* extract, *E. coli* lysate containing recombinant mycoHSP65, and recombinant mycoHSP65 produced in *E. coli*, and *E. coli* lysate without the mycoHSP65 insert. The conclusion must be that T-cells from inflammatory sites are more reactive and that this enhanced reactivity is inflammation rather than disease-specific. This is also the conclusion of Res et al. [32].

Closer inspection of the published work reveals the rather disappointing picture that there is no consensus as to whether T-cells from patients with RA exhibited increased responsiveness to mycobacterial antigens and/or bacterial HSP [31–41] or not [32, 36, 37, 42–46]. There are several explanations for this failure. The first is that the underlying hypothesis may be wrong. We believe the evidence is most compatible with this conclusion. Other reasons for failure are mainly technical and methodological. Thus, it is not always clear whether the recombinant proteins used have been purified from other contaminating *E. coli* proteins, including *E. coli* HSP, or whether they contain endotoxin and, if so, the quantity. The culture conditions, methods for cell separation, the cell concentration and duration of cell culture are also different between the different studies. The duration of culture is crucial as culture times equal to or less than seven days are looking at antigen recall responses, the relevant response in this context. By contrast, culture times of greater than seven days may be eliciting a primary immune response *in vitro*. This is clearly not relevant to the underlying hypothesis.

This is one reason why we have not focused on work in which T-cell lines or clones have been generated as it is difficult to determine whether the cells thus generated came originally from a recall or a primary immune response [47, 48]. Celis et al. [49] made T-cell clones from RA PB and SF that were T-cell receptor (TCR) αβ positive and preferentially recognised peptides 1–170 and 303–540 of mycoHSP65. However, there was no evidence that such T-cells were expanded *in vivo* in the patients. Crucially for the hypothesis, these T-cell clones did not cross-react with huHSP60. However, for the sake of balance, it must be pointed out that the investigators made no effort to determine whether these mycoHSP-specific T-cell clones were able to recognise the homologous huHSP60 peptides as altered peptide ligands with a different cytokine secreting profile and lower proliferative rate. The predominant anti-mycoHSP T-cell clones were TH1 with high IFN-γ production although some produced IL-10. One clone was of the TH0 type as it produced equal amounts of IFN-γ, IL-10 and IL-4 [50].

Some of the experiments, however, give insights into how antigen-responsive T-cells may contribute and promote the pathogenesis of RA, as the processes uncovered are almost certainly not antigen-specific but rather mechanism-specific. Some of the mechanisms uncovered during work on HSP T-cell responses in RA are fascinating. These include lysis of macrophages by $CD4^+$ and $CD8^+$ T-cells [51]; single-cell explants from RA SM organise into pannus-like tissue if T-cells and mycobacterial antigens are added to the cultures [52]; and suppression of cartilage proteoglycan synthesis *in vitro via* the secretion of IL-4 and TNF-α [35]. The relevance of these observations to the pathogenesis of RA is clear. What is clear, from the analysis of the published evidence, is that HSP are not driving antigen responses in the rheumatoid joint. The nature of the antigens that do, remain to be determined.

Finally, we must briefly comment on the role, if any, of γδ T-cells in the pathogenesis of RA. Holoshitz [40] has recently reviewed this controversial area. T-cells responding to various mycobacterial antigens were found to be $CD8^+$ in the RA SF with a high proportion of Vδ1 while, by contrast, in the PB the γδ T-cells were $Ti/\gamma A^+/BB3^+$ [38]. The relevance of these observations to the pathogenesis of RA is not at all clear but the SF $\gamma\delta^+$ T-cells could cause joint damage by cytotoxic mechanisms [51]. It would be interesting and important if the $\gamma\delta^+$ T-cells found in the rheumatoid joint could be shown to be exerting regulatory influences within the RA SM, albeit inadequate for the degree of inflammation present.

Summary

There is no consensus that mycobacterial/bacterial HSP and the cross-reacting human HSP are significant antigens in the pathogenesis of RA.

Many of the studies suffer from methodological and technical flaws or use different experimental systems so that direct comparisons are difficult to make.

Some of the studies provide useful insights into possible pathogenetic mechanisms initiated and maintained by T-cells at play in the rheumatoid joint.

Perhaps the only test of the veracity of the hypothesis, that molecular mimicry between mycobacterial/bacterial HSP and human HSP is an important and relevant factor in the pathogenesis of RA, would be a successful therapeutic intervention based on such a hypothesis.

Molecular mimicry between the third hypervariable region of HLA-DRB1 and bacterial heat shock proteins

Having considered the role of molecular mimicry between mycobacterial/bacterial HSP and the homologous human HSP in the pathogenesis of RA *via* the stimulation of T-cells, we shall now consider the very special proposal by Roudier and his colleagues (reviewed in [53]) that the pathogenesis of RA is linked to molecular mimicry between the third hypervariable region (3HVR) of the HLA-DRB1 alleles linked to RA and molecular mimicry to the 40kD HSP from *E. coli*, DnaJ. DnaJ has an 11-

amino acid stretch of homology, particularly the segment spanning residues 61–65, with the 3HVR of HLA DRBI* 0401, encompassing residues 70–74. This suggests that molecular mimicry between these molecules could trigger autoimmunity or be a factor contributing to chronicity of rheumatoid arthritis (RA). Albani and colleagues [54] immunised rabbits with recombinant DnaJ (rDnaJ) protein and produced antibodies that reacted specifically with HLA DRBI*0401 B-lymphocytes. Conversely, a rabbit antibody raised against the 3HVR of HLA DRBI*0401 recognised rDnaJ in ELISA. The anti-rDnaJ antibody recognised native, conformational HLA DRBI*0401 present on the surface of B-lymphocytes. The gene for DnaJ is located immediately downstream of the DnaK gene, which codes for the *E. coli* HSP70. The DnaJ and DnaK proteins constitute a heat shock-inducible operon whose function includes protein folding and assembly. The human homologue of the *E. coli* HSP70, huHSP70, maps in the human MHC between the HLA-DR and HLA-B loci. It is not known whether there is a human homologue to DnaJ. Infection with *E. coli* is universal and most human sera contain anti-rDnaJ antibodies. Yet not all humans who are HLA DRBI*0401 positive have RA; clearly there is more to developing the disease than cross-reactivity at the antibody level between RA-associated DRB1 alleles and bacterial HSP.

A more relevant investigation would be to determine T-cell reactivity to the shared epitope and homologues of the shared epitope found in bacteria. Salvat et al., from the same group that carried out the foregoing investigation, looked at T-cell proliferation to the 3HVR peptide [55]. They used 15-mer peptides from the 3HVR of *0401, *0402, *0404, *0301, 1601, 0701 and 0801. Proliferation to self-peptides encompassing amino acids 65–79 from the 3HVR regions of seven different DRBI alleles was a common phenomenon. By contrast, lymphocytes from DRBI 0401 healthy control or patients with RA did not respond to DRBI *0401 3HVR peptide. Having excluded several confounding factors, including failure to present the peptide, they concluded that this was due to tolerance to the self-peptide. The favoured interpretation of this finding is that the RA-associated shared epitope may shape the T-cell repertoire during T-cell development in the thymus. However, McColl and colleagues [56] did not find increased T-cell reactivity to shared epitope peptides from DRB1*0401 or from *E. coli* peptides, both for proliferation and for production of IFN-γ, in patients with early onset RA. Interestingly, these patients did show T-cell reactivity to articular Type II collagen indicating that the failure to react to the peptides was not due to a general failure of lymphocyte responsiveness. Interestingly, T-cells from patients with RA do not proliferate to *E. coli* DnaK [37] but this is not predicted by Roudier's hypothesis, as this bacterial HSP does not contain the shared epitope.

Since the QKRAA sequence is found also on bacterial HSP, which are able to bind to and to transport other proteins, the shared epitope may promote strong protein-protein interaction. Direct confirmation of this hypothesis has been provided by Auger et al. [57]. Affinity columns of different 3HVR DR peptides were used to screen protein extracts from *E. coli.* Those columns carrying peptides containing the shared epitope QKRAA or the conservatively substituted RRRAA bound the 70 Kd HSP, DnaK. This finding would support the proposal, put forward above, that the QKRAA motif may promote protein interactions. Of additional, and possibly greater, relevance was the finding that in B-lymphoblastoid cell lines, homozygous for these same DRBI alleles, the constitutive 70Kd HSP, HSP73, co-precipitated with the HLA DRBI chains. HSP73 targets HLA DRBI*0401 to liposomes and thereby influence the processing of this and other proteins into antigenic peptides for presentation by antigen presenting cells. What is puzzling, and remains unexplained, is why the HSP73/HLA-DRBI interaction was not seen with another RA-susceptibility sequence, QRRAA.

In experiments similar to the foregoing, Auger and Roudier [58] showed that the QKRAA motif mediated the binding of DnaJ to DnaK, the 70 Kd *E. coli* HSP. Thus, the shared epitope does promote protein-protein interactions. They propose that it is possible that after exposure to Enterobacteriae in HLA-DRB*0401 individuals, bacterial DnaK proteins may bind the QKRAA motif on HLA-DR, triggering strong T-cell responses to HSP70s and, by unknown mechanisms, triggering RA. Plainly many other genes and, most likely, environmental factors, are involved in the triggering event in order to explain the relative rarity of RA compared to the frequency of alleles containing the shared epitope in the population.

However, the preceding experiments still left some doubts about the precise mechanisms involved in the interactions between HSP73, the human homologue of DnaK the 70Kd HSP of *E. coli* and HLA-DRBI alleles associated or not associated with rheumatoid arthritis. Auger and colleagues [59] attacked this problem head on by developing a quantitative precipitation assay and a direct binding assay. The development of these assays was necessary in order to refute criticisms of their previous by Rich et al. [60]. They showed that RA–associated HLA-DRBI alleles bound HSP73 better than did HLA-DRBI alleles that were not associated with RA. HLA-DRBI*0401 was the best HSP73 binder.

E. coli HSP70, which is the molecular chaperone DnaK, interacts with its co-chaperone DnaJ in an ATP-controlled cycle of polypeptide binding and release to mediate protein folding [61]. The bindings of DnaJ to DnaK can also be mediated by possession of QKRAA motif in DnaJ as shown in Figure 1, modified from Auger and Roudier (1997) [58]. Very strong support for this model has been provided by Roth et al. [62]. Since HSP73 is known to transport peptides for MHC Class I presentation [63] and also displays chaperone activity in MHC Class II presentation [64], it could be that the privileged interactions between HSP73 and HLA-DRBI*0401 could cause more efficient processing in the Class II pathway and either induce the development of autoimmunity or perpetuate T-cell responses during the course of joint inflammation. This mechanism could be operating during T-cell education in the thymus as well as within the periphery, during peptide presentation.

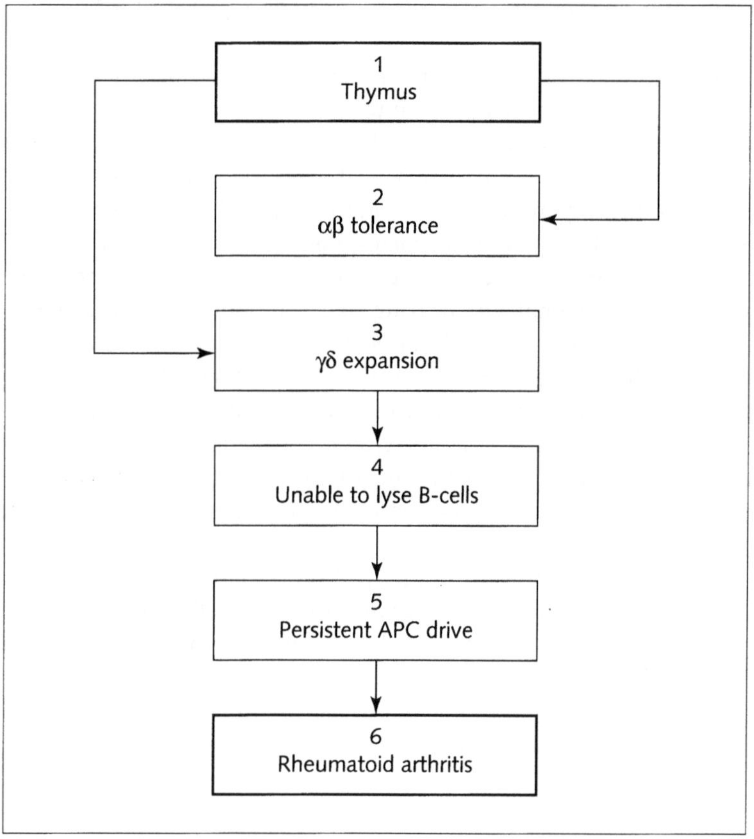

Figure 1
During thymic education (1) there is tolerance induction of αβ⁺ T-cells to the "shared epitope" (2), according to the hypothesis of Roudier [94].
Holoshitz [40] proposes that this leads to the expansion of γδ⁺ T-cells (3) that are unable to lyse B-cells (4) As a consequence, B-cells accumulate in the rheumatoid synovial membrane and efficiently present antigens, which drive the rheumatoid process (5), to disease specific T-cells, thus causing rheumatoid inflammation (6).

Maier et al. [65] have provided partial confirmation of these findings which showed, using a sensitive binding technique, that peptides covering the 3HVR from RA-susceptibility and non-susceptible alleles bound DnaK. Similar binding specificities were found for the constitutively expressed mammalian HSP, Hsc73, and the inducible mammalian Hsp72. Of great interest however was the finding that peptides containing the amino acid sequence DERAA, found in HLA-DR alleles strongly associated from protection from RA, did not bind any HSP70.

Unification of Roudier concept with T-cell responses to HSP in RA

Holoshitz has proposed the hypothesis that a possible explanation for the 3HVR association with RA is that it confers αβ T-cell tolerance so that it allows γδ T-cells with autoimmune potential to dominate [66]. The αβ T-cell tolerance is, of course, the corner stone of the Roudier hypothesis reviewed and discussed above. Holoshitz notes that there are significantly more γδ T-cells in the synovial membranes and synovial fluids of patients with RA than in the corresponding blood. However, there is controversy as to which subsets are expanded. On balance it would appear that there is expansion of an oligoclonal Vγ9-bearing γδ T-cell population in the peripheral blood and joints of patients with RA. Holoshitz has shown that early in the course of RA there is a high frequency of Vγ9/Vδ2 T-cell clones. Interestingly, the frequency decreases with progression of the disease. The implication from these observations is that in established disease polyclonally activated γδ T-cells predominate. The natural mycobacterial ligands that may be responsible for the expansion of Vγ9/Vδ2 cells have been identified as isopentenyl pyrophosphate and the related prenyl pyrophosphate derivative. The relevance of these observations to the pathogenesis of RA is not clear. The T-cells bearing Vγ9/Vδ2 recognise the same epitope, 180–188, within mycobacterial HSP65 recognised by αβ T-cells. These cells have the ability to lyse a large number of tumour cell lines but not lymphoblastoid B-cell lines from RA patients. It is of interest to note in this context that ablation of B cells with an anti-CD20 monoclonal antibody improves disease activity in RA [67]. These observations can be put together into a single unified concept as is shown in Figure 1.

BiP as an autoantigen and immuno-modulator in RA

We [68] and Blass and colleagues [69] have proposed that BiP is an autoantigen in RA on the basis of auto-antibody production and T-cell proliferation. There is a significantly increased concentration of anti-BiP antibodies in the serum of patients as compared to patients with other inflammatory joint diseases or healthy controls [70, 71]. Interestingly, patients with primary Sjøgren's Syndrome also have increased levels of anti-BiP antibodies not significantly different from those found in RA. Thus, patients with RA or primary Sjøgren's Syndrome share two auto-antibody systems, namely, rheumatoid factor and antibodies to BiP. There is no correlation between the two antibodies. The mechanism linking the co-expression of these two antibody systems is presently unknown.

Both groups have shown that the proliferation to BiP of PB T-cells is low whilst we have found that the proliferation of SF T-cells is significantly higher than in the corresponding PB and only found in patients with RA [68]. It is of interest that, despite the low proliferation rate of these RA PB and SF T-cells, we have been able

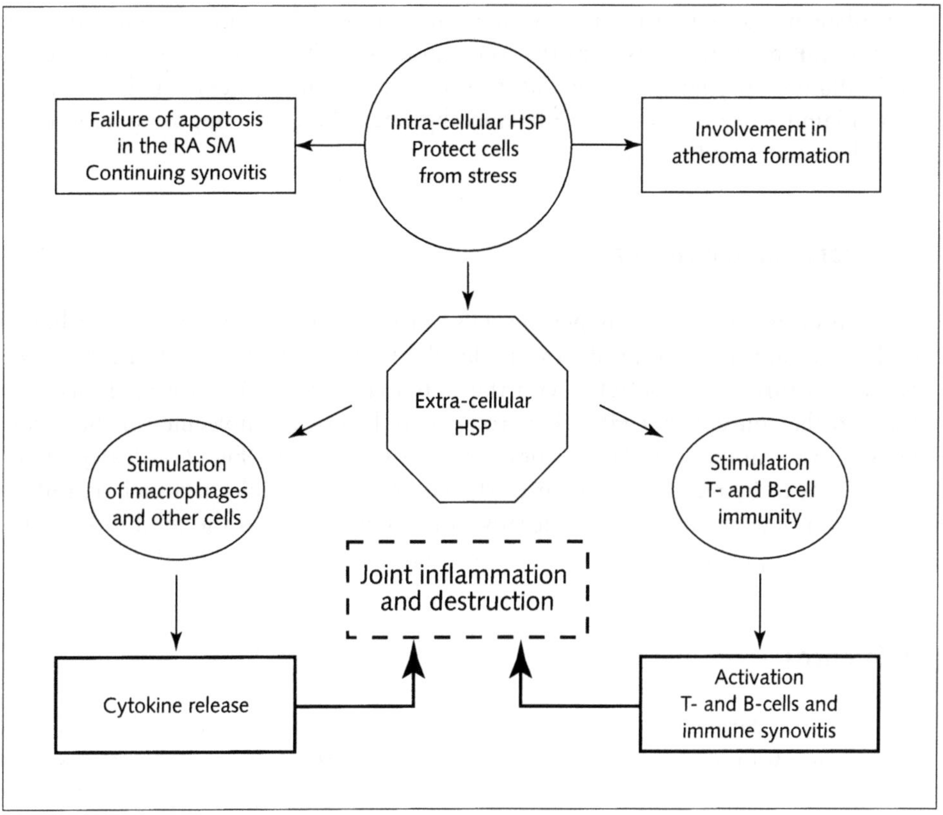

Figure 2
The multi-faceted involvement of HSP/chaperones in the pathogenesis of rheumatoid arthritis. The up-regulation of HSP in the rheumatoid synovium will lead to failure of apoptosis and hence, contribute to chronicity of inflammation by survival of cells such as the synovial fibroblasts. Systemic over-expression may also contribute to atheroma formation – a leading cause of death in RA. Extra-cellular cytokines may have pro- and anti-inflammatory effects depending on the nature of the cytokines released and the type of T-cells activated.

to generate both CD4 and CD8 T-cell clones from the SF or PB of patients with RA and from the PB of normal, healthy individuals [72–74]. These clones secrete IL-10 and are presently being investigated for their immunomodulatory potential.

BiP may have two additional immunomodulatory functions. The first is based on our observation that BiP stimulates anti-inflammatory cytokines, such as IL-10, from monocytes [75]. This suggests that BiP may have therapeutic potential for the treatment of RA. Indeed, this is borne out by our finding that intravenous (Myers L., unpublished) or subcutaneous (Thompson S.J., unpublished) BiP will treat ongoing collagen-induced arthritis. Thus one can propose the hypothesis that some chaperones/HSP may be released in the joint under conditions of stress and induce an anti-inflammatory/immunomodulatory programme while others may be inducing a pro-inflammatory response. The balance between these two opposing actions may well determine the final outcome.

Summary and conclusion

Our views of the role of chaperones/HSP in the pathogenesis of RA will have to undergo a radical reappraisal from molecules that may be involved in the perpetuation of arthritis to molecules with the potential to treat the disease. Indeed, this leads to the concept that HSP/chaperones may be critical molecules in the balance between pro- and anti-inflammatory forces in the RA SM (Fig. 2). Clearly the balance needs to be pushed in favour of the latter. Perhaps molecules such as BiP may usher in a new era of biologic therapy not based simply on the inhibition or neutralisation of pro-inflammatory cytokines such as IL-1, TNF-α or IL-6.

References

1 Dupuy C, Buzoni-Gatel D, Touze A, Bout D, Coursaget P (1999) Nasal immunization of mice with human papillomavirus type 16 (HPV-16) virus-like particles or with the HPV-16 L1 gene elicits specific cytotoxic T lymphocytes in vaginal draining lymph nodes. *J Virol* 73: 9063–9071

2 Klavinskis LS, Gao L, Barnfield C, Lehner T, Parker S (1997) Mucosal immunization with DNA-liposome complexes. *Vaccine* 15: 818–820

3 Tone M, Powell MJ, ToneY, Thompson SA, Waldmann H (2000) IL-10 gene expression is controlled by the transcription factors Sp1 and Sp3. *J Immunol* 165: 286–291.

4 Ma Y, Thornton S, Duwel LE, Boivin GP, Giannini EH, Leiden JM, Bluestone JA, Hirsch R (1998) Inhibition of collagen-induced arthritis in mice by viral IL-10 gene transfer. *J Immunol* 161: 1516–1524

5 Hajeer AH, Lazarus M, Turner D, Mageed RA, Vencovsky J, Sinnott P, Hutchinson IV, Ollie WE (1998) IL-10 gene promoter polymorphisms in rheumatoid arthritis. *Scand J Rheumatol* 27: 142–145

6 Raziuddin S, Bahabri S, Al Dalaan A, Siraj AK, Al Sedairy S (1998) A mixed Th1/Th2 cell cytokine response predominates in systemic onset juvenile rheumatoid arthritis: immune-regulatory IL-10 function. *Clin Immunol Immunopathol* 86: 192–198

7 Bucht A, Larsson P, Weisbrot L, ThorneC, Pisa P, Smedegard G, Keystone EC, Gronberg A (1996) Expression of interferon-gamma (IFN-gamma), IL-10, IL-12 and transforming growth factor-beta (TGF-beta) mRNA in synovial fluid cells from patients in the early and late phases of rheumatoid arthritis (RA). *Clin Exp Immunol* 103: 357–367

8 Kawakami A, Eguchi K, Matsuoka N, Tsuboi M, Urayama S, Kawabe Y, Aoyagi T, Maeda K, Nagataki S (1997) Inhibitory effects of interleukin-10 on synovial cells of rheumatoid arthritis. *Immunology* 91: 252–259

9 Kasama T, Strieter RM, Lukacs NW, Lincoln PM, Burdick MD, Kunkel SL. (1995) Interleukin-10 expression and chemokine regulation during the evolution of murine type II collagen-induced arthritis. *J Clin Invest* 95: 2868–2876

10 Herfarth, HH, Mohanty SP, Rath HC, Tonkonogy S, Sartor RB (1996) Interleukin 10 suppresses experimental chronic, granulomatous inflammation induced by bacterial cell wall polymers. *Gut* 39: 836–845

11 van Roon JA, van Roy JL, Gmelig-Meyling FH, Lafeber FP, Bijlsma JW (1996) Prevention and reversal of cartilage degradation in rheumatoid arthritis by interleukin-10 and interleukin-4. *Arthritis Rheum* 39: 829–835

12 Isomaki P, Luukkainen R, Saario R, Toivanen P, Punnonen J (1996) Interleukin-10 functions as an anti-inflammatory cytokine in rheumatoid synovium. *Arthritis Rheum* 39: 386–395

13 Dasgupta B, Dolan AL, Panayi GS, Fernandes L (1998) An initially double-blind controlled 96 week trial of depot methylprednisolone against oral prednisolone in the treatment of polymyalgia rheumatica. *Br J Rheumatol* 37: 189–195

14 Bergroth V, Kontinnen YT, Nykanen P, Von Essen R, Koota K (1985) Proliferating cells in the synovial fluid in rheumatic disease. An analysis with autoradiography-immunoperoxidase double staining. *Scand J Immunol* 22: 383–388

15 Panayi GS (1997) Clinical improvement and radiological deterioration in rheumatoid arthritis. *Br J Rheumatol* 36: 820–821

16 Mertz PM, DeWitt DL, Stetler-Stevenson WG, Wahl LM (1994) Interleukin 10 suppression of monocyte prostaglandin H synthase-2. Mechanism of inhibition of prostaglandin-dependent matrix metalloproteinase production. *J Biol Chem* 269: 21322–21329

17 Katsikis PD, Chu CQ, Brennan FM, Maini RN, Feldmann M (1994) Immunoregulatory role of interleukin 10 in rheumatoid arthritis. *J Exp Med* 179: 1517–1527

18 Pitzalis C, Pipitone N, Bajocchi G, Hall M, Goulding N, Lee A, Kingsley G, Lanchbury J, Panayi G (1997) Corticosteroids inhibit lymphocyte binding to endothelium and intercellular adhesion: an additional mechanism for their anti-inflammatory and immunosuppressive effect. *J Immunol* 158: 5007–5016

19 Akbar AN, Terry L, Timms AM, Beverley PCL, Janossy G (1988) Loss of CD45R and gain of UCHL1 reactivity is a feature of primed T cells. *J Immunol* 140: 2171–2178

20 Panayi GS (1997) T-cell-dependent pathways in rheumatoid arthritis. *Curr Opin Rheumatol* 9: 236–240

21 Hirata D, Hirai I, Iwamoto M, Yoshio T, Takeda A, Masuyama JI, Mimori A, Kano S, Minota S (1997) Preferential binding with *Escherichia coli* hsp60 of antibodies prevalent in sera from patients with rheumatoid arthritis. *Clin Immunol Immunopathol* 82: 141–148

22 Handley HH, Yu J, Yu DT, Singh B, Gupta RS, Vaughan JH (1996) Autoantibodies to

human heat shock protein (hsp)60 may be induced by *Escherichia coli* groEL. *Clin Exp Immunol* 103: 429–435

23 Krenn V, Vollmers HP, von Landenberg P, Schmausser B, Rupp M, Roggenkamp A, Muller-Hermelink HK (1996) Immortalized B-lymphocytes from rheumatoid synovial tissue show specificity for bacterial HSP 60. *Virchows Arch* 427: 511–518

24 Handley HH, Ngyuen MD, Yu DT, Gupta RS, Vaughan JH (1995) Purification of recombinant human Hsp60: use of a GroEL-free preparation to assess autoimmunity in rheumatoid arthritis. *J Autoimmun* 8: 659–673

25 Jarjour WN, Jeffries BD, Davis JS, Welch WJ, Mimura T, Winfield JB (1991) Autoantibodies to human stress proteins. A survey of various rheumatic and other inflammatory diseases. *Arthritis Rheum* 34: 1133–1138

26 Panchapakesan J, Daglis M, Gatenby P (1992) Antibodies to 65 kDa and 70 kDa heat shock proteins in rheumatoid arthritis and systemic lupus erythematosus. *Immunol Cell Biol* 70 (Pt 5): 295–300

27 Tsoulfa G, Rook GA, Bahr GM, Sattar MA, Behbehani K, Young DB, Mehlert A, Van Embden JD, Hay FC, Isenberg DA et al (1989) Elevated IgG antibody levels to the mycobacterial 65-kDa heat shock protein are characteristic of patients with rheumatoid arthritis. *Scand J Immunol* 30: 519–527

28 Kalabay L, Fekete B, Czirjak L, Horvath L, Daha MR, Veres A, Fonyad G, Horvath A, Viczian A, Singh M et al (2002) *Helicobacter pylori* infection in connective tissue disorders is associated with high levels of antibodies to mycobacterial hsp65 but not to human hsp60. *Helicobacter* 7: 250–256

29 Worthington J, Rigby AS, MacGregor AJ, Silman AJ, Carthy D, Ollier WE (1993) Lack of association of increased antibody levels to mycobacterial hsp65 with rheumatoid arthritis: results from a study of disease discordant twin pairs. *Ann Rheum Dis* 52: 542–544

30 Panayi GS, Corrigall VM, Pitzalis C (2001) Pathogenesis of rheumatoid arthritis. The role of T cells and other beasts. *Rheum Dis Clin North Am* 27: 317–334

31 Res PC, Telgt D, van Laar JM, Pool MO, Breedveld FC, de Vries RR (1990) High antigen reactivity in mononuclear cells from sites of chronic inflammation. *Lancet* 336: 1406–1408

32 Res PC, Schaar CG, Breedveld FC, van Eden W, Van Embden JD, Cohen IR, de Vries RR (1988) Synovial fluid T cell reactivity against 65 kD heat shock protein of mycobacteria in early chronic arthritis. *Lancet* 2: 478–480

33 van Roon JA, van Roy JL, Duits A, Lafeber FP, Bijlsma JW (1995) Proinflammatory cytokine production and cartilage damage due to rheumatoid synovial T helper-1 activation is inhibited by interleukin-4. *Ann Rheum Dis* 54: 836–840

34 Kogure A, Miyata M, Nishimaki T, Kasukawa R (1994) Proliferative response of synovial fluid mononuclear cells of patients with rheumatoid arthritis to mycobacterial 65 kDa heat shock protein and its association with HLA-DR^{+}gamma delta^{+} T cells. *J Rheumatol* 21: 1403–1408

35 Wilbrink B, Holewijn M, Bijlsma JW, van Roy JL, den Otter W, van Eden W (1993) Sup-

pression of human cartilage proteoglycan synthesis by rheumatoid synovial fluid mononuclear cells activated with mycobacterial 60-kd heat-shock protein. *Arthritis Rheum* 36: 514–518
36 Pope RM, Lovis RM, Gupta RS (1992) Activation of synovial fluid T lymphocytes by 60-kd heat-shock proteins in patients with inflammatory synovitis. *Arthritis Rheum* 35: 43–48
37 Pope RM, Wallis RS, Sailer D, Buchanan TM, Pahlavani MA (1991) T cell activation by mycobacterial antigens in inflammatory synovitis. *Cell Immunol* 133: 95–108
38 Soderstrom K, Halapi E, Nilsson E, Gronberg A, van Embden J, Klareskog L, Kiessling R (1990) Synovial cells responding to a 65-kDa mycobacterial heat shock protein have a high proportion of a TcR gamma delta subtype uncommon in peripheral blood. *Scand J Immunol* 32: 503–515
39 Zou J, Rudwaleit M, Thiel A, Lauster R, Braun J, Sieper J (2002) T cell response to human HSP60 and yersinia 19 kDa in ankylosing spondylitis and rheumatoid arthritis: no evidence for a causal role of these antigens in the pathogenesis. *Ann Rheum Dis* 61: 473–474
40 Holoshitz J (1999) Activation of gammadelta T cells by mycobacterial antigens in rheumatoid arthritis. *Microbes Infect* 1: 197–202
41 MacHt LM, Elson CJ, Kirwan JR, Gaston JS, Lamont AG, Thompson JM, Thompson SJ (2000) Relationship between disease severity and responses by blood mononuclear cells from patients with rheumatoid arthritis to human heat-shock protein 60. *Immunology* 99: 208–214
42 Doran MC, Goodstone NJ, Hobbs RN, Ashton BA (1995) Cellular immunity to cartilage link protein in patients with inflammatory arthritis and non-arthritic controls. *Ann Rheum Dis* 54: 466–470
43 Crick FD, Gatenby PA (1992) Limiting-dilution analysis of T cell reactivity to mycobacterial antigens in peripheral blood and synovium from rheumatoid arthritis patients. *Clin Exp Immunol* 88: 424–429
44 Fischer HP, Sharrock CE, Colston MJ, Panayi GS (1991) Limiting dilution analysis of proliferative T cell responses to mycobacterial 65-kDa heat-shock protein fails to show significant frequency differences between synovial fluid and peripheral blood of patients with rheumatoid arthritis. *Eur J Immunol* 21: 2937–2941
45 Burmester GR, Altstidl U, Kalden JR, Emmrich F (1991) Stimulatory response towards the 65 kDa heat shock protein and other mycobacterial antigens in patients with rheumatoid arthritis. *J Rheumatol* 18: 171–176
46 Ramage JM, Gaston JS (1999) Depressed proliferative responses by peripheral blood mononuclear cells from early arthritis patients to mycobacterial heat shock protein 60. *Rheumatology (Oxford)* 38: 631–635
47 Quayle AJ, Wilson KB, Li SG, Kjeldsen-Kragh J, Oftung F, Shinnick T, Sioud M, Forre O, Capra JD, Natvig JB (1992) Peptide recognition, T cell receptor usage and HLA restriction elements of human heat-shock protein (hsp) 60 and mycobacterial 65-kDa

hsp-reactive T cell clones from rheumatoid synovial fluid. *Eur J Immunol* 22: 1315–1322
48 Melchers I, Jooss-Rudiger J, Peter HH (1997) Reactivity patterns of synovial T-cell lines derived from a patient with rheumatoid arthritis. I. Reactions with defined antigens and auto-antigens suggest the existence of multi-reactive T-cell clones. *Scand J Immunol* 46: 187–194
49 Celis L, Vandevyver C, Geusens P, Dequeker J, Raus J, Zhang J (1997) Clonal expansion of mycobacterial heat-shock protein-reactive T lymphocytes in the synovial fluid and blood of rheumatoid arthritis patients. *Arthritis Rheum* 40: 510–519
50 Quayle AJ, Chomarat P, Miossec P, Kjeldsen-Kragh J, Forre O, Natvig JB (1993) Rheumatoid inflammatory T-cell clones express mostly Th1 but also Th2 and mixed (Th0-like) cytokine patterns. *Scand J Immunol* 38: 75–82
51 Li SG, Quayle AJ, Shen Y, Kjeldsen-Kragh J, Oftung F, Gupta RS, Natvig JB, Forre OT (1992) Mycobacteria and human heat shock protein-specific cytotoxic T lymphocytes in rheumatoid synovial inflammation. *Arthritis Rheum* 35: 270–281
52 Holoshitz J, Kosek J, Sibley R, Brown DA, Strober S (1991) T lymphocyte-synovial fibroblast interactions induced by mycobacterial proteins in rheumatoid arthritis. *Arthritis Rheum* 34: 679–686
53 Auger I, Toussirot E, Roudier J (1998) HLA-DRB1 motifs and heat shock proteins in rheumatoid arthritis. *Int Rev Immunol* 17: 263–271
54 Albani S, Tuckwell JE, Esparza L, Carson DA, Roudier J (1992) The susceptibility sequence to rheumatoid arthritis is a cross-reactive B cell epitope shared by the *Escherichia coli* heat shock protein dnaJ and the histocompatibility leukocyte antigen DRB10401 molecule. *J Clin Invest* 89: 327–331
55 Salvat S, Auger I, Rochelle L, Begovich A, Geburher L, Sette A, Roudier J (1994) Tolerance to a self-peptide from the third hyper-variable region of HLA DRB1*0401 in rheumatoid arthritis patients and normal subjects. *J Immunol* 153: 5321–5329
56 McColl GJ, Hammer J, Harrison LC (1997) Absence of peripheral blood T cell responses to 'shared epitope' containing peptides in recent onset rheumatoid arthritis. *Ann Rheum Dis* 56: 240–246
57 Auger I, Escola JM, Gorvel JP, Roudier J (1996) HLA-DR4 and HLA-DR10 motifs that carry susceptibility to rheumatoid arthritis bind 70-kD heat shock proteins. *Nat Med* 2: 306–310
58 Auger I, Roudier J (1997) A function for the QKRAA amino acid motif: mediating binding of DnaJ to DnaK. Implications for the association of rheumatoid arthritis with HLA-DR4. *J Clin Invest* 99: 1818–1822
59 Auger I, Lepecuchel L, Roudier J (2002) Interaction between heat-shock protein 73 and HLA-DRB1 alleles associated or not with rheumatoid arthritis. *Arthritis Rheum* 46: 929–933
60 Rich T, Gruneberg U, Trowsdale J (1998) Heat shock proteins, HLA-DR and rheumatoid arthritis. *Nat Med* 4: 1210–1211
61 Suh WC, Lu CZ, Gross CA (1999) Structural features required for the interaction of the

Hsp70 molecular chaperone DnaK with its cochaperone DnaJ. *J Biol Chem* 274: 30534–30539

62 Roth S, Willcox N, Rzepka R, Mayer MP, Melchers I (2002) Major differences in antigen-processing correlate with a single Arg71<-- >Lys substitution in HLA-DR molecules predisposing to rheumatoid arthritis and with their selective interactions with 70 kDa heat shock protein chaperones. *J Immunol* 169: 3015–3020

63 Castellino F, Boucher PE, Eichelberg K, Mayhew M, Rothman JE, Houghton AN, Germain RN (2000) Receptor-mediated uptake of antigen/heat shock protein complexes results in major histocompatibility complex class I antigen presentation *via* two distinct processing pathways. *J Exp Med* 191: 1957–1964

64 Panjwani N, Akbari O, Garcia S, Brazil M, Stockinger B (1999) The HSC73 molecular chaperone: involvement in MHC class II antigen presentation. *J Immunol* 163: 1936–1942

65 Maier JT, Haug M, Foll JL, Beck H, Kalbacher H, Rammensee HG, Dannecker GE (2002) Possible association of non-binding of HSP70 to HLA-DRB1 peptide sequences and protection from rheumatoid arthritis. *Immunogenetics* 54: 67–73

66 Holoshitz J (1990) Potential role of gamma delta T cells in autoimmune diseases. *Res Immunol* 141: 651–657

67 Leandro MJ, Edwards JC, Cambridge G (2002) Clinical outcome in 22 patients with rheumatoid arthritis treated with B lymphocyte depletion. *Ann Rheum Dis* 61: 883–888

68 Corrigall VM, Bodman-Smith MD, Fife MS, Canas B, Myers LK, Wooley P, Soh C, Staines NA, Pappin DJ, Berlo SE et al (2001) The human endoplasmic reticulum molecular chaperone BiP is an autoantigen for rheumatoid arthritis and prevents the induction of experimental arthritis. *J Immunol* 166: 1492–1498

69 Blass S, Union A, Raymackers J, Schumann F, Ungethum U, Muller-Steinbach S, De Keyser F, Engel JM, Burmester GR (2001) The stress protein BiP is overexpressed and is a major B and T cell target in rheumatoid arthritis. *Arthritis Rheum* 44: 761–771

70 Bodman-Smith MD, Corrigall VM, Chan C, Panayi GS (2002) Anti-BiP antibodies in the serum of patients with autoimmune disease. *Immunology* 107 (Suppl): S1 OP203

71 Bodman-Smith MD, Corrigall VM, Chan C, Panayi GS (2003) Anti-BiP antibodies in the serum of patients with autoimmune disease. *Rheumatology (Oxford)* 42 (Suppl): 44

72 Bodman-Smith MD, Corrigall VM, Panayi GS (2001) The molecular chaperone BiP preferentially stimulates IL-10 producing T cell clones from normal individuals. *FASEB J* 15: 5 A1215

73 Bodman-Smith MD, Corrigall VM, Kemeny DM, Panayi GS (2002) The human chaperone BiP stimulates interleukin (IL) 10 producing CD8 T cells: implications for rheumatoid arthritis. *Arthritis Res* 4: S1 A21

74 Bodman-Smith MD, Corrigall VM, Kemeny DM, Panayi GS (2003) The putative RA autoantigen BiP stimulates interleukin-10 producing $CD8^+$ cells. *Rheumatology (Oxford)* 42 (Suppl)

75 Bodman-Smith MD, Corrigall VM, Kemeny DM, Panayi GS (2003) BiP, a putative

autoantigen in rheumatoid arthritis, stimulates IL-10 producing CD8-positive T cells from normal individuals. *Rheumatology* 42: 1–8

76 Corrigall VM, Bodman-Smith MD, Panayi GS (2002) BiP may have a regulatory function mediated through IL-10 production. *Arthritis Res* 4: S1 A20

77 Karlsson-Parra A, Soderstrom K, Ferm M, Ivanyi J, Kiessling R, Klareskog L (1990) Presence of human 65 kD heat shock protein (hsp) in inflamed joints and subcutaneous nodules of RA patients. *Scand J Immunol* 31: 283–288

78 Evans DJ, Norton P, Ivanyi J (1990) Distribution in tissue sections of the human groEL stress-protein homologue. *APMIS* 98: 437–441

79 Sharif M, Worrall JG, Singh B, Gupta RS, Lydyard PM, Lambert C, McCulloch J, Rook GA (1992) The development of monoclonal antibodies to the human mitochondrial 60 kd heat-shock protein, and their use in studying the expression of the protein in rheumatoid arthritis. *Arthritis Rheum* 35: 1427–1433

80 Kodama E, Kasukawa R, Miyata M, Shigeta S, Ito M (1995) Analysis of human 65 kD heat shock protein mRNA using polymerase chain reaction in synovia of rheumatoid arthritis patients. *Fukushima J Med Sci* 41: 95–102

81 Sato H, Miyata M, Kasukawa R (1996) Expression of heat shock protein on lymphocytes in peripheral blood and synovial fluid from patients with rheumatoid arthritis. *J Rheumatol* 23: 2027–2032

82 Kurzik-Dumke U, Schick C, Rzepka R, Melchers I (1999) Over-expression of human homologs of the bacterial DnaJ chaperone in the synovial tissue of patients with rheumatoid arthritis. *Arthritis Rheum* 42: 210–220

83 Fang SM, Lin CS, Lyon V (1977) Progesterone retention by rat uterus I. Pharmacokinetics after uterine intraluminal instillation. *J Pharm Sci* 66: 1744–1748

84 Giacomello A, Salerno C, Ferrari M, Giartosio A, Fasella P 1979 A collagen film for microdetermination of collagenase activity. *Physiol Chem Phys* 11: 169–173

85 Oda A, Miyata M, Kodama E, Satoh H, Sato Y, Nishimaki T, Nomaguchi H, Kasukawa R (1994) Antibodies to 65 Kd heat-shock protein were elevated in rheumatoid arthritis. *Clin Rheumatol* 13: 261–264

86 Adebajo AO, Williams DG, Hazleman BL, Maini RN (1995) Antibodies to the 65 kDa mycobacterial stress protein in west Africans with rheumatoid arthritis, tuberculosis and malaria. *Br J Rheumatol* 34: 352–354

87 Lai NS, Lan JL, Yu CL, Lin RH (1995) Antibody to Mycobacterium tuberculosis 65 kDa heat shock protein in patients with rheumatoid arthritis – a survey of antigen-specific antibody isotypes and subclasses in an endemic area of previous tuberculosis infection. *Ann Rheum Dis* 54: 225–228

88 Hayem G, De Bandt M, Palazzo E, Roux S, Combe B, Eliaou JF, Sany J, Kahn MF, Meyer O (1999) Anti-heat shock protein 70 kDa and 90 kDa antibodies in serum of patients with rheumatoid arthritis. *Ann Rheum Dis* 58: 291–296

89 Hattori T, Takahash K, Yutani Y, Fujisawa T, Nakanishi T, Takigawa M (2000) Rheumatoid arthritis-related antigen 47kDa (RA-A47) is a product of colligin-2 and acts as a human HSP47. *J Bone Miner Metab* 18: 328–334

90 Yokota SI, Hirata D, Minota S, Higashiyama T, Kurimoto M, Yanagi H, Yura T, Kubota H (2000) Autoantibodies against chaperonin CCT in human sera with rheumatic autoimmune diseases: comparison with antibodies against other Hsp60 family proteins. *Cell Stress Chaperones* 5: 337–346

91 Yoshida A, Nakano Y, Yamashita Y, Oho T, Ito H, Kondo M, Ohishi M, Koga T (2001) Immunodominant region of Actinobacillus actinomycetemcomitans 40-kilodalton heat shock protein in patients with rheumatoid arthritis. *J Dent Res* 80: 346–350

92 Horvath L, Czirjak L, Fekete B, Jakab L, Prohaszka Z, Cervenak L, Romics L, Singh M, Daha MR, Fust G (2001) Levels of antibodies against C1q and 60 kDa family of heat shock proteins in the sera of patients with various autoimmune diseases. *Immunol Lett* 75: 103–109

93 van Roon JA, van Eden W, van Roy JL, Lafeber FJ, Bijlsma JW (1997) Stimulation of suppressive T cell responses by human but not bacterial 60 kD heat-shock protein in synovial fluid of patients with rheumatoid arthritis. *J Clin Invest* 100: 459–463

94 Roudier J (2000) Association of MHC and rheumatoid arthritis. Association of RA with HLA-DR4: the role of repertoire selection. *Arthritis Res* 2: 217–220

95 Gaston JS, Life PF, van Der ZR, Jenner PJ, Colston MJ, Tonks S, Bacon PA (1991) Epitope specificity and MHC restriction of rheumatoid arthritis synovial T cell clones which recognize a mycobacterial 65 kDa heat shock protein. *Int Immunol* 3: 965–972

Heat shock proteins for immunotherapy of rheumatoid arthritis

Gisella L. Puga Yung[1,2], *Tho D. Le*[1], *Sarah Roord*[3], *Berent Prakken*[3,4] *and Salvatore Albani*[1,4,5]

[1]Departments of Medicine and Pediatrics, University of California San Diego, 9500 Gilman Drive, La Jolla, CA 92093-0663, USA; [2]Instituto de Ciencias Biomédicas (ICBM), Facultad de Medicina Norte, Universidad de Chile, Av. Independencia 1027, Casilla 13898, Santiago 653499, Chile; [3]Wilhelmina Children's Hospital, University Medical Center, P.O. Box 85090, 3508 AB Utrecht, The Netherlands; [4]IACOPO Institute for Translational Medicine, 9500 Gilman Drive, La Jolla, CA 92093-0663, USA; [5]Androclus Therapeutics, 4204 Sorrento Valley Blvd., Ste A-C, San Diego, CA 92121, USA

Introduction

Rheumatoid arthritis (RA) is an inflammatory disease that primarily involves the joints and has a worldwide prevalence of about one percent, with a female to male ratio of 3:1. This chapter aims to summarize some of the recent progresses in molecular immunology, and to discuss the application of this new knowledge for therapeutic purposes. We will focus on recent experiences from us and others in modulation of antigen specific responses as a tool for manipulating autoimmune inflammation. Particular emphasis will be given to the concept of exploiting for therapeutic purposes a natural mechanism of immune regulation. This mechanism is based on sequential cross recognition of bacterial and human derived heat shock protein peptides.

Rheumatoid arthritis

Rheumatoid arthritis is a chronic systemic inflammatory disease which occurs when the synovial membranes in the joints are damaged by infiltrating mononuclear phagocytes, lymphocytes, and neutrophils [1]. Even though the course of RA is variable, patients tend to develop a progressive loss of cartilage and bone around the joints, resulting in highly painful and impaired mobility.

Even if the cause and pathogenesis of this disease is complex, involving both genetic and environmental factors [2], several of its features are suggestive of an autoimmune etiology. The pathology of arthritic joints is suggestive of a T-cell-mediated chronic inflammatory reaction [3, 4].

Heat Shock Proteins and Inflammation, edited by Willem van Eden

In recent years, much research has been focused on the infiltrating T-cells and mononuclear phagocytes. Research from several groups has tried to identify different populations of T-cells that may be involved in the pathogenic process. These studies have demonstrated that the T-cell repertoire in RA is relatively restricted, mimicking the characteristics of an antigen driven, yet polymorphic population. A pool of antigens may exist with the capability of triggering Th1 delayed hypersensitivity responses. If such conditions in the synovial micro-environment are allowed to persist, such responses would be expected to lead to joint damage [2, 5–15]. Depletion of T-cells by thoracic duct drainage [16] or by immunosuppressive drugs, such as cyclosporin, has resulted in marked improvement, implying the importance of T-cells in the pathogenesis of RA [17].

Susceptibility to RA on the genetic level has been found to be associated with genes located in the Human Class II Major Histocompatibility complex (HLA) encoding the β-chains of HLA-DR molecules [18, 19]. A positive association is observed with several alleles of the DRB1 gene. Upon comparison of protein sequences in the third hyper-variable region of these disease associated alleles, it has been observed that they share a five-amino acid stretch QKRAA, called the "shared epitope sequence" [20], thereby suggesting a role for auto antigen presentation [21].

Evolving therapy: From pharmacologic to immuno-mediated suppression

Current therapy for rheumatoid arthritis includes a wide spectrum of medications that can be grouped according to their characteristics and to the pathophysiologic pathways which they target (Tab. 1). A common trait within the plethora of different medications available is their generic lack of specificity. Most available drugs target, in fact, one or more pathways of pathophysiologic inflammation. This leads to clinically detectable positive effects, such as, for instance, reduction of pain or control of the erosive process. However, shutting down entirely a given pathway, which has many physiologic roles, leads often to problems, which may range from tolerable side effects to grave induces immunodeficiencies. The balancing act between disease control and undesirable effects of the treatment is becoming increasingly more difficult. In fact, the recent dramatic advances in molecular immunology have brought to the market a new generation of potent biological agents, eminently interfering with a cytokine which is part of pro-inflammatory pathways. Efficacy for these drugs is often remarkable, to the point that some, particularly those interfering with TNF-α-mediated inflammation, are gaining ground as first line agents in some aggressive therapeutic protocols. Aside for the cost, the limitations of use for these agents reside in increased risks of transitory immunodeficiency, with some reported cases of tuberculosis infections. Moreover the duration of the therapeutic effects for these drugs is often associated with continuous administration. In fact, if the drug is withdrawn there is often a relapse in disease activity.

Table 1 - Traditional rheumatoid arthritis treatments

Drug	Physiological effects (process influenced)
NSAIDs	Prostaglandin synthesis, leukotriene synthesis, superoxide radical production, superoxide scavenging, lysosomal enzyme release, cell membrane activities; enzymes NAPDDH oxidase and phospholipase; transmembrane ion transport; uptake of prostaglandin precursor, neutrophil aggregation and adhesion, lymphocyte function, rheumatoid factor production, cytokine production, cartilage metabolism, syntesis of oxide nitric.
Analgesics	Central level
Glucocorticoids	Reduction of: inflammation caused by cytokines, NO, prostaglandins, leukotrienes; reduced emigration of leukocytes from vessels, induction of apoptosis in lymphocytes and eosinophils.
DMARDs	Interferes with adenine and guanine ribonucleosides, crosslink DNA, suppresses IL2 synthesis and release, suppresses T-cell response and interaction, inhibition of dihydrofolate reductase, thymidylate synthetase and phosphoribosylaminoimidazolecarboxamide transformylase activity, IL1 and IL2 suppression, metalloproteinase inhibition: possible effects on PMN and lymphocyte function.
Biologic response modifiers	Neutralization of pro-inflammatory cytokines, blockage of pro-inflammatory cytokine signaling.

Other biological agents are currently being developed and tested. Some target small molecules involved in signaling, and are still faraway from large scale trials. Other efforts have focused on eliminating indiscriminately, by using monoclonal antibodies, one or more subsets of immune cells involved in the pathogenic process. These approaches have often shown transitory results with a high degree of side effects. A new generation of monoclonal antibodies is being developed, which capitalizes on the previous experience in the hopes of ameliorating the balance between efficacy and safety. Other therapeutic attempts have been aimed, by employing either peptides or antibodies, at T- and B-cell idiotype/antidiotype networks with yet unconvincing results.

Altogether, the next generation of immune therapy drugs will face the challenge of matching the efficacy of currently available biologics while reducing costs and side effects. This challenge can probably be met only by evolving the target focus from non-specific to disease related pathophysiologic mechanisms. These mechanisms include the likely multiple antigens which contribute to generation and perpetuation of the inflammatory process. Unfortunately, this approach has been hampered by the attempt to identify the "one" antigen responsible for the disease. This almost holistic search has, to date, been unsuccessful, neither have been the correlated therapeutic attempt aimed at one specific antigen considered as the main culprit. The mixed results with tolerization to collagen and the recent negative outcome of Organon's gp39 Phase II trial are among the recent examples. Reasons for these unsatisfactory outcomes can be several, starting from inappropriate patients' selection to lack of measurement of biological efficacy of the treatment. Conceptually, however, antigen-specific T-cell therapy could have a better chance to succeed if the antigen chosen played a specific role in pathogenic rather than in the etiologic process.

Immune basis of physiologic tolerance

Understanding the role that T-cells play in the immune system is important in understanding how autoimmunity occurs. During the maturation process, T-cells that have a specificity for foreign antigens are selected while those that recognize autoantigens are deleted, tolerated, down-regulated [22], or may simply ignore the target protein [23, 24]. Autoimmunity occurs when this system is altered and these T-cells become activated upon encounter with the body's own proteins. This can occur for several reasons; when auto-antigens that are normally not exposed to the peripheral environment become exposed, or it can occur due to cross recognition between foreign and self-antigens carrying similar epitopes.

The vertebrate immune system is comprised of two systems: innate (or natural) and adaptive (or acquired) immunity. Innate immunity provides for a rapid antimicrobial response that precedes a more specific adaptive immune response. The innate immune system employs different cell populations and soluble components (e.g., antibacterial peptide, complement, tumor necrosis factor (TNF), IL-1, IL-12 and IL-18). The early host defense also has an additional role in determining the nature of downstream adaptive immune responses [25]. Adaptive immunity involves the use of lymphocytes to remove foreign invaders in an antigen-specific fashion.

Inflammation, which is closely associated with both arms of the immune system, is another mechanism for clearance of foreign antigens. However, inflammation is also involved in driving autoimmune reactions. It is clear that the difference between autoimmune protection and autoimmune disease is a matter of the intensity and the timing of the autoimmune inflammation, because the autoimmune T-cells may be the same [26]. Destructive autoimmunity is the final consequence of a complex multi-step process, and is strongly supported by inflammation. Multiple steps that accomplish the inflammatory process are described in Figure 1. The first inflamma-

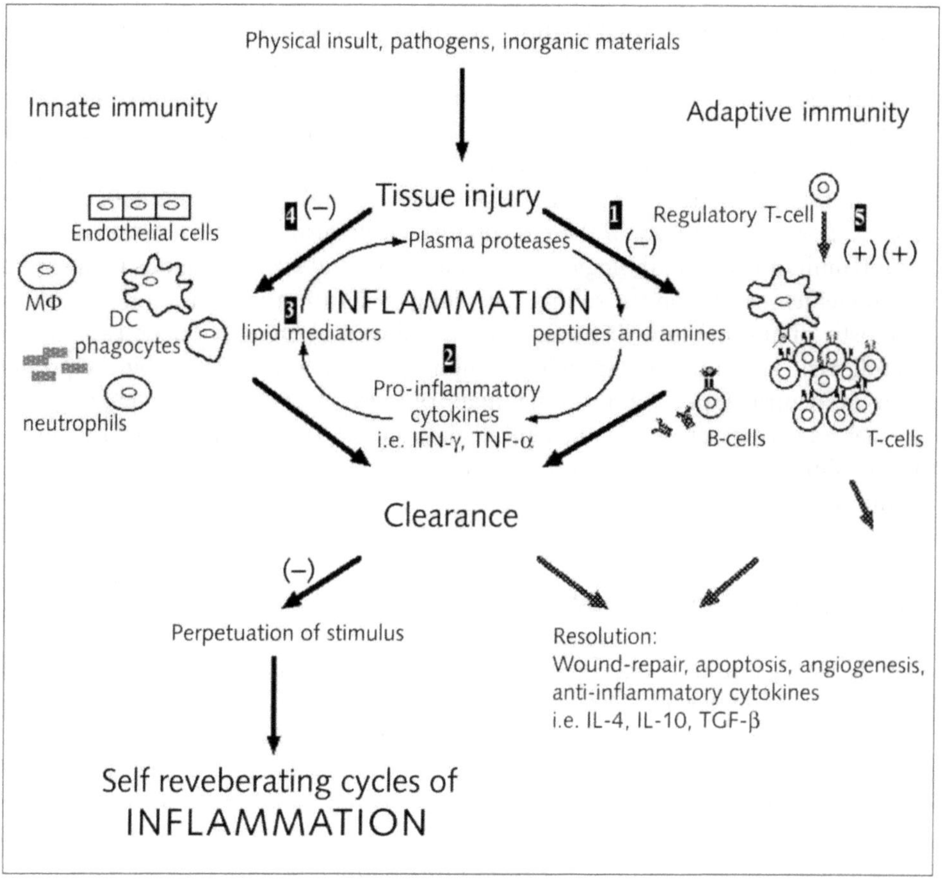

Figure 1
Inflammatory process involves both the innate and adaptive arms of immunity.
After damage, for instance by physical insult, pathogens, organic materials, tissue injury follows. Many mediators are produced such as plasma proteases, peptides and amines, pro-inflammatory cytokines, and lipid mediators. These compounds are produced by several different kinds of cells: endothelial cells, macrophages (MΦ), dendritic cells (DC), phagocytes, neutrophils, NK, NKT, T-, and B-cells. The main purpose of inflammation is the quick death and clearance of foreign antigen as well as the damage repair. Clearance and resolution may take the form of wound repair, apoptosis, angiogenesis, and/or anti-inflammatory cytokines. There is also a possibility of stimulus perpetuation thus producing self reverberating cycles of inflammation that can lead to autoimmune diseases. T-cells are involved in the process to turn off the initial inflammation by the production of regulatory cytokines as well as the down-regulation of their T-cell receptors, or by going into clonal deletion or anergy. Traditional therapy in RA focuses on steps 1 through to 4 but disease outcome would be better controlled through the modulation of the immune system in step 5.

tory events may be generated by different ways as physical insult, pathogens or inorganic materials. Innate as much as adaptive immunity is activated, thus releasing several agents into the micro-environment. Under normal conditions, the inflammatory process is turned off as soon as possible to avoid its deleterious effect. Mechanisms involving regulatory T-cells and anti-inflammatory cytokines are secreted later in the inflammatory process; helping the organism to stop the inflammation. Absence or low efficiency of these regulatory mechanisms may lead to perpetuation of inflammation and pathologic autoimmunity. Several steps along this pathway can be implemented to stop the self-reverberating cycles of inflammation that are characteristic of autoimmune diseases. These are shown in Figure 1 (steps 1–5) and are as follows: (1) inhibit non-specific T-lymphocyte activation, (2) inhibit the production or neutralization of pro-inflammatory cytokines, (3) reduce inflammatory mediators such as prostaglandins and leukotrienes, (4) intervene in innate immunity by inhibiting nitric oxide synthase and adhesion molecules, and (5) T-cell specific intervention.

Strategy behind the therapy

T-cells play a role in the events initiating and propagating autoimmune inflammation in RA. Modulation of such T-cell driven, epitope-specific pathways would represent a major addition to the available therapeutic options. This method would be specific, non-toxic, and possibly permanent. Modulation of T-cells in an antigen-specific fashion may lead to control of autoimmune inflammation, with possible clinical improvement. In order to not repeat some of the mistakes of the past, however, several things need to be changed. In particular:

1) a clear understanding of the mechanism that leads to epitope specific immune deviation is essential prior to venturing into larger scale clinical application;
2) the characteristics of the epitope of choice need to be relevant to immune pathogenesis;
3) the characteristics of the route chosen for immune deviation need to be fully exploited.

The route – mucosal tolerance

Tolerance could be defined as any mechanism by which recognition of an antigen is not followed by an immune response devoted at its elimination. As such, tolerance is an important component of a healthy immune system. Autoreactive cells, such as those reacting with brain antigens, thyroglobulin, serum albumin, collagen and others autoantigens are, in fact, not deleted and are present in all individuals

[27]. They not only remain harmless under normal conditions, but autoreactive cells may have an important function in maintaining tissue homeostasis and may be differentially focused depending on the tissue and the autoantigen [28]. The basis of immunologic tolerance does not appear to be simply distinguishing between self and non-self, but reacting to danger signals that confront the immune system [29]. Thus, immunological tolerance cannot rely solely on neonatal deletional events, but requires and active process that functions during the entire life of the organism.

Mucosal tolerance is an immunological mechanism designed to deal with external antigens such as foods that gain access to the body. Tolerance is an active immunologic process and is mediated by more than one mechanism. The administration of low doses of antigen favors the induction of active cellular regulation [30], whereas higher doses favor the induction of anergy or deletion [31, 32]. Thus, mucosal tolerance is a complex process that involves suppression/modulation of some immune responses and the induction of others. Antigen-specific peripheral T-cell tolerance can be induced through oral or nasal administration of a relevant antigen [33, 34]. In several experimental autoimmune models, oral administration of a disease triggering autoantigen has led to considerable suppression of disease activity [35–37].

Successful experiments in RA rat models (adjuvant induced arthritis (AIA)) have shown promising results using heat shock protein (HSP) peptides to modulate the induced disease. Such examples include HSP60 [38], HSP65 [39], HSP10 [40], HSP70 [41], and various other examples reported in the literature. For instance, in rat AIA it demonstrated that using an altered peptide ligand of HSP60, it was possible to induce highly effective protection against AIA through the generation of regulatory cells that produce IL-4, TGF-β, and IL-10. Furthermore, it has been reported that this induced tolerance is driven mainly by production of IL10 [38]. Also Wendling et al., has shown that mucosal tolerance, with a peptide like a HSP70 peptide, the intrinsic capacity of a conserved bacterial HSP to trigger self-HSP cross-reactive T-cells with the potential to down regulate arthritis *via* IL10 [41].

Pan-HLA DR binding heat shock protein peptides as immunoregulatory agents

The approach which we are testing is based on the concept that epitope-specific T-cell therapy should target immune responses directly involved in the pathogenesis rather than in the etiology of autoimmune disease. It is therefore possible that some pathways are shared among different diseases. It should also be considered that T-cell mediated inflammation may be driven by several different antigens, which should have in common the following characteristics in order to be candidates as immunomodulatory agents:

1) Be part of proteins with documented strong antigenic potential.
2) Induce the production of cytokines with either stimulating or regulatory function in inflammation.
3) Be part of proteins that are readily available and possibly over expressed at the site of inflammation.
4) Possibly be part of proteins that contain domains conserved across species. This latter characteristic would be important in the context of abnormalities in immune regulation induced by cross-reactive recognition (i.e., molecular mimicry).

Heat shock proteins are uniquely positioned in this context. Heat shock proteins are ubiquitous, being expressed by virtually every living cell, from prokaryotes to eukaryotes. They are up-regulated under stress-type stimuli, including inflammatory processes occurring during normal immune responses [42]. Heat shock proteins are highly conserved, with significant interspecies homologies [43]. HSPs are strongly immunogenic molecules as well, having the capability of modulating autoimmune processes, as shown in several systems [7, 44–47].

More than ten years of experiments have been devoted to studying the potential role of bacterial heat shock proteins in the pathogenesis of arthritis, first in animal models, then in people [46, 48–50]. Heat shock proteins are major bacterial antigens. For example, antibodies and T-cells reactive with HSP65 and dnaJ class of HSPs are abundant in synovial fluids of RA patients [44, 48, 50–56].

Sequence similarities between self-and exogenous proteins are common. For proteins such as enzymes, this might simply reflect a conservation of functional sites as with HSPs. However, this might also have additional implications for proteins in the immune system. Molecular mimicry might play an important part in the creation of the immune repertoire; it might also represent a way for a pathogen to escape the host's immune surveillance. Molecular mimicry might be the trigger for abnormal cross reactive responses that lead, in some instances, to autoimmune responses [57–59].

Recent studies done by several groups including ourselves have found that immunological responses to bacterial HSPs are implicated in the pathogenesis of arthritis in animals and in humans. The role of HSPs has been widely described and provide a promising avenue of intervention. HSPs are present at the site of inflammation and have been described as relevant targets of T-cell responses in immune mediated diseases such as arthritis. Bacterial HSPs have human counterparts with a high degree of homology. Hence it has been suggested that potentially pathogenic responses are initially triggered by the encounter of the immune system of the host with proteins of bacterial origin, and subsequently perpetuated by recognition on self-homologues. This cross reactivity may be part of the chronic inflammation process. We have found (Kampuis, Albani and Prakken, submitted, and Massa, Prakken and Albani, manuscript submitted) in several inflammatory diseases that

E. coli dnaJp1 peptide sequence	**QKRAA**YDQYGHAAFE
S. typhi dnaJ	**QKRAA**YDRFGHAAFE
V. cholerae dnaJ	**QKRAA**YDRYGHAAFE
Brucella ovis dnaJ	**QKRAA**YDRFGHAAFE
R. fredii dnaJ	**QKRAA**YDRYGHAAFE
H. influenzae dnaJ	**QKRAA**YDQYGHAAFE
C. crescentus dnaJ	**QKRAA**YDRFGHAGVN
Lactobacillus lactis dnaJ	**QKRAA**YDQYGEAGAN
HLA DRB1*0401	**QKRAA**VDTYCRHNYG
	shared epitope

Figure 2
*Aligned amino acid sequences of the dnaJ proteins that contain the "shared epitope". Also included is the sequence of the third variable region of HLA DRB1*0401.*

recognition of bacterial HSP-derived peptides is associated with an up-regulation of the inflammatory process. Conversely, recognition of human-derived HSP peptides leads to generation of regulatory mechanisms which may be central in physiologic regulation of inflammation but may be altered in autoimmune conditions, such as rheumatoid arthritis. Our objective is therefore to induce a recalibration of this "molecular dimmer" for therapeutic purposes. The idea is to exploit and expand the regulatory mechanisms by either inducing tolerance to pro-inflammatory epitopes or by increasing T-cells specific for self-derived epitopes that possess naturally regulatory properties.

The dnaJ class of HSP from *E. coli* and several other pathogens contain the common QKRAA sequence called the "shared epitope" (Fig. 2).

In preliminary experiments, cellular and humoral immune responses to the region of *E. coli* dnaJ that includes the QKRAA sequence in adult patients with early RA have strong immune responses to the epitope that resembles the RA sequence motif, and not in normal subjects or patients with other autoimmune diseases [7]. A very good correlation exists between the HLA-binding avidity of a peptide derived from dnaJp, dnaJp1, and T-cell responsiveness in patients with RA [7]. The "shared epitope" sequence, when placed at the N-terminal of the peptide (dnaJp1), was necessary both for T-cell activation and the observed effects on HLA binding. Indeed, individual amino acid substitutions or sequence frame shifts resulted in dramatic reductions in HLA-binding avidity and T-cell responsiveness. HLA-DR is the presenting molecule for the dnaJp1 peptide [7].

We have also found that T-cell recognition of the dnaJp1 peptide, is associated with RA and not simply a consequence of DRB1*0401 expression [15]. Early RA patients also have enhanced humoral and cellular responses to antigen from the

Epstein-Barr virus, *Brucella bovis*, and *Lactobacillus lactis* that also contain the QKRAA sequence [14].

There are several hypotheses that have been proposed to explain why this reactivity is preferentially observed in patients with RA. The first hypothesis assumes that these amino acids are essential for the formation of a special binding pocket [60, 61]. Another hypothesis proposed that the "shared epitope" also exists in foreign antigens, e.g., micro-organisms. Within a different framework, the same amino acid sequence may function as an epitope for the immune system and eventually break the tolerance of the system toward peptides derived from its own DRB chains (a variation of the "molecular mimicry" hypothesis) [15, 62]. Indeed, the QKRAA sequence was identified within the *E. coli* dnaJ chaperone; in rabbit antisera, immunologic cross reactions between these foreign antigens and the DR1 chain have been described [13]. Also, another possible hypothesis is that this QKRAA stretch may be an important sequence motif with a biologic function different from antigen presentation (the "functional motif" hypothesis) [63–65].

We propose that in RA, interplay between HLA and dnaJp-derived peptides maintains and stimulates T-cells, which participate in autoimmune inflammation [14, 15, 66, 67]. DnaJp1 belongs to a group of HSP-derived peptides which share many functional aspects that are found in different inflammatory diseases. These peptides may act as natural dimmers of inflammation and have been designed in order to contain HLA binding motifs enabling binding to the most commonly represented haplotypes. Hence, the approach which we developed for dnaJp1 in RA could be theoretically extended to various autoimmune diseases bypassing haplotypic differences.

DnaJp1 has several interesting characteristics that make it a good candidate for T-cell epitope specific immunotherapy:

1) it belongs to an HSP that is highly antigenic;
2) it induces cytokine production during the inflammation process;
3) dnaJ human homologues are expressed in the inflamed synovium tissues [51] and may be part of the "immunological dimmer" mechanism;
4) as with other HSPs, it is highly conserved across species;
5) it has the "shared epitope" sequence;
6) it is also able triggers pro-inflammatory responses in T-cells from RA patients in a disease specific fashion [7]. This is a unique approach that allows a large field of possibilities.

A peptide as "molecular dimmer"

Based on the route of administration and the antigen, we propose to gradually modulate T-cell responses to the target epitope. Expansion of regulatory mechanisms can be achieved by either inducing tolerance to pro-inflammatory epitopes or by increasing T-cells specific for self-derived epitopes, which possess natural regulatory properties acting through bystander suppression. In essence, we are using a conserved domain of *E. coli* HSP that is also a homolog of human antigens to exploit the normal regulatory cells with reactivity against such epitopes to re-establish the balance between autoimmunity and immunity.

As discussed before, autoreactive T-cells upon cross recognition between pathogenic and self-homologues may become activated and driven toward a pro-inflammatory state (Fig. 3 upper). On the other hand, the same autoreactive T-cells may encounter the same antigen (dnaJp1 peptide), but in a different environment, such as the mucosal tissue; thereby producing an entirely different result. Due to the characteristic of the mucosa as discussed before, the immune response assembled will drive T autoreactive cells towards a regulatory pathway. Now these cells will be able to modulate and turn off the inflammatory process by the secretion of anti-inflammatory cytokines such as IL-10 and TGF-β. These T-cell populations are able to deviate into regulatory T-cells and thus carry out their function by specific or through bystander suppression mechanisms (Fig. 3 lower).

Mucosal modulation of immune response to HSP in RA

A Phase I clinical trial was performed using 15 patients with early rheumatoid arthritis that were treated for six months with dnaJp1. The objective of this project was to modulate immune responses to dnaJp1 by inducing a qualitative switch of antigen-specific T-cell responses from a pro-inflammatory phenotype to a more tolerogenic one. Inclusion criteria for the study required diagnosis of clinically active disease and immune responsiveness to dnaJp1. Patients were monitored monthly for clinical and immunological effects of the treatment. *In vitro* T-cell responses to dnaJp1 were monitored by measuring T-cell proliferation and cytokine production. Data obtained from this completed Phase I clinical trial has yielded interesting observations:

1) Treatment with dnaJp1 is safe. No side effects have been reported to date by the patients enrolled. This confirms the general consensus regarding the proven safety of oral tolerization approaches in humans.
2) *In vitro* responses to dnaJp1 are accompanied, in patients with RA before the beginning of the treatment, by production of pro-inflammatory cytokines such as IFN-γ, and interestingly, also TNF-α (Tab. 2). This latter finding may underscore the central role played by HSP-derived peptides in chronic inflammatory pathways related to autoimmunity.
3) By employing a novel technique (T-cell capture, TCC) for the detection of epitope specific T-cells [68], we could estimate dnaJp1-specific T-cell numbers pre-

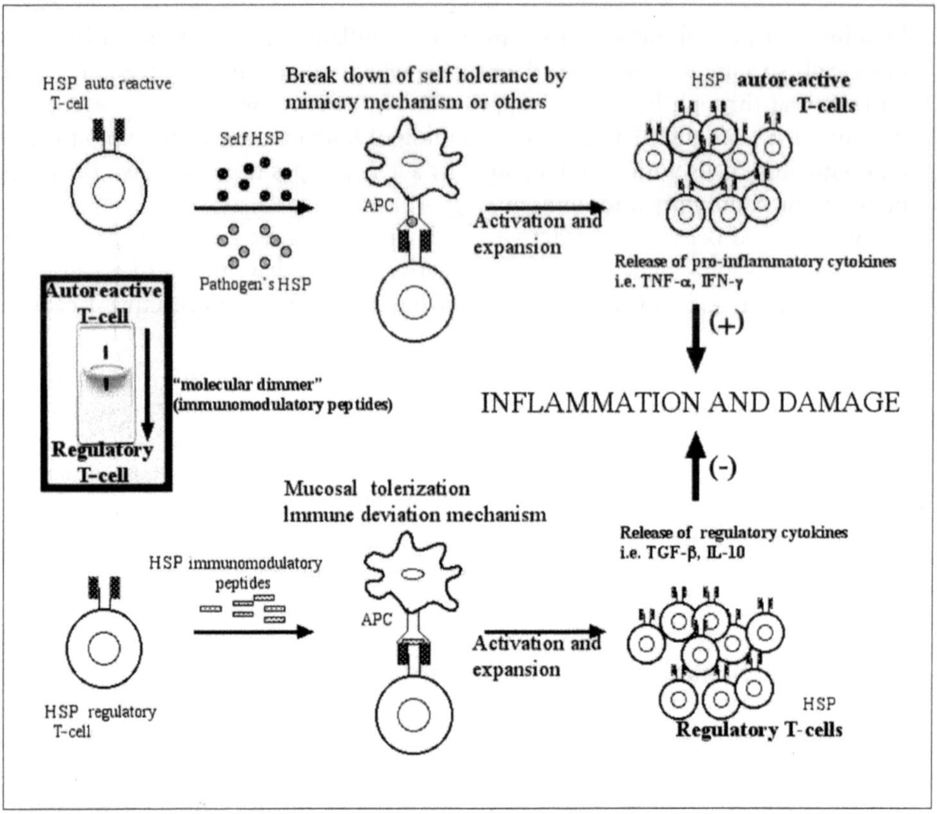

Figure 3
"Molecular dimmer" model.
Upper panel shows how peptide epitopes are able to induce inflammation. These peptides may belong to foreign or self antigens, but under an inflammatory environment they are able to be activated into effector T-cells and produce more inflammation and damage. On the other hand, the same or homologous peptides may also be able to induce tolerogenic behaviors (lower panel) depending on the route that they are presented (i.e., mucosal tolerization). These immunomodulatory peptides share common characteristics because they are recognized by regulatory cells and are able to stimulate those regulatory cells to maintain immune balance through secretion of regulatory cytokines like IL-10 and TGF-β.

sent in PBMC as 1% of $CD3^+$ cells specific for the peptide in short-term cell lines stimulated with dnaJp1. This number is significantly lower than the number of $CD3^+$ cells producing, upon stimulation with dnaJp1 in the same culture, IFN-γ and TNF-α. Hence, bystander activation effects are implicated with reactivity to

Table 2 - DnaJp1 modulation of in vitro *immune response*

Day of treatment	***In vitro* proliferation % of reduction**	**IL-2 % of reduction**	**IFN-γ % of reduction**	**TNF-α % of reduction**
14	0%	0%	0%	0%
28	9.1%	98.2%	99.4%	25%
56	−27%	46.5%	94%	90%
84	9.1%	98.2%	97.6%	100%
112	36.4%	98.2%	99.8%	100%
140	−27%	98.2%	100%	100%

Cytokine data represent intracellular staining of $CD3^+$ gated PBMC after incubation with dnaJp1. Results are expressed as % change as compared to day 14. In vitro *proliferation performed by ^3H uptake using freshly isolated PBMC. Results of proliferation assays are expressed as % change as compared to day 14.*

dnaJp1 in RA. The modulation of this pattern of reactivity alone may represent a valid therapeutical approach; complementary to other immunologic or pharmaceutical strategies.

4) Oral treatment with dnaJp1 reduces T-cell proliferative response, as well as production of IL-2, IFN-γ and TNF-α by both specific and bystander cells, as seen in Table 2.
5) Production of IL-4, occurs upon treatment. The number of cells producing IL-4 largely exceeds the stimulated number of dnaJp1-specific T-cells. This effect, which is the consequence of an epitope-specific treatment and which is presumably tolerogenic is therefore, affecting a large population of bystander cells [69].
6) We found by TCC that dnaJp1-specific T-cell numbers did not change, while the majority of TNF-α-producing cells were replaced by IL-10 and IL-4 producing cells, thus indicating true immune deviation rather than clonal deletion. The success of this trial has been followed by a Phase II trial that is currently running.

In summary, we have evidence of a treatment-induced switch from a pro-inflammatory to a tolerogenic pattern of cytokine production from T-cells in RA patients. This qualitative change is not a consequence random cyclic relapses of RA, as shown by the fact that changes in T-cell responses to control antigens and mitogens changed randomly during the course of the treatment.

Table 3 - Statistical analysis of clinical outcome data

Treatment groups	p-values *versus* no treatment on day 23	Area under the curve (AUC)	p-values *versus* no treatment (AUC)
Etanercept (3 doses)	0.004	42.639	0.03
Etanercept (1 dose)	0.299	48.858	0.1
180–188 peptide (4 doses)	0.069	54.726	0.26
Etanercept (1 dose) + 180–188 peptide (4 doses)	0.0004	40.369	0.02
No treatment		77.748	

Combination treatment with Etanercept and mycobacterial HSP60 peptide 180–188 considerably suppresses adjuvant induced arthritis (AIA), statistical analysis of clinical outcome data. AIA was induced in susceptible Lewis rats on day 0 with Mycobacterium tuberculosis *in IFA. From day seven, arthritis scores were assessed. On day nine rats were randomly divided into four treatment groups (n = 15), receiving:*

(1) 3 doses of Etanercept s.c. (day 9, 11, 13);

(2) 1 dose of Etanercept s.c.;

(3) 4 doses of HSP60 peptide 180–188, nasally;

(4) 1 dose of Etanercept s.c. (day 9) followed by four doses of 180–188 nasally (day 10, 13, 16, 19) and

(5) no treatment.

Combination of epitope-specific and anti-cytokine therapy as a novel therapeutic approach

In studies in AIA [70] we have explored whether the combination of epitope-specific T-cell therapy and anti-TNF-α therapy would enable disease control while allowing for a significant reduction of anti TNF-α dose. Our studies have shown that single dose of Etanercept, a potent anti-inflammatory agent, followed by mucosal tolerance induction to the arthritogenic epitope (HSP60 peptide 180–188), led to a significant suppression of arthritis to the same extent as a full course of Etanercept therapy in adjuvant induced arthritis (AIA) rats (Tab. 3).

This combination therapy led to the induction of antigen specific regulatory cells, characterized by the production of IL-10, IL-4 and TGB-β as well as to the expression of CTLA4 in the mandibular lymphonode cells (MNL) and to a decrease of IFN-γ production. This indicates that this tolerance might be directed by IL-10 producing regulatory T-cells.

This finding may, in our view, offer exciting prospects for immunotherapy of various autoimmune diseases.

Conclusion

We have found that *E. coli* HSP derived dnaJp1 peptide is only one of several possible antigens able to modulate the immune system in RA. We were able to show that using an HSP derived peptide it was possible to modulate the immune behavior of T-cells with clinically positive effects on RA patients. Effective immunotherapy of RA in the near future will most likely comprise of a combination of several approaches, given the complexity of pathogenic mechanisms, the likely multiplicity of the antigenic triggers, and the central role played by bystander, antigen non-specific mechanisms. Immunomodulation of the inflammatory process by one or more combined HSP through mucosal tolerization is one of the enticing tools which current research is offering. In this context, it will be important to emphasize also that trial design must take into account measurement of changes induced by the treatment on relevant immune responses, therefore dissecting to a molecular level the mechanisms which the therapy is targeting. Reliance on clinical outcomes alone may in fact be deceiving. This may be a crucial point to differentiate future attempts from previous failures.

References

1 Palmer DG (1995) The anatomy of the rheumatoid lesion. *Br Med Bull* 51: 286–295
2 Harris ED Jr (1990) Rheumatoid arthritis. Pathophysiology and implications for therapy. *N Engl J Med* 322: 1277–1289
3 Salmon M, Gaston JS (1995) The role of T-lymphocytes in rheumatoid arthritis. *Br Med Bull* 51: 332–345
4 De Keyser F, Elewaut D, Vermeersch J, De Wever N, Cuvelier C, Veys EM (1995) The role of T cells in rheumatoid arthritis. *Clin Rheumatol* 14 (Suppl) 2: 5–9
5 Feldmann M, Brennan FM, Maini RN (1996) Rheumatoid arthritis. *Cell* 85: 307–310
6 Fox D A (1997) The role of T cells in the immunopathogenesis of rheumatoid arthritis: new perspectives. *Arthritis Rheum* 40: 598–609
7 Albani S, Keystone EC, Nelson JL, Ollier WE, La Cava A, Montemayor AC, Weber DA, Montecucco C, Martini A, Carson DA (1995) Positive selection in autoimmunity: abnormal immune responses to a bacterial dnaJ antigenic determinant in patients with early rheumatoid arthritis. *Nat Med* 1: 448–452

8 Weyand CM, Goronzy JJ (1997) The molecular basis of rheumatoid arthritis. *J Mol Med* 75: 772–785
9 Kohsaka H, Nanki T, Ollier WE, Miyasaka N, Carson DA (1996) Influence of the rheumatoid arthritis-associated shared epitope on T-cell receptor repertoire formation. *Proc Assoc Am Physicians* 108: 323–328
10 Walser-Kuntz DR, Weyand CM, Weaver AJ, O'Fallon WM, Goronzy JJ (1995) Mechanisms underlying the formation of the T cell receptor repertoire in rheumatoid arthritis. *Immunity* 2: 597–605
11 Hingorani R, Monteiro J, Furie R, Chartash E, Navarrete C, Pergolizzi R, Gregersen PK (1996) Oligoclonality of V beta 3 TCR 20 chains in the CD8^{+} T cell population of rheumatoid arthritis patients. *J Immunol* 156: 852–858
12 Weyand CM, McCarthy TG, Goronzy JJ (1995) Correlation between disease phenotype and genetic heterogeneity in rheumatoid arthritis. *J Clin Invest* 95: 2120–2126
13 Albani S, Tuckwell JE, Esparza L, Carson D A, Roudier J (1992) The susceptibility sequence to rheumatoid arthritis is a cross-reactive B cell epitope shared by the Escherichia coli heat shock protein dnaJ and the histocompatibility leukocyte antigen DRB10401 molecule. *J Clin Invest* 89: 327–331
14 La Cava A, Nelson JL, Ollier WE, MacGregor A, Keystone EC, Thorne JC, Scavulli JF, Berry CC, Carson DA, Albani S (1997) Genetic bias in immune responses to a cassette shared by different microorganisms in patients with rheumatoid arthritis. *J Clin Invest* 100: 658–663
15 Albani S, Carson DA (1996) A multistep molecular mimicry hypothesis for the pathogenesis of rheumatoid arthritis. *Immunol Today* 17: 466–470
16 Vaughan JH, Fox RI, Abresch RJ, Tsoukas CD, Curd JG, Carson DA (1984) Thoracic duct drainage in rheumatoid arthritis. *Clin Exp Immunol* 58: 645–653
17 Sany J (1990) Immunological treatment of rheumatoid arthritis. *Clin Exp Rheumatol* 8 Suppl 5: 81–88
18 Ollier W, Thomson W (1992) Population genetics of rheumatoid arthritis. *Rheum Dis Clin North Am* 18: 741–759
19 Wordsworth BP, Lanchbury JS, Sakkas LI, Welsh KI, Panayi GS, Bell JI (1989) HLA-DR4 subtype frequencies in rheumatoid arthritis indicate that DRB1 is the major susceptibility locus within the HLA class II region. *Proc Natl Acad Sci USA* 86: 10049–10053
20 Rich T, Gruneberg U, Trowsdale J (1998) Heat shock proteins, HLA-DR and rheumatoid arthritis. *Nat Med* 4: 1210–1211
21 Winchester RJ, Gregersen PK (1988) The molecular basis of susceptibility to rheumatoid arthritis: the conformational equivalence hypothesis. *Springer Semin Immunopathol* 10: 119–139
22 Hammerling GJ, Schonrich G, Ferber I, Arnold B (1993) Peripheral tolerance as a multistep mechanism. *Immunol Rev* 133: 93–104
23 Ohashi PS, Oehen S, Buerki K, Pircher H, Ohashi CT, Odermatt B, Malissen B, Zinker-

nagel RM, Hengartner H (1991) Ablation of "tolerance" and induction of diabetes by virus infection in viral antigen transgenic mice. *Cell* 65: 305–317
24 Oldstone MB, Nerenberg M, Southern P, Price J, Lewicki H (1991) Virus infection triggers insulin-dependent diabetes mellitus in a transgenic model: role of anti-self (virus) immune response. *Cell* 65: 319–331
25 Fearon DT, Locksley RM (1996) The instructive role of innate immunity in the acquired immune response. *Science* 272: 50–53
26 Cohen IR (2000) Discrimination and dialogue in the immune system. *Semin Immunol* 12: 215–219; discussion 257–344
27 Zhang J, Markovic-Plese S, Lacet B, Raus J, Weiner HL, Hafler DA (1994) Increased frequency of interleukin 2-responsive T cells specific for myelin basic protein and proteolipid protein in peripheral blood and cerebrospinal fluid of patients with multiple sclerosis. *J Exp Med* 179: 973–984
28 Cohen IR, Young DB (1991) Autoimmunity, microbial immunity and the immunological homunculus. *Immunol Today* 12: 105–110
29 Matzinger P (1994) Tolerance, danger, and the extended family. *Annu Rev Immunol* 12: 991–1045
30 Chen Y, Kuchroo VK, Inobe J, Hafler DA, Weiner HL (1994) Regulatory T cell clones induced by oral tolerance: suppression of autoimmune encephalomyelitis. *Science* 265: 1237–1240
31 Whitacre CC, Gienapp IE, Orosz CG, Bitar DM (1991) Oral tolerance in experimental autoimmune encephalomyelitis. III. Evidence for clonal anergy. *J Immunol* 147: 2155–2163
32 Chen Y, Inobe J, Marks R, Gonnella P, Kuchroo VK, Weiner HL. (1995) Peripheral deletion of antigen-reactive T cells in oral tolerance. *Nature* 376: 177–180
33 Metzler B, Wraith DC (1993) Inhibition of experimental autoimmune encephalomyelitis by inhalation but not oral administration of the encephalitogenic peptide: influence of MHC binding affinity. *Int Immunol* 5: 1159–1165
34 Hoyne GF, O'Hehir RE, Wraith DC, Thomas WR, Lamb JR (1993) Inhibition of T cell and antibody responses to house dust mite allergen by inhalation of the dominant T cell epitope in naive and sensitized mice. *J Exp Med* 178: 1783–1788
35 Higgins PJ, Weiner HL (1988) Suppression of experimental autoimmune encephalomyelitis by oral administration of myelin basic protein and its fragments. *J Immunol* 140: 440–445
36 Zhang ZJ, Davidson L, Eisenbarth G, Weiner HL (1991) Suppression of diabetes in nonobese diabetic mice by oral administration of porcine insulin. *Proc Natl Acad Sci USA* 88: 10252–10256
37 Thompson HS, Staines NA (1986) Gastric administration of type II collagen delays the onset and severity of collagen-induced arthritis in rats. *Clin Exp Immunol* 64: 581–586
38 Prakken BJ, Roord S, van Kooten PJ, Wagenaar JP, van Eden W, Albani S, Wauben MH (2002) Inhibition of adjuvant-induced arthritis by interleukin-10-driven regulatory cells

induced *via* nasal administration of a peptide analog of an arthritis-related heat-shock protein 60 T cell epitope. *Arthritis Rheum* 46: 1937–1946
39 Cobelens PM, Kavelaars A, van der Zee R, van Eden W, Heijnen CJ (2002) Dynamics of mycobacterial HSP65-induced T-cell cytokine expression during oral tolerance induction in adjuvant arthritis. *Rheumatology (Oxford)* 41: 775–779
40 Agnello D, Scanziani E, Di GM, Leoni F, Modena D, Mascagni P, Introna M, Ghezzi P, Villa P (2002) Preventive administration of *Mycobacterium tuberculosis* 10-kDa heat shock protein (hsp10) suppresses adjuvant arthritis in Lewis rats. *Int Immunopharmacol* 2: 463–474
41 Wendling U, Paul L, van der Zee R, Prakken B, Singh M, van Eden W (2000) A conserved mycobacterial heat shock protein (hsp) 70 sequence prevents adjuvant arthritis upon nasal administration and induces IL-10-producing T cells that crossreact with the mammalian self-hsp70 homologue. *J Immunol* 164: 2711–2717
42 Dubois P (1989) Heat shock proteins and immunity. *Res Immunol* 140: 653–659
43 Jones DB, Coulson AF, Duff GW (1993) Sequence homologies between hsp60 and autoantigens. *Immunol Today* 14: 115–118
44 Albani S, Ravelli A, Massa M, De Benedetti F, Andree G, Roudier J, Martini A, Carson DA (1994) Immune responses to the *Escherichia coli* dnaJ heat shock protein in juvenile rheumatoid arthritis and their correlation with disease activity. *J Pediatr* 124: 561–565
45 Windhagen A, Nicholson LB, Weiner HL, Kuchroo VK, Hafler DA (1996) Role of Th1 and Th2 cells in neurologic disorders. *Chem Immunol* 63: 171–186
46 van Eden W, Thole JE, van der Zee R, Noordzij A, van Embden JD, Hensen EJ, Cohen IR (1988) Cloning of the mycobacterial 24 epitope recognized by T lymphocytes in adjuvant arthritis. *Nature* 331: 171–173
47 van Eden W (1991) Heat-shock proteins as immunogenic bacterial antigens with the potential to induce and regulate autoimmune arthritis. *Immunol Rev* 121: 5–28
48 Winfield JB (1989) Stress proteins, arthritis, and autoimmunity. *Arthritis Rheum* 32: 1497–1504
49 Shinnick TM, Vodkin MH, Williams JC (1988) The Mycobacterium tuberculosis 65-kilodalton antigen is a heat shock protein which corresponds to common antigen and to the *Escherichia coli* GroEL protein. *Infect Immun* 56: 446–451
50 Holoshitz J, Klajman A, Drucker I, Lapidot Z, Yaretzky A, Frenkel A, van Eden W, Cohen IR (1986) T lymphocytes of rheumatoid arthritis patients show augmented reactivity to a fraction of mycobacteria cross-reactive with cartilage. *Lancet* 2: 305–309
51 Kurzik-Dumke U, Schick C, Rzepka R, Melchers I (1999) Overexpression of human homologs of the bacterial DnaJ chaperone in the synovial tissue of patients with rheumatoid arthritis. *Arthritis Rheum* 42: 210–220
52 Gaston JS, Life PF, Bailey LC, Bacon PA (1989) *In vitro* responses to a 65-kilodalton mycobacterial protein by synovial T cells from inflammatory arthritis patients. *J Immunol* 143: 2494–2500
53 Res PC, Schaar CG, Breedveld FC, van Eden W, van Embden JD, Cohen IR, de Vries RR

(1988) Synovial fluid T cell reactivity against 65 kD heat shock protein of mycobacteria in early chronic arthritis. *Lancet* 2: 478–480

54 Gaston JS, Life PF, Jenner PJ, Colston MJ, Bacon PA (1990) Recognition of a mycobacteria-specific epitope in the 65-kDa heat-shock protein by synovial fluidderived T cell clones. *J Exp Med* 171: 831–841

55 Quayle AJ, Wilson KB, Li SG, Kjeldsen-Kragh J, Oftung F, Shinnick T, Sioud M, Forre O, Capra JD, Natvig JB (1992) Peptide recognition, T cell receptor usage and HLA restriction elements of human heat-shock protein (hsp) 60 and mycobacterial 65-kDa hsp-reactive T cell clones from rheumatoid synovial fluid. *Eur J Immunol* 22: 1315–1322

56 Henwood J, Loveridge J, Bell JI, Gaston JS (1993) Restricted T cell receptor expression by human T cell clones specific for mycobacterial 65-kDa heat-shock protein: selective *in vivo* expansion of T cells bearing defined receptors. *Eur J Immunol* 23: 1256–1265

57 Baum H, Wilson C, Tiwana H, Ahmadi K, Ebringer A (1995) HLA association with autoimmune disease: restricted binding or T-cell selection? *Lancet* 346: 1042–1043

58 Baum H, Brusic V, Choudhuri K, Cunningham P, Vergani D, Peakman M (1995) MHC molecular mimicry in diabetes. *Nat Med* 1: 388

59 Baum H, Davies H, Peakman M (1996) Molecular mimicry in the MHC: hidden clues to autoimmunity? *Immunol Today* 17: 64–70

60 Stern LJ, Brown JH, Jardetzky TS, Gorga JC, Urban RG, Strominger JL, Wiley DC (1994) Crystal structure of the human class II MHC protein HLA-DR1 complexed with an influenza virus peptide. *Nature* 368: 215–221

61 Weyand CM, Goronzy JJ (1995) Inherited and noninherited risk factors in rheumatoid arthritis. *Curr Opin Rheumatol* 7: 206–213

62 Oldstone MB (1987) Molecular mimicry and autoimmune disease. *Cell* 50: 819–820

63 Auger I, Escola JM, Gorvel JP, Roudier J (1996) HLA-DR4 and HLA-DR10 motifs that carry susceptibility to rheumatoid arthritis bind 70-kD heat shock proteins. *Nat Med* 2: 306–310

64 Roudier C, Auger I, Roudier J (1996) Molecular mimicry reflected through database screening: serendipity or survival strategy? *Immunol Today* 17: 357–358

65 Auger I, Roudier J (1997) A function for the QKRAA amino acid motif: mediating binding of DnaJ to DnaK. Implications for the association of rheumatoid arthritis with HLA-DR4. *J Clin Invest* 99: 1818–1822

66 Nepom GT, Byers P, Seyfried C, Healey LA, Wilske KR, Stage D, Nepom BS (1989) HLA genes associated with rheumatoid arthritis. Identification of susceptibility alleles using specific oligonucleotide probes. *Arthritis Rheum* 32: 15–21

67 Albert LJ, Inman RD (1999) Molecular mimicry and autoimmunity. *N Engl J Med* 341: 2068–2074

68 Prakken B, Wauben M, Genini D, Samodal R, Barnett J, Mendevil A, Leoni L, Albani S (2000) Artificial antigen-presenting cells as a tool to exploit the immune ‘synapse’. *Nat Med* 6: 1406–1410

69 Bonnin D, Albani S (1998) Mucosal modulation of immune responses to heat shock proteins in autoimmune arthritis. *Biotherapy* 10: 213–221
70 Roord S, Le T, Koffeman E, Cox E, Puga Yung G, Canavese M, Kuis W, Prakken B, Albani S (2002) Combination of epitope specific immunotherapy and antiinflammatory therapy supressess arthritis by T cell tolerization and provides a tool for *ex vivo* modulation in the treatment of autoimmune diseases. ACR/ARHP Annual Scientific Meeting, New Orleans, Louisiana, Oct 24–29, Poster presentation

Immunity to heat shock proteins and atherosclerosis

Michael Knoflach[1], Bruno Mayrl[1], Mahavir Singh[2] and Georg Wick[1]

[1]Institute of Pathophysiology, University of Innsbruck, Medical School, Fritz-Pregl-Str. 3, 6020 Innsbruck, Austria; [2]Lionex Diagnostics and Therapeutics GmbH, Mascheroder Weg 1B, 38124 Braunschweig, Germany

Introduction

As documented elsewhere in this book, heat shock proteins (HSPs) clearly play a central role in activation and modulation of unspecific as well as specific immune responses. In this chapter, we will describe early atherosclerosis as an autoimmune disorder and heat shock protein 60 (HSP60) as the culprit-autoantigen. After a brief review of experimental *in vitro* data and results from animal experiments, we will focus on clinical data confirming our pathophysiological concept that early atherosclerosis is the price we pay for protective immunity against microbial HSP60 or *bona fide* autoimmunity to eliminate biochemically-altered autologous HSP60.

Atherosclerosis

Cardiovascular diseases (CVD) are the leading cause of death in developed countries. Pathologic conditions summarized under this heading include coronary heart disease, congestive heart failure, and hypertensive heart and renal diseases. All these conditions show a possible direct link to atherosclerosis as a risk factor or underlying cause.

Until the early 1990s, most atherosclerosis research focused exclusively on lipid metabolism, but for histopathologists, the role of the immune system in early atherogenesis was clearly documented by previous microscopic studies. As early as 1840, the German pathologist von Virchow demonstrated that during atherogenesis, the first cells to infiltrate the innermost layer of the arterial vessel wall – the intima – are mononuclear inflammatory cells [1], later shown to be T lymphocytes [2]. In 1908, Osler proposed a correlation between inflammation and "arterial diseases" [3], and the first experimental proof that infection had a role in atherogenesis was provided by Fabricant and colleges in the late 1970s [4]. These authors showed that infecting chickens with Mareks' disease virus (MDV), an avian herpes virus, not only induced neurolymphomatosis, but also led to the development of atherosclerosis. The late 1980s brought clinical evidence of significant correlations between inflammation and myocardial infarction [5], and numerous studies and reviews

Heat Shock Proteins and Inflammation, edited by Willem van Eden

since then have explored the connection between inflammation and different manifestations or markers of atherosclerosis [6–8].

Atherosclerosis is a chronic, multi-factorial disease wherein the effect of various risk factors becomes manifest on appropriate genetic backgrounds. The disease begins in early childhood with the appearance of mononuclear cells in the vessel wall at certain predilection sites, such as arterial branching points, which are known to be subject to major turbulent haemodynamic stress. With progression of the lesions, smooth muscle cells migrate from the media into the intima, where they proliferate and lead to the deposition of extracellular matrix proteins, most notably collagen fibers. This sequence of events then leads to thickening and hardening of the artery (arteriosclerosis). This early stage can be readily demonstrated *in vivo* with high resolution ultrasound. Atherosclerosis is characterized by the additional formation of foam cells, i.e., macrophages and smooth muscle cells that have taken up chemically-modified (e.g., oxidized) low density lipoproteins *via* non-saturable scavenger receptors. This leads to lipid overload in these cells, and eventual deposition of extracellular cholesterol crystals. These cells form the so-called 'fatty streaks' comprised primarily of foam cells that progress slowly into more severe, rupture-prone, often exulcerated and even calcified lesions, called the atherosclerotic plaque. The thickened vessel walls narrow the lumen and, once they rupture, lead to a rapid occlusion of the vessel by the formation of thrombi that manifest in a variety of ways, depending on the affected vessel. Usually decades pass between the early inflammatory infiltration and the eventual myocardial infarction (narrowing of the coronary arteries), stroke (narrowing of the brain vessels or the carotid arteries) or intermittent claudication (narrowing of the femoral or popliteal arteries).

In addition to this standard process of atherogenesis, other non-classical forms of the disease are observed. These include genetically-determined hypercholesterinaemia, e.g., in patients or knock-out mice lacking the classical low density lipoprotein (LDL) receptor (so-called familial hypercholesterolemia) or arthrosclerosis emerging in transplant recipients (transplant atherosclerosis). These atypical forms will not be further discussed in the present context.

In this chapter, we will focus on the role of the HSP60 family, which has been identified as a culprit autoantigen in early atherogenesis. Later, other authors have provided evidence that different HSPs might also be involved in atherosclerosis – for example, HSP40 [9] or HSP70 [10] – but in contrast to the situation with HSP60, there has been no evidence that atherosclerosis can be induced by injection of these HSPs into experimental animals.

Experimental evidence

Inducing a specific immune reaction to HSP60s by immunizing normocholesterolemic rabbits with recombinant mycobacterial HSP65 (mHSP65), results in lesion formation without foam cells [11]. In the absence of a cholesterol-rich diet, the vessel wall is infiltrated by mononuclear cells – predominantly T-cells – at predilection sites for atherosclerosis. These early lesions resolve themselves in the absence of further immunization or other atherosclerosis risk factors. Vaccination combined with a high cholesterol uptake in these early lesions show foam cells in addition to the mononuclear infiltration, and the lesions are no longer reversible [12]. Notably, T-cell lines derived from lesions of non-immunized animals showed a higher reactivity to mHSP65 than peripheral blood lymphocytes [2].

Wild-type mice are notoriously resistant to induction of atherosclerosis, yet the mild lesions that develop in C57BL/6J mice receiving a high cholesterol diet can be aggravated by immunization with mHSP65 [13].

It is appropriate to mention that oxidized LDLs have also been suggested as a potential autoantigen in atherogenesis, but immunisation of Apo $E^{-/-}$ or LDL-receptor-deficient mice with biochemically-altered LDL (malondealdehyde LDL) turned out to be protective rather than atherogenic [14, 15].

In vitro, different stressors, such as heat, mechanical stress, tumour necrosis factor (TNF)-α, or oxygen radicals, induce HSP60 production by endothelial cells (EC), together with up-regulation of (among other proteins) intercellular adhesion molecule-1 (ICAM-1), and vascular cell adhesion molecule-1 (VCAM-1), which are important for lymphocyte adhesion *in vivo* [16]. Interestingly, venous and arterial endothelial cells, which are indistinguishable once detached from the vessel, still react differently to various stressors, i.e., arterial ECs express HSP60 molecules more promptly than venous ECs upon exposure to classical atherosclerosis risk factors, notably oxidized LDL, probably due to the pre-stress exerted in the arterial system due to the higher arterial blood pressure [16].

Although HSP60 is usually localized in mitochondria, it can also be found on the surface of EC and macrophages under special stress conditions *in vitro* [17, 18]. Affinity-chromatography-purified human antibodies cross-reactive between mycobacterial HSP65 and human HSP60 are able to lyse stressed, but not unstressed, human EC [19] and macrophages [20] in a complement-mediated fashion or by antibody-dependent cellular cytotoxicity (ADCC). This observation proves that anti-HSP60 antibodies are not only of secondary importance, but can be considered to play a possible primary pathogenic role.

The infiltrating mononuclear cells were further characterized on histologic sections of early atherosclerotic lesions. As expected, a considerable percentage of the T-cells (CD3-positive) are activated (CD25-positive). Compared with peripheral blood, the ratio of $\alpha\beta$ to $\gamma\delta$ T-cell-receptor (TCR)-bearing CD3-positive cells in the lesion is shifted, as in the local lymphatic system of the gut (mucosa associated lymphoid tissue – MALT), in favour of the $\gamma\delta$ T-cells. This fact, together with the demonstration of a network of dendritic cells (DC) in the intima of stressed parts of the arterial tree, led to the description of the so-called vascular associated lymphoid tissue (VALT) (reviewed in [21]).

Table 1 - The role of HSPs in atherosclerosis

Important *in vitro* data	**Key refs.**
- Expression of HSP60 in endothelial cells under different stress conditions on human umbilical venous endothelial cells (HUVEC) and rat aortic endothelium	[36, 37]
- Different forms of stress induce up-regulation of HSP60 and adhesion molecules in endothelial cells (protein- and RNA-level)	[16]
- Lysis of stressed, but not unstressed, human endothelial cells and macrophages by cross-reactive HSP60/65 antibodies	[19, 20]
Important *in vivo* animal data	**Key refs.**
- Expression of HSP60 in vessel walls of experimental animals models for atherosclerosis (rats, mice)	[38, 39]
- Induction of atherosclerosis in rabbits or mice by immunization with mycobacterial heat shock protein 65 (mHSP65)	[11, 13]
- Atherosclerotic changes induced in rabbits by immunization with mHSP65, but not by additional cholesterol-rich diet, are reversible	[12]
- T-cells from lesions induced by immunization with HSP65 as well as by cholesterol-rich diet show higher reactivity to mHSP65 than T-cells from the peripheral blood	[2]
- Tolerance induced by nasal or oral application of recombinant mHSP65 decreases atherosclerotic lesions in LDL-receptor-deficient mice	[40, 41]
Important *in vivo* human data	**Key refs.**
- Correlation between the occurrence and titers of antibodies against HSP60, 65, 70 and cardiovascular disease	[30, 31, 42–56]
- Soluble HSP60 (sHSP60) and atherosclerosis	[26]
- HSP60-reactive T-cells are present in late atherosclerotic lesions	[57, 58];[a]
- T-cell reactivity against HSP60s correlates with early atherosclerotic lesions in young males, but not in the elderly	[58a]
Reviews on inflammation and atherosclerosis	**suggested refs.**
- HSP and atherosclerosis – the autoimmune concept	[59–61]
- Immune mechanisms in atherosclerosis	[62]
- Histomorphology of early atherosclerotic plaques	[21, 34]
- Inflammation and atherosclerosis	[6, 8]

[a]*Rossmann et al., unpublished data.*

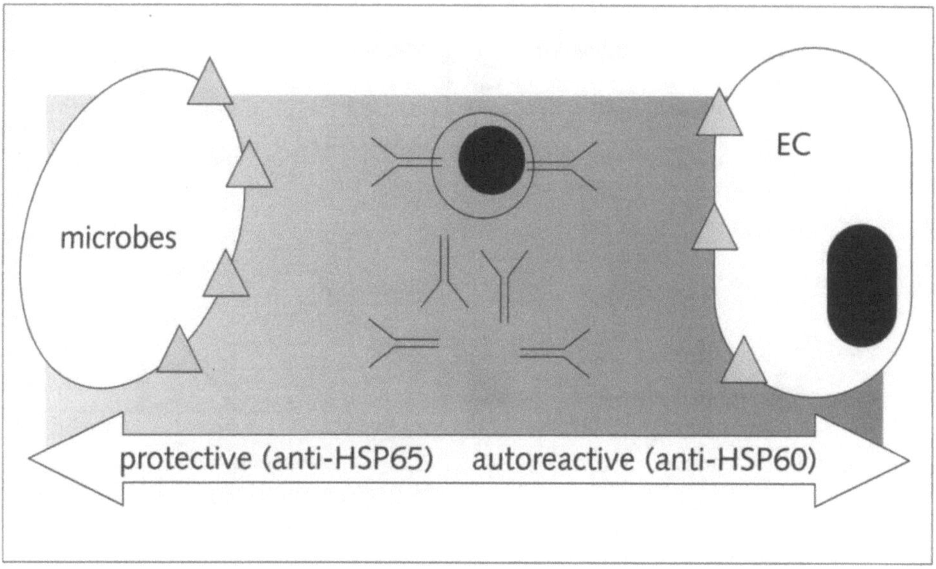

Figure 1
HSPs, as important microbial antigens, are recognized by the immune system. Due to high sequence homology between microbial (in this example mycobacterial HSP65) and human HSP60, cross-reactive antibodies (Ab) and T-cells emerge that can trigger an autoimmune reaction. EC, endothelial cell; grey triangles symbolise HSP60s.

The autoimmune hypothesis of early atherogenesis

Based on the data summarized above and in Table 1, we formulated a hypothesis describing early atherogenesis as an inflammatory autoimmune disease with HSP60 as the autoantigen, and postulate the following mechanism: Upon microbial infection, B- and T-cells, specific for different microbial antigens with HSPs as a major component, proliferate. As with all HSPs, the members of the HSP60 family (mammalian HSP60, mycobacterial HSP65 and the *Escherichia coli* homolog GroEL) show high sequence homology. Thus, various bacterial HSP60s show more than 98% homology on the amino acid and DNA levels [22]. Bacterial and human HSP60s still share greater than 55% of the overall amino acid sequence. Due to this sequence homology, protective immune reactivity may have to be "paid for" by the risk of cross-reactivity with autologous HSP60 (Fig. 1). Another possibility is that autologous chemically-altered HSP60 can activate the immune system, thus leading to *bona fide* autoimmunity, as suggested by our group in a different context earlier [23].

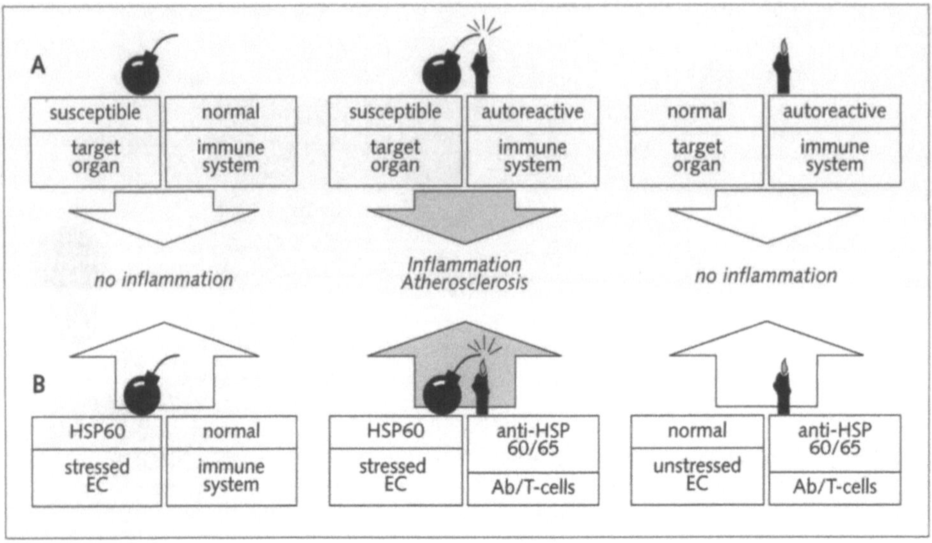

Figure 2
In order for an autoimmune disorder to occur, at least two essential genetically encoded components must be present simultaneously: A susceptible target organ structure and an autoreactive immune system (A) [23].
In the special case of atherogenesis, antibodies (Ab) and T-cells, both cross-reactive between human HSP60 and bacterial HSP65 (anti-HSP60/65), recognise the HSP60 expressed on stressed – but not on unstressed – endothelial target cells (EC) (B).

In autoimmune diseases, at least two essential genetic prerequisites have to be present simultaneously – an autoreactive immune system and a susceptible target organ structure [23] (Fig. 2A). In the case of atherosclerosis, the alteration of the immune system is induced by cross-reactivity due to repeated bacterial infections or *bona fide* autoimmunity, while the target organ, the endothelium, is altered by different forms of stress, i.e., classical atherosclerosis risk factors, and induced to express HSP60 on the surface [17, 18] (Fig. 2B). Apart from the antibody-dependent damage, and thus inherent further stress [19], the adhesion of HSP60 specific lymphocytes is facilitated by the simultaneous expression of adhesion molecules on EC [16]. With the migration of the lymphocytes into the vessel wall, the initial inflammatory stage of atherosclerosis has begun.

The two main factors mentioned above, i.e., altered immune system and target organ susceptibility, are influenced by general modulators of the immune system such as cytokines, hormones (e.g., glucocorticoids), T-regulator cells etc., or of the target organ, i.e., iodine concentration in Hashimoto's thyroditis [24]. Toll-like

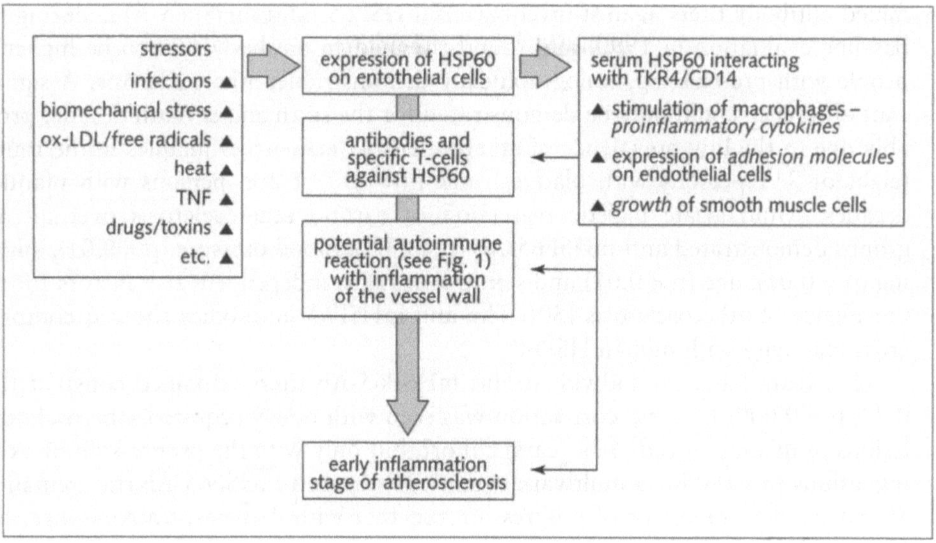

Figure 3
The role of HSP60 in early atherogenesis.

receptors (TLR), known to bind microbial HSPs and other microbial peptides, play a crucial role in innate immune defence and have recently also been shown to bind human HSP60 [25]. This, together with the facts that serum levels of soluble HSP60 correlate with the severity of sonographically-demonstrable atherosclerosis [26], and that TLR4-polymorphisms influence atherosclerosis [27], underlie our concept of a pathogenetic network for early atherosclerosis, as pictured in Figure 3.

Immunity to HSP60 in a longitudinal prospective atherosclerosis prevention study (Bruneck Study)

To extend our hypothesis into the human arena, we have participated in the Bruneck Study; a population-based longitudinal epidemiologic study to assess risk factors for, and the natural course of, human atherosclerosis. The baseline evaluation was performed in 1990, with subsequent re-evaluations in 1995 and 2000. Originally, 93.6% of the 1,000 invited 40–79 year-old inhabitants of the city of Bruneck (Bolzano Province, Italy) participated, and the follow-ups were 96.5% complete. Atherosclerosis was assessed by high resolution ultrasound of the carotid and femoral arteries [28, 29].

As described above, we postulated a humoral and cellular cross-reaction of the immune system between human and bacterial HSP60. Therefore, we first determined antibody titers against mycobacterial HSP65 (anti-mHSP65-Abs) during the baseline evaluation in 1990, and found the median antibody titer to be higher in people with pre-existing, sonographically demonstrable, atherosclerosis. A significant difference could only be demonstrated for the sixth and seventh decade, probably due to the low prevalence of atherosclerotic (echogenic) plaques in the fourth (eight of 211 persons with plaques) and fifth (30 of 204 persons with plaques) decades. Multivariate logistic regression of carotid atherosclerosis over all age groups demonstrated anti-mHSP65-Ab ($p < 0.01$), blood pressure ($p = 0.01$), smoking ($p = 0.01$), age ($p < 0.01$) and sex ($p < 0.01$) as independent risk factors for the prevalence of atherosclerosis [30]. The anti-mHSP65 antibodies showed complete cross-reactivity with human HSP60.

Five years later, the individual anti mHSP65-Ab titers remained constant ($r = 0.78$; $p < 0.0001$), yet no correlation was seen with newly acquired atherosclerotic lesions in this rather old (50+ years) cohort, but only with the progression of existing lesions ($p < 0.05$). In multivariate analysis, the anti-mHSP65-Ab titer remained an independent predictor of progression together with diabetes, smoking, age, fibrinogen, Lp(a) and ApoB [31]. Baseline antibody-titers not only correlated with morbidity but also with mortality between 1990 and 1995 ($p < 0.001$) [31].

The data strongly indicated that anti-HSP60/65-antibodies are involved in early plaque formation and progression of those early lesions, but are less important in late-onset atherosclerosis of the aged.

Measurements of soluble HSP60 (sHSP60) in the sera of the Bruneck study participants revealed a significant correlation between prevalence ($p < 0.05$) and progression ($p < 0.01$) of atherosclerotic plaques and vessel wall thickness ($p < 0.05$). Statistical analysis was performed in a multivariate manner, indicating a role of sHSP60 independent of classical risk factors [26]. We do not yet have *in vivo* data on immune complex formation between sHSP60 and cross-reactive anti mHSP65-Ab, and further studies are required to confirm and further define the role of sHSP60 in atherogenesis, i.e., as inducers of *bona fide* autoimmunity, cross-reactive antigen in immune complexes or activation of EC *via* binding to TLR.

Next, we approached the possible role of the cellular immune system in atherogenesis. During the 2000 evaluation of the Bruneck-Study, we determined T-cell reactivity against human, mycobacterial, chlamydial and *E. coli* HSP60. As expected, a high cross-reactivity between all HSP60 on the cellular level could be demonstrated, with the best correlation between T-cell reactivity to human *versus* chlamydial HSP60 and GroEL (*E. coli* HSP60) *versus* mycobacterial HSP65. However, none of those T-cell reactions proved to be predictors of atherosclerosis. Initially, this might seem surprising, but previous studies in other late-stage autoimmune diseases have shown that autoreactive T-cells are no longer found in the peripheral blood at this stage, but rather accumulate in the target organ [23]. As mentioned above, this concept also seems to hold for atherosclerosis, since our animal studies showed a higher immune reactivity against mHSP65 in T-cells

Table 2 - The role of antibodies and T-cells specific for mycobacterial HSP65 and human HSP60 in atherosclerosis of young males and elderly persons

	aged (50+)	**young males (17–18 years) [58a]**
Antibodies against mycobacterial HSP65	$p < 0.01$ [31]	$p = 0.0514$
Antibodies against human HSP60	$p < 0.01$ [55]	no data
T-cells specific against mycobacterial HSP65	no correlation to atherosclerosis	$p = 0.0566$
T-cells specific against human HSP60	no correlation to atherosclerosis	$p < 0.01$

derived from plaques than those from peripheral blood in non-immunized cholesterol fed rabbits [2].

HSP60 in the ARMY (Atherosclerosis Risk factors in Male Youth) Study

Data from studies on humoral and cellular immune reactions against HSP60 indicate a role for both in early atherogenesis. Therefore, we extended our investigations to younger individuals. From the PDAY (Pathobiological Determinants of Atherosclerosis in Youth) and the Bogalusa Studies [32, 33], as well as from our own pathohistological investigations, within the framework of the former [34], we know that atherosclerosis can be found in children and young adults. Together with the Austrian Ministry of Defence, we examined 141 17–18 year-old males (ARMY Study) for classical and immunological risk factors for atherosclerosis. Vessel wall thickness was also measured at eight different sites using high resolution ultrasound, a surrogate marker for early atherosclerosis, as used in the Bruneck or ARIC (Atherosclerosis Risk in Communities) Studies [35]. Early atherosclerosis was defined as sonographically-detectable thickening of the innermost layers of the vessel wall (intima-media thickness, IMT) on at least one of the eight vessel sites measured.

Even in these clinically-healthy young males, peripheral blood T-cell reactivity against human HSP60 was an independent predictor of early atherosclerosis ($p < 0.01$), together with smoking status, diastolic blood pressure, pulmonary function, HDL-cholesterol and alcohol intake. Due to the exploratory nature of this assessment and the relatively small number of persons enrolled, we added the anti mHSP65-antibody titer to the multivariate risk model with a borderline significance of $p = 0.0514$. When T-cell reactivity to mycobacterial HSP65 was included in the model instead of the reaction to human HSP60, the chosen set of variables stayed

Table 3 - Immune reaction against HSPs in participants in the Bruneck study

- Antibody-titer against HSP60/65 significantly correlated with:
  - prevalence of atherosclerotic plaques in 1990, but not in 1995
  - progression of pre-existing atherosclerosis from 1990 to 1995
  - mortality between 1990 and 1995
- Antibody-titer against HSP60/65 highly conserved for most individuals between 1990 and 1995
- Further correlations with atherosclerosis:
  - soluble HSP60
- No correlation with atherosclerosis:
  - cellular immune reaction against human, mycobacterial, chlamydial or *E. coli* HSP60
  - antibody-titer against HSP70

essentially the same. A simultaneous multivariate analysis was prevented by the high cellular cross-reactivity between the two HSPs [58a]. Results are summarized in Tables 2 and 3.

In conclusion, it seems that both the humoral and cellular immune reactions against HSP60s play an important role in early-onset atherogenesis. By the time cardiovascular prevention programmes start, which is generally after the first myocardial infarction, a vast number of irreversible atherosclerotic plaques have already been formed. Thus, prevention requires identification of persons at risk in the early – and still reversible – stages of inflammatory vessel wall infiltration (arteriosclerosis) or fatty streak development in order to set more stringent criteria for other risk factors, like smoking, blood pressure and lipid levels. In the future, hyposensibilisation, as used in the therapy of allergies, against HSP60 or special atherogenic epitopes, might offer a causal preventive therapy for atherosclerosis.

Acknowledgements
The work of the authors has been supported by the Austrian Research Fund (project # 14741), the Austrian Ministry of Defence (ARMY-Study) and the Merkur insurance.

References

1 von Virchow R (1971) *Cellular pathology as based upon physiological and pathological histology*. (English translation of second German edition), JB Lippincott, Philadelphia
2 Xu Q, Kleindienst R, Waitz W, Dietrich H, Wick G (1993) Increased expression of heat

shock protein 65 coincides with a population of infiltrating T lymphocytes in atherosclerotic lesions of rabbits specifically responding to heat shock protein 65. *J Clin Invest* 91: 2693–2702

3 Osler W (1908) Diseases of the arteries. In: W Osler, T MacCrae (eds): *Modern medicine. Its theory and practice in original contributions by Americans and foreign authors.* Vol 4. Lea & Fabiger, Philadelphia, 426–447

4 Fabricant CG, Fabricant J, Litrenta MM, Minick CR (1978) Virus-induced atherosclerosis. *J Exp Med* 148: 335–340

5 Mattila KJ (1989) Viral and bacterial infections in patients with acute myocardial infarction. *J Intern Med* 225: 293–296

6 Ross R (1999) Atherosclerosis – an inflammatory disease. *N Engl J Med* 340: 115–126

7 Wick G, Xu Q (1999) Atherosclerosis – an autoimmune disease. *Exp Gerontol* 34: 559–566

8 Libby P (2002) Inflammation in atherosclerosis. *Nature* 420: 868–874

9 Nguyen TQ, Jaramillo A, Thompson RW, Dintzis S, Oppat WF, Allen BT, Sicard GA, Mohanakumar T (2001) Increased expression of HDJ-2 (hsp40) in carotid artery atherosclerosis: a novel heat shock protein associated with luminal stenosis and plaque ulceration. *J Vasc Surg* 33: 1065–1071

10 Kirby LB, Mondy JS, Brophy CM (1999) Balloon angioplasty induces heat shock protein 70 in human blood vessels. *Ann Vasc Surg* 13: 475–479

11 Xu Q, Dietrich H, Steiner HJ, Gown AM, Schoel B, Mikuz G, Kaufmann SH, Wick G (1992) Induction of arteriosclerosis in normo-cholesterolemic rabbits by immunization with heat shock protein 65. *Arterioscler Thromb* 12: 789–799

12 Xu Q, Kleindienst R, Schett G, Waitz W, Jindal S, Gupta RS, Dietrich H, Wick G (1996) Regression of arteriosclerotic lesions induced by immunization with heat shock protein 65-containing material in normocholesterolemic, but not hypercholesterolemic, rabbits. *Atherosclerosis* 123: 145–155

13 George J, Shoenfeld Y, Afek A, Gilburd B, Keren P, Shaish A, Kopolovic J, Wick G, Harats D (1999) Enhanced fatty streak formation in C57BL/6J mice by immunization with heat shock protein-65. *Arterioscler Thromb Vasc Biol* 19: 505–510

14 Palinski W, Miller E, Witztum JL (1995) Immunization of low density lipoprotein (LDL) receptor-deficient rabbits with homologous malondialdehyde-modified LDL reduces atherogenesis. *Proc Natl Acad Sci USA* 92: 821–825

15 George J, Afek A, Gilburd B, Levkovitz H, Shaish A, Goldberg I, Kopolovic Y, Wick G, Shoenfeld Y, Harats D (1998) Hyperimmunization of apo-E-deficient mice with homologous malondialdehyde low-density lipoprotein suppresses early atherogenesis. *Atherosclerosis* 138: 147–152

16 Amberger A, Maczek C, Jurgens G, Michaelis D, Schett G, Trieb K, Eberl T, Jindal S, Xu Q, Wick G (1997) Co-expression of ICAM-1, VCAM-1, ELAM-1 and Hsp60 in human arterial and venous endothelial cells in response to cytokines and oxidized low-density lipoproteins. *Cell Stress Chaperones* 2: 94–103

17 Xu Q, Schett G, Seitz CS, Hu Y, Gupta RS, Wick G (1994) Surface staining and cyto-

toxic activity of heat-shock protein 60 antibody in stressed aortic endothelial cells. *Circ Res* 75: 1078–1085
18 Soltys BJ, Gupta RS (1997) Cell surface localization of the 60 kDa heat shock chaperonin protein (hsp60) in mammalian cells. *Cell Biol Int* 21: 315–320
19 Schett G, Xu Q, Amberger A, Van der ZR, Recheis H, Willeit J, Wick G (1995) Autoantibodies against heat shock protein 60 mediate endothelial cytotoxicity. *J Clin Invest* 96: 2569–2577
20 Schett G, Metzler B, Mayr M, Amberger A, Niederwieser D, Gupta RS, Mizzen L, Xu Q, Wick G (1997) Macrophage-lysis mediated by autoantibodies to heat shock protein 65/60. *Atherosclerosis* 128: 27–38
21 Millonig G, Schwentner C, Mueller P, Mayerl C, Wick G (2001) The vascular-associated lymphoid tissue: A new site of local immunity. *Curr Opin Lipidol* 12: 547–553
22 Jones DB, Coulson AF, Duff GW (1993) Sequence homologies between hsp60 and autoantigens. *Immunol Today* 14: 115–118
23 Wick G, Kroemer G, Neu N, Faessler R, Ziemiecki A, Muller RG, Ginzel M, Beladi I, Kuehr T, Hala K (1987) The multi-factorial pathogenesis of autoimmune disease. *Immunol Lett* 16: 249–257
24 Wick G, Hala K, Wolf H, Ziemiecki A, Sundick RS, Stoeffler-Meilicke M, DeBaets M (1986) The role of genetically-determined primary alterations of the target organ in the development of spontaneous autoimmune thyroiditis in obese strain (OS) chickens. *Immunol Rev* 94: 113–136
25 Habich C, Baumgart K, Kolb H, Burkart V (2002) The receptor for heat shock protein 60 on macrophages is saturable, specific, and distinct from receptors for other heat shock proteins. *J Immunol* 168: 569–576
26 Xu Q, Schett G, Perschinka H, Mayr M, Egger G, Oberhollenzer F, Willeit J, Kiechl S, Wick G (2000) Serum soluble heat shock protein 60 is elevated in subjects with atherosclerosis in a general population. *Circulation* 102: 14–20
27 Kiechl S, Lorenz E, Reindl M, Wiedermann CJ, Oberhollenzer F, Bonora E, Willeit J, Schwartz DA (2002) Toll-like receptor 4 polymorphisms and atherogenesis. *N Engl J Med* 347: 185–192
28 Kiechl S, Willeit J (1999) The natural course of atherosclerosis. Part I: incidence and progression. *Arterioscler Thromb Vasc Biol* 19: 1484–1490
29 Kiechl S, Willeit J (1999) The natural course of atherosclerosis. Part II: vascular remodeling. Bruneck study group. *Arterioscler Thromb Vasc Biol* 19: 1491–1498
30 Xu Q, Willeit J, Marosi M, Kleindienst R, Oberhollenzer F, Kiechl S, Stulnig T, Luef G, Wick G (1993) Association of serum antibodies to heat-shock protein 65 with carotid atherosclerosis. *Lancet* 341: 255–259
31 Xu Q, Kiechl S, Mayr M, Metzler B, Egger G, Oberhollenzer F, Willeit J, Wick G (1999) Association of serum antibodies to heat-shock protein 65 with carotid atherosclerosis : Clinical significance determined in a follow-up study. *Circulation* 100: 1169–1174
32 McGill HC Jr, McMahan CA, Zieske AW, Sloop GD, Walcott JV, Troxclair DA, Malcom GT, Tracy RE, Oalmann MC, Strong JP (2000) Associations of coronary heart dis-

ease risk factors with the intermediate lesion of atherosclerosis in youth. The Pathobiological Determinants of Atherosclerosis in Youth (PDAY) Research Group. *Arterioscler Thromb Vasc Biol* 20: 1998–2004

33 Berenson GS, Srinivasan SR, Bao W, Newman WP III, Tracy RE, Wattigney WA (1998) Association between multiple cardiovascular risk factors and atherosclerosis in children and young adults. The Bogalusa Heart Study. *N Engl J Med* 338: 1650–1656

34 Millonig G, Malcom GT, Wick G (2002) Early inflammatory-immunological lesions in juvenile atherosclerosis from the Pathobiological Determinants of Atherosclerosis in Youth (PDAY)-study. *Atherosclerosis* 160: 441–448

35 Heiss G, Sharrett AR, Barnes R, Chambless LE, Szklo M, Alzola C (1991) Carotid atherosclerosis measured by B-mode ultrasound in populations: associations with cardiovascular risk factors in the ARIC study. *Am J Epidemiol* 134: 250–256

36 Mayr M, Metzler B, Kiechl S, Willeit J, Schett G, Xu Q, Wick G (1999) Endothelial cytotoxicity mediated by serum antibodies to heat shock proteins of *Escherichia coli* and *Chlamydia pneumoniae*: immune reactions to heat shock proteins as a possible link between infection and atherosclerosis. *Circulation* 99: 1560–1566

37 Seitz CS, Kleindienst R, Xu Q, Wick G (1996) Co-expression of heat-shock protein 60 and intercellular-adhesion molecule-1 is related to increased adhesion of monocytes and T cells to aortic endothelium of rats in response to endotoxin. *Lab Invest* 74: 241–252

38 Hochleitner BW, Hochleitner EO, Obrist P, Eberl T, Amberger A, Xu Q, Margreiter R, Wick G (2000) Fluid shear stress induces heat shock protein 60 expression in endothelial cells *in vitro* and *in vivo*. *Arterioscler Thromb Vasc Biol* 20: 617–623

39 Kanwar RK, Kanwar JR, Wang D, Ormrod DJ, Krissansen GW (2001) Temporal expression of heat shock proteins 60 and 70 at lesion-prone sites during atherogenesis in ApoE-deficient mice. *Arterioscler Thromb Vasc Biol* 21: 1991–1997

40 Harats D, Yacov N, Gilburd B, Shoenfeld Y, George J (2002) Oral tolerance with heat shock protein 65 attenuates Mycobacterium tuberculosis-induced and high-fat-diet-driven atherosclerotic lesions. *J Am Coll Cardiol* 40: 1333–1338

41 Maron R, Sukhova G, Faria AM, Hoffmann E, Mach F, Libby P, Weiner HL (2002) Mucosal administration of heat shock protein-65 decreases atherosclerosis and inflammation in aortic arch of low-density lipoprotein receptor-deficient mice. *Circulation* 106: 1708–1715

42 Birnie DH, Holme ER, McKay IC, Hood S, McColl KE, Hillis WS (1998) Association between antibodies to heat shock protein 65 and coronary atherosclerosis. Possible mechanism of action of *Helicobacter pylori* and other bacterial infections in increasing cardiovascular risk. *Eur Heart J* 19: 387–394

43 Burian K, Kis Z, Virok D, Endresz V, Prohaszka Z, Duba J, Berencsi K, Boda K, Horvath L, Romics L, et al (2001) Independent and joint effects of antibodies to human heat-shock protein 60 and *Chlamydia pneumoniae* infection in the development of coronary atherosclerosis. *Circulation* 103: 1503–1508

44 Zhu J, Quyyumi AA, Rott D, Csako G, Wu H, Halcox J, Epstein SE (2001) Antibodies to human heat-shock protein 60 are associated with the presence and severity of coro-

nary artery disease: evidence for an autoimmune component of atherogenesis. *Circulation* 103: 1071–1075

45 Huittinen T, Leinonen M, Tenkanen L, Manttari M, Virkkunen H, Pitkanen T, Wahlstrom E, Palosuo T, Manninen V, Saikku P (2002) Autoimmunity to human heat shock protein 60, Chlamydia pneumoniae infection, and inflammation in predicting coronary risk. *Arterioscler Thromb Vasc Biol* 22: 431–437

46 Gruber R, Lederer S, Bechtel U, Lob S, Riethmuller G, Feucht HE (1996) Increased antibody titers against mycobacterial heat-shock protein 65 in patients with vasculitis and arteriosclerosis. *Int Arch Allergy Immunol* 110: 95–98

47 Hoppichler F, Lechleitner M, Traweger C, Schett G, Dzien A, Sturm W, Xu Q (1996) Changes of serum antibodies to heat-shock protein 65 in coronary heart disease and acute myocardial infarction. *Atherosclerosis* 126: 333–338

48 Hoppichler F, Koch T, Dzien A, Gschwandtner G, Lechleitner M (2000) Prognostic value of antibody titer to heat-shock protein 65 on cardiovascular events. *Cardiology* 94: 220–223

49 Frostegard J, Lemne C, Andersson B, Van der ZR, Kiessling R, de Faire U (1997) Association of serum antibodies to heat-shock protein 65 with borderline hypertension. *Hypertension* 29: 40–44

50 Pockley AG, Wu R, Lemne C, Kiessling R, de Faire U, Frostegard J (2000) Circulating heat shock protein 60 is associated with early cardiovascular disease. *Hypertension* 36: 303–307

51 Mukherjee M, De Benedictis C, Jewitt D, Kakkar VV (1996) Association of antibodies to heat-shock protein-65 with percutaneous transluminal coronary angioplasty and subsequent restenosis. *Thromb Haemost* 75: 258–260

52 Chan YC, Shukla N, Abdus-Samee M, Berwanger CS, Stanford J, Singh M, Mansfield AO, Stansby G (1999) Anti-heat-shock protein 70 kDa antibodies in vascular patients. *Eur J Vasc Endovasc Surg* 18: 381–385

53 Gromadzka G, Zielinska J, Ryglewicz D, Fiszer U, Czlonkowska A (2001) Elevated levels of anti-heat shock protein antibodies in patients with cerebral ischemia. *Cerebrovasc Dis* 12: 235–239

54 Prohaszka Z, Duba J, Lakos G, Kiss E, Varga L, Janoskuti L, Csaszar A, Karadi I, Nagy K, Singh M et al (1999) Antibodies against human heat-shock protein (hsp) 60 and mycobacterial hsp65 differ in their antigen specificity and complement-activating ability. *Int Immunol* 11: 1363–1370

55 Prohaszka Z, Duba J, Horvath L, Csaszar A, Karadi I, Szebeni A, Singh M, Fekete B, Romics L, Fust G (2001) Comparative study on antibodies to human and bacterial 60 kDa heat shock proteins in a large cohort of patients with coronary heart disease and healthy subjects. *Eur J Clin Invest* 31: 285–292

56 Ciervo A, Visca P, Petrucca A, Biasucci LM, Maseri A, Cassone A (2002) Antibodies to 60-kilodalton heat shock protein and outer membrane protein 2 of Chlamydia pneumoniae in patients with coronary heart disease. *Clin Diagn Lab Immunol* 9: 66–74

57 Mosorin M, Surcel HM, Laurila A, Lehtinen M, Karttunen R, Juvonen J, Paavonen J,

Morrison RP, Saikku P, Juvonen T (2000) Detection of Chlamydia pneumoniae-reactive T lymphocytes in human atherosclerotic plaques of carotid artery. *Arterioscler Thromb Vasc Biol* 20: 1061–1067
58 Curry AJ, Portig I, Goodall JC, Kirkpatrick PJ, Gaston JS (2000) T lymphocyte lines isolated from atheromatous plaque contain cells capable of responding to *Chlamydia* antigens. *Clin Exp Immunol* 121: 261–269
58a Knoflach M, Kiechl S, Kind M, Said M, Sief R, Gisinger M, van der Zee R, Gaston H, Jarosch E, Willeit J, Wick G (2003) Cardiovascular risk factors and atherosclerosis in young males. *Circulation* 108
59 Wick G, Kleindienst R, Dietrich H, Xu Q (1992) Is atherosclerosis an autoimmune disease? *Trends Food Sci Technol* 3: 114–119
60 Wick G, Schett G, Amberger A, Kleindienst R, Xu Q (1995) Is atherosclerosis an immunologically mediated disease? *Immunol Today* 16: 27–33
61 Wick G, Perschinka H, Millonig G (2001) Atherosclerosis as an autoimmune disease: an update. *Trends Immunol* 22: 665–669
62 Hansson GK (2001) Immune mechanisms in atherosclerosis. *Arterioscler Thromb Vasc Biol* 21: 1876–1890

Chaperonins: Chameleon proteins that influence myeloid cells

Brian Henderson

Cellular Microbiology Research Group, Eastman Dental Institute, University College London, 256 Gray's Inn Road, London WC1X 8LD, UK

Introduction

The term chaperonin was introduced by John Ellis and colleagues [1] and now refers to two distinct subfamilies – the GroE subfamily found in prokaryotes, plastids and mitochondria and the TCP-1 subfamily found in eukaryotes and the *Archaeae*. This chapter will focus on the GroE subfamily which consists of two members: chaperonin (Cpn)60 and the co-chaperone – Cpn10. Both proteins form oligomeric structures required for protein folding [2] and the mechanism of folding has largely been delineated [3]. These proteins show significant sequence homology [4] which is normally assumed to underpin functional "homology". While for some cell biologists this information would seem to be the end-of-the-line for the chaperonins, as subjects of scientific interest, it is emerging from various laboratories around the world that the chaperonins have other activities which suggest a very much larger role for these proteins in physiological homeostasis. An emerging concept in protein biochemistry is of the "moonlighting" protein [5]. These are proteins, which have more than one function and which may participate in one process, say within the cell, and with another process in some other body compartment. A fascinating example of a moonlighting protein is the glycolytic enzyme, phosphoglucose isomerase (PGI). Indeed, a growing number of glycolytic enzymes "moonlight". It is now established that in addition to participating in glycolysis, PGI is also:

1) an autocrine motility factor, which stimulates cell migration;
2) a neuroleukin, which is both a nerve growth factor and B-cell maturation factor, and;
3) a differentiation and maturation mediator that induces human myeloid cell differentiation [6].

Another example, and perhaps one most closely resembling chaperonin 60, is the other glycolytic enzyme, glyceraldehyde 3-phosphate dehydrogenase (GAPD). It is now recognised that GAPD proteins from bacteria, in spite of their significant sequence homology, can have a wide range of biological actions in addition to their

Heat Shock Proteins and Inflammation, edited by Willem van Eden

role in glycolysis. This includes being a receptor for epidermal growth factor in *Mycobacterium tuberculosis* [7] and a receptor for plasmin in *Streptococci* [8]. In this article the hypothesis being propounded is that chaperonin 60, and possibly also chaperonin 10, are moonlighting proteins with a very extensive range of moonlighting abilities. It is this range of biological activities that the chaperonins can perform that has suggested their resemblance to chameleons.

Chaperonins unfold their signalling activity

By the early 1990s enormous advances had been made in our understanding of the role of molecular chaperones in normal and aberrant cell functioning and the mechanism of action of the prototypic protein, the chaperonin 60 of *E. coli*, GroEL, was being established [9]. It was at this time that a paper appeared from St George's Hospital Medical School in London announcing the finding that the chaperonin 60.2 protein of *Mycobacterium tuberculosis*, better known as HSP65 [10], stimulated human monocytic cells to produce pro-inflammatory cytokines [11]. This was the beginning of a novel area of research that is still, at the time of writing, somewhat controversial. This initial finding that one of the key immunogens of *M. tuberculosis* had the capacity to activate myeloid cells was quickly confirmed by Van Furth's group who found that this protein would stimulate human monocyte cytokine synthesis without causing other signs of macrophage activation [12], and by Retzlaff and co-workers who found activation of murine monocytes by chaperonin 60 [13]. These studies indicated that, in addition to protein folding, the chaperonin 60 proteins of bacteria could also act as cell-to-cell signals. Of course the possibility existed that these effects are as a result of contamination of the chaperonin 60 preparations with known cytokine-stimulating agonists such as lipopolysaccharide (LPS), lipoarabinomannan (LAM) etc.

Chaperonins stimulate bone resorption *via* modulation of myeloid cell differentiation

In a study completely unrelated to that described above, the author of this article was searching for the potently osteolytic protein released by the oral bacterium, *Actinobacillus actinomycetemcomitans*. This oral bacterium is the causative agent of a very severe form of periodontal disease in which inflammation of the gums is associated with destruction (resorption) of the alveolar bone of the jaw. When this osteolytic protein was isolated and sequenced it turned out to be the chaperonin 60 of *A. actinomycetemcomitans* [14]. Immunogold labelling demonstrated the presence of this chaperonin on the surface of the bacterium and on the proteinaceous fibrillar material secreted by this bacterium (Henderson, unpublished data). This surface location has subsequently been confirmed and the *A. actinomycetemcomitans* chaperonin 60 has been reported to both act as a mitogen for some cells and as an apoptotic signal for others [15, 16]. Recombinant *E. coli* GroEL also potently stimulated bone resorption but, surprisingly, neither the chaperonin 60 proteins from *M. tuberculosis* or *M. leprae* had any significant bone resorbing activity [14]. As a control for LPS contamination, calvaria from the LPS-insensitive C3H/HeJ mouse were used. This murine strain has a mutation in the gene encoding Toll-like receptor (TLR) 4 such that this receptor does not recognise activating ligands such as LPS [17]. The *A. actinomycetemcomitans* chaperonin 60 was capable of stimulating the calvarial bone from this murine strain to resorb. In contrast, and as expected, LPS was inactive [14]. This established that LPS contamination in the chaperonin 60 preparation was not causing the bone resorption. It also suggested that activation of bone is not *via* TLR4 and suggested some other receptor must be involved. An even more intriguing finding was that in spite of the significant sequence homology between the Gram-negative chaperonin 60 proteins and the mycobacterial chaperonin 60 proteins the latter lacked the osteolytic activity of the former. This was the first indication that chaperonin 60 proteins did not represent a unitary cell-cell signalling activity and that variation in the activity of this family of proteins was possible.

To understand the mechanism of action of the Gram-negative chaperonin 60 proteins on bone it is important to understand the cellular basis of bone remodelling. Bone is a fibre-reinforced matrix containing two major cell lineages. The osteoblast is a mesenchymal cell responsible for the formation of the bone matrix and for recognising most of the signals controlling bone remodelling. The osteoclast is a multinucleate myeloid cell whose function is to remove damaged matrix components. The osteoblast and osteoclast populations of bone interact by cell-cell contact and in recent years the major controlling network in bone remodelling has been elucidated. This involves three TNF receptor-like proteins:

1) receptor activator of NF-κB (RANK);
2) RANK ligand (RANKL), and;
3) osteoprotegerin (OPG) (Fig 1).

RANK is found on osteoclast precursor cells. RANKL is an inducible protein found on mesenchymal cells, principally osteoblasts and also on activated T-cells. OPG is also an inducible protein made by mesenchymal cells. In normal bone remodelling, the binding of RANKL to RANK stimulates myeloid cell differentiation to induce osteoclast formation and promote bone resorption. Induction of OPG results in the OPG competing with the RANK for binding to RANKL thus switching off myeloid cell differentiation and halting bone resorption [18].

Having identified that Gram-negative chaperonin 60 proteins induce bone resorption in the murine calvarial bone resorption assay the mechanism of action

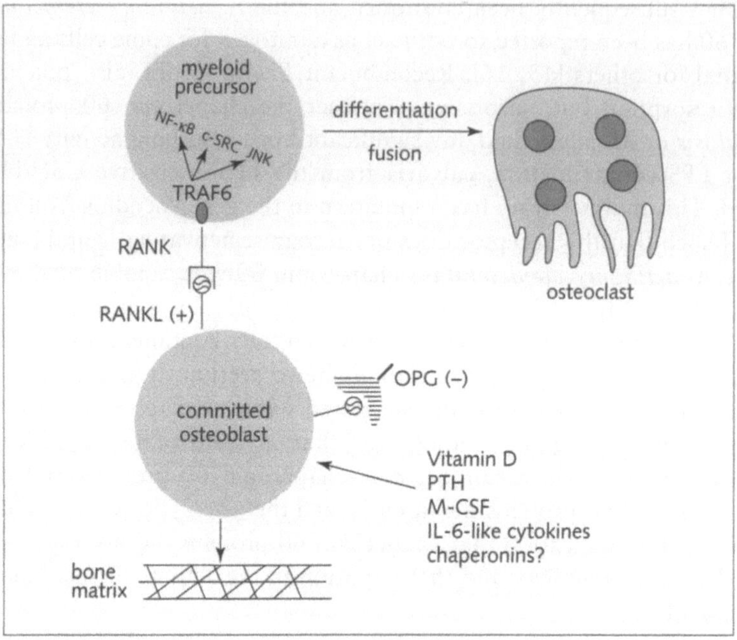

Figure 1
The interaction of the mesenchymal osteoblast population with the myeloid osteoclast population via the TNF receptor-like proteins RANKL, RANK and OPG.
RANKL is an inducible protein that acts as the ligand for binding to RANK on osteoclast precursors. Such activation results in the maturation of the latter cells into osteoclasts. Osteoprotegerin (OPG) is another inducible protein that competes with RANK for binding to RANKL and is effectively a soluble form of RANK that can inhibit osteoclastogenesis. It appears that chaperonins can modulate this system in a number of ways.

needed to be addressed. Three possibilities existed. The chaperonin 60 proteins could either:

1) activate preformed osteoclasts;
2) induce the formation of osteoclasts, or;
3) inhibit the matrix forming activity of the osteoblast cell population.

Analysis of recombinant GroEL in a variety of bone and bone cell assays revealed that this protein was a potent inducer of osteoclast formation and that it could also stimulate preformed osteoclasts to promote bone resorption [19]. GroEL appeared to be able to activate cells *via* the RANK receptor to induce osteoclast maturation.

However, the mechanism was not established. Other molecular chaperones, including HSP27, HSP70, HSP90, but not HSP47, were also found to be capable of stimulating resorption of murine calvaria [20].

These findings were interesting but potentially confusing. It is well known that *M. tuberculosis* infects bone causing an osteomyelitis associated with significant bone resorption [21]. The finding that chaperonin 60 from Gram-negative bacteria could promote bone resorption, but that the chaperonin 60 of mycobacterial species was inactive, was surprising. An obvious explanation emerged when it was realised that in fact *M. tuberculosis* has two chaperonin 60 genes encoding what are termed chaperonin 60.1 [22] and chaperonin 60.2 better known as HSP65. It was the latter that had been tested in earlier studies [14]. The key osteolytic component of *M. tuberculosis* was then identified [23]. Surprisingly, it was not chaperonin 60.1 but the co-chaperone of chaperonin 60 – chaperonin 10. It was possible to completely block the osteolytic activity of sonicates of *M. tuberculosis* by use of a blocking monoclonal antibody to *M. tuberculosis* chaperonin 10 [23]. This implied that both chaperonin 60 proteins of *M. tuberculosis* were without osteolytic activity. Using synthetic *M. tuberculosis* chaperonin 10 peptides established that the osteolytic activity of this bacterial protein resided in a large flexible loop region and in a small conserved loop around position 70 of the chaperonin 10 protein [23].

Recombinant human chaperonin 60 has also been tested in the murine calvarial bone resorption assay and is a reasonably potent inducer of bone resorption. In collaboration with Professor Hill Gaston, University of Cambridge, a number of human chaperonin 60 truncation mutants have been tested. Only the N/terminal/C-terminal truncation mutant Δ1–26 Δ466–573 was inactive while N-terminal mutants (e.g., Δ1–137) were fully active [24].

Gram-negative bacterial and mitochondrial chaperonin 60 proteins have the ability to induce cellular changes in bone which result in its breakdown. The mitochondria are presumed to have evolved from the Proteobacteria [25], which includes Gram-negative bacteria such as *E. coli*. This evolutionary relationship may explain why these apparently diverse proteins have similar biological actions.

It was still not clear what biological functions the mycobacterial chaperonin 60 proteins played. To address this in more detail all three *M. tuberculosis* chaperonins were cloned and expressed [23, 26]. These proteins were individually tested to see if they had any activity against adjuvant arthritis in the rat [27]. This is a much-studied model of rheumatoid arthritis that is used as a test-bed for anti-inflammatory compounds. It has been shown in previous studies that *M. tuberculosis* chaperonin 60.2 (HSP65) has immunomodulatory effects in this model [28]. Intradermal administration of mycobacterial chaperonin 60.1, 60.2 or 10 to rats revealed that only chaperonin 10 produced any anti-inflammatory effects. However, it was surprising to find that while chaperonin 60.1 had absolutely no effect on the joint swelling there was an almost complete suppression of the enormous amount of osteoclastic bone remodelling that accompanies joint inflammation in adjuvant arthritis [29]. This was a completely unexpected finding. In this respect the *M. tuberculosis* chaperonin 60.1 produces an identical effect to osteoprotegerin [30] or to the c-Jun amino-terminal kinase (JNK) inhibitor, SP600125 [31]. JNK is believed to be responsible for transducing the signalling between RANKL and RANK. To determine if the inhibition of bone resorption was a direct effect on bone, as opposed to some modulation of immunity (as activated T-cells can stimulate bone resorption [32]), murine calvaria were stimulated to resorb by LPS in the presence or absence of recombinant *M. tuberculosis* chaperonin 60 proteins or OPG. Both OPG and *M. tuberculosis* chaperonin 60.1, but not the 60.2 protein, dose-dependently inhibited bone resorption (Henderson and Meghji, unpublished data). The mechanism of action of the *M. tuberculosis* chaperonin 60.1 is not defined. It is possible that this protein can selectively induce the formation of OPG or it may mimic the actions of OPG. Alternatively, it may compete with RANKL for binding to RANK. The end result of any of these mechanisms would be the inhibition of osteoclast formation and the cessation of bone remodelling.

To conclude this section, chaperonin 60 proteins from bacteria can both stimulate and inhibit the complex process of myeloid differentiation that leads to the generation of activated osteoclasts capable of driving bone resorption. The simplest explanation is that these different chaperonins act either as agonists or antagonists of a common receptor. The nature of the receptor remains to be defined. The next section describes the findings that chaperonin 60 proteins can stimulate myeloid cells and other cell populations to produce cytokines and provides further evidence for the hypothesis that in spite of the conservation of sequence chaperonin 60 proteins can exhibit markedly different actions with myeloid cells.

Chaperonin 60 proteins can be cytokine-inducing agonists – but not always

The first evidence that chaperonin 60 proteins could have cell signalling properties arose from the finding that they could induce myeloid cells to synthesise cytokines [11–13]. As this author knows from the referees of both papers and grant applications there is still a hard core of individuals out there who do not believe that molecular chaperones have direct cell signalling activity. Two good reasons for this belief with the chaperonin 60 proteins is the fact that recombinant chaperonin 60 proteins are contaminated with LPS. In the author's experience, commercial preparations of chaperonin 60 can contain large quantities of LPS which contribute to the majority of any cytokine inducing activity recorded. The second problem that workers with chaperonin 60 proteins have to contend with is the large number of contaminating proteins that co-purify even with recombinant histidine tagged fusion proteins [33, 34]. The author and his colleagues have recently described methodology for the purification of recombinant 6His-tagged chaperonin 60 proteins which minimises both LPS contamination and extraneous protein contamination [35]. In this protocol the LPS is removed, not by passing the recombinant protein over a polymyxin B column, as would normally be done, but by washing the recombinant protein bound to the nickel column with polymyxin B. Our early experience with cleaning up recombinant GroEL on commercially available polymyxin B columns (Detoxigel columns: Pierce) was that most of the protein bound avidly to the column and was difficult to recover. Washing the column with polymyxin B reduces LPS contamination to very low levels and then the polymyxin B can be removed by further washing. Once the LPS is removed the chaperonin 60 is eluted and desalted and contaminating proteins are removed by applying the chaperonin 60 to a reactive red column and treating with ATP to prime the chaperonin 60 for the release of bound proteins. The final recombinant chaperonin 60 proteins have very low levels of LPS contamination and of protein contaminants [35].

This clean-up methodology was first applied to recombinant GroEL and the removal of both the LPS and extraneous proteins verified [36]. Further checks of LPS contamination involved use of polymyxin B which had no effect and the addition of monoclonal antibodies to CD14 which could block the activity of enteric LPS. Again, such antibodies did not inhibit the capacity of GroEL to stimulate human monocyte cytokine synthesis. An additional control involved treatment of the GroEL with trypsin. In spite of complete proteolysis of the GroEL into its 60-odd tryptic peptides there was no loss of monocyte-stimulating activity [36]. This was a completely unexpected finding and one that was repeated on many, many occasions before it was believed. This suggested that the activity of GroEL was not due to the whole protein but to one or more of the tryptic peptides from which the protein was constituted. Analysis of the tryptic peptides by reverse phase HPLC and bioassay identified four active peptides. These peptides were identified by a combination of Edman degradation and MALDI-TOF mass spectrometry (Tab. I). When these were modelled onto the GroEL crystal structure they were all found on the surface of the protein. To confirm the activity of these peptides they were synthesised and tested for monocyte activating activity. All peptides proved very difficult to synthesise. When they were tested for their ability to activate monocytes none demonstrated the expected activity. It is not clear why this was the case and this work was not followed up.

A number of groups worldwide have confirmed that chaperonin 60 proteins from both bacteria and mitochondria can activate mammalian myeloid cells (Tab. 2). There is controversy with respect to the mechanism by which chaperonin 60 proteins from different sources interact and activate myeloid cells. Work with the human chaperonin 60 protein (HSP60) and with the chaperonin 60 of *Chlamydia trachomatis* [37–39] has suggested that these proteins interact with the CD14/TLR4 receptor. It has also been proposed that HSP60 interacts with TLR2 and TLR4 and requires endocytosis for cell activation [39, 40].

Table 1 - Identification of the bioactive tryptic peptides of GroEL

Fraction number	Peptide sequence	Position of peptide in groEL	IL-6 pg/ml	IL-8 pg/ml
17	VEDALHATR	395–401	1478 ± 331	10866 ± 926
24	AAVEEGVVAGGGALIR	405–421	2585 ± 15	9550 ± 1835
28	GGDGNYGYNAATEEYGNMID MGILDPTK	471–498	1709 ± 551	10055 ± 2920
30/31	EGVITVEDGTGLQDELDVVE GMQFDR	171–196	1491 ± 551	27023 ± 5004

IL-6 and IL-8 synthesis measured by ELISA.

However, GroEL activation of human monocytes was not inhibited by monoclonal antibodies to CD14 that could block the biological activity of enteric LPS [36] and this chaperonin would activate the CD14-negative cell line LR9 (Lewthwaite and Henderson, unpublished). When the two highly conserved chaperonin 60 proteins of *M. tuberculosis* were compared for their capacity to activate human monocyte cytokine synthesis there were a number of surprises [26]. Firstly, they demonstrated different potencies, with the chaperonin 60.1 being 10–100 times more potent than chaperonin 60.2. Secondly, the biological activity of chaperonin 60.2 was not blocked by neutralising anti-CD14 antibodies while that of chaperonin 60.1 was completely inhibited by high concentrations of such antibodies [26]. Both chaperonins were capable of inducing human monocytes to synthesise a wide range of pro-inflammatory and anti-inflammatory (IL-10) cytokines. However, they did not induce the synthesis of interferon (IFN)-γ. Analysis of the biological activity of synthetic mycobacterial chaperonin 60 peptides, designed as putative T-cell ligands, revealed that one chaperonin 60.1 peptide (195–219) was biologically active. Interestingly, this peptide also induced the synthesis of IFN-γ and activity was blocked by antibodies to CD14. The same peptide sequences in *M. tuberculosis* chaperonin 60.2 and in GroEL were inactive. It is speculated that this is because they contain two and three proline residues respectively, compared with no prolines in the peptide from *M. tuberculosis* chaperonin 60.1. Prolines significantly interfere with peptide structures such as α-helices. These findings suggest that there are at least two sites in the *M. tuberculosis* chaperonin 60.1 able to interact with host cells in a CD14-dependent manner and activate cytokine gene transcription. Another surprise was the thermal stability of both mycobacterial chaperonin 60 proteins. Boiling these proteins had almost no influence on their ability to induce cytokine synthesis. It required autoclaving to inhibit activity and even then complete inhibition of activ-

Table 2 - Chaperonin 60 proteins stimulating cytokine synthesis

Cpn60 homologue	Cell population	Response induced	Refs.
Mycobacterium bovis, *M. leprae*, *E. coli*	Macrophages	TNF-α, IL-1β, IL-6, GM-CSF	[13]
M. tuberculosis Cpn60.2		TNF-α, IL-1β, IL-2	[63, 64]
Human		IL-6	[65]
		Cyclooxygenase	[66]
Chlamydia trachomatis		IL-6	[65]
M. tuberculosis Cpn60.1	Monocytes	IL-1β, IL-6, IL-8, IL-10, IL-12, TNF-α, GM-CSF, IFN-γ	[26]
M. tuberculosis Cpn60.2		IL-1β, IL-6, IL-8, IL-10, IL-12, TNF-α, GM-CSF, IFN-γ	[12, 26]
H. pylori		IFN-γ, IL-10	[67]
E. coli		IL-1, IL-6, TNF-α, IL-10, GM-CSF	[36, 68]
Rh. leguminosarum Cpn60.3		IL-1, IL-6, TNF-α	[43]
M. tuberculosis Cpn60.2 , *E. coli*	HUVECs	ICAM-1, VCAM-1, E-selectin	[68, 69]
Human		ICAM-1, VCAM-1, E-selectin	[65]
Chl. trachomatis		ICAM-1, VCAM-1, E-selectin. NF-κB	[39, 65]
E. coli	Keratinocytes	TNF-α, IL-1, IL-6, ICAM-1	[70]
Human	Smooth muscle cells	IL-6	[65]
Campylobacter rectus	Fibroblasts	IL-6, IL-8	[71]
H. pylori	Epithelial cells	IL-8	[72]
A. actinomycetemcomitans	Bone	Bone breakdown/cytokines	[14, 20]

ity was not achieved. In contrast, autoclaving had no effect on the bioactivity of *E. coli* LPS [26]. This suggests that the mycobacterial chaperonin 60 proteins could be extremely long-lived bacterial signals if released from sites of infection.

Another bacterium that encodes more than one chaperonin 60 is *Rhizobium leguminosarum*. This is a bacterium normally found in association with leguminous plants that have nitrogen-fixing root nodules. It has three chaperonin 60 proteins: chaperonin 60.1, 60.2 and 60.3 [41]. Although this organism does not come into contact with *Homo sapiens* it was of interest to see if the chaperonin 60 proteins of this bacterium had the capacity to activate human monocytes. Recombinant purified *Rh. leguminosarum* chaperonin 60.1 and 60.3 proteins, which demonstrate >70% sequence identity, were investigated. Both proteins exhibited ATP-ase activity and were active in folding assays showing that they themselves were properly folded and biologically active [42]. When incubated with human peripheral blood monocytes the chaperonin 60.3 exhibited cytokine-stimulating activity, which could be blocked by LPS-neutralising anti-CD14 monoclonal antibodies. In contrast, the highly homologous chaperonin 60.1 was completely inactive even at very high concentrations [43].

Categorisation of chaperonin 60 proteins

Thus it appears that chaperonin 60 proteins can be divided into a number of categories depending on their possession or absence of particular cell-cell signalling abilities (Tab. 3). There are those proteins that can induce bone resorption and cytokine synthesis and those can not induce bone resorption but can induce cytokine synthesis. There are those proteins that induce cytokine synthesis *via* interaction with CD14 and downstream TLRs and those that do not interact with CD14. Finally there is the *Rh. leguminosarum* chaperonin 60 protein which is unable to activate human monocytes to synthesise cytokines. This suggests that there is sufficient plasticity in the sequence/structure of chaperonin 60 proteins, in spite of sequence conservation, for different bioactivities to be exhibited by these molecules.

What is the receptor for chaperonin 60?

The paradigm for explaining the activation of a cell by an extracellular ligand is that the binding of the ligand to a particular cell surface receptor results in selective signal transduction and the activation of a particular pattern of cellular changes and/or gene transcription. Most extracellular ligands bind with high affinity to single cell surface receptors. However, there are certain ligands that appear to have a promiscuous affair with the plasma membrane of host cells. The most prominent of these

Table 3 - Categorisation of chaperonin 60 proteins

Chaperonin	CD14-dependent	CD14-independent	Inactive	Effect on bone
Human	+	–	–	+
GroEL	–	+	–	+
Chl. Trachomatis	+	–	–	?
*Aa**	–	+	–	+
M. tuberculosis 60.1	+	–	–	–
M. tuberculosis 60.2	–	+	–	–
Rh. leguminosarum 60.1	–	–	+	?
Rh. leguminosarum 60.3	+	–	–	?

**Actinobacillus actinomycetemcomitans.*

is LPS which, over the years, has been proposed to bind to a very wide variety of receptors. The work on the LPS receptor up until 1997 has been reviewed in [44]. In the last five years a growing number of receptors for LPS and other pattern-associated molecular patterns (PAMPs), including the Toll-like receptors, have been identified. These include the nucleotide-binding site leucine-rich repeat (NBS-LRR) family [45], trigger receptor expressed on myeloid cells (TREM) [46], moesin [47] and a cell surface complex composed of HSPs70 and 90, the chemokine receptor 4 (CXCR4), and growth differentiation factor 5 [48]. Mammalian cells are clearly replete with receptors able to recognise constituents of bacteria, and in particular LPS. Moreover, it is now emerging that LPS, depending on its source, and presumably reflecting its structure, can act either as an agonist or an antagonist with TLR4 [49].

What receptors bind chaperonin 60? The evidence, such as it is, would suggest that the answer to this will be somewhat similar to the story of LPS. The author has demonstrated that fluorescein-labelled GroEL binds to around 40% of the CD14-positive monocytes in human blood samples and that overnight culture of such monocytes on plastic, a process which promotes macrophage maturation, results in the majority of cells binding to GroEL. The obvious interpretation is that CD14 is not the receptor for GroEL and that whatever the receptor is it is increased in density in macrophages compared with monocytes. Of approximately 30 samples of blood from different individuals tested, one was found not to respond to GroEL in terms of cytokine synthesis. The cells from this normal individual did not bind to fluorescein-labelled GroEL. This could be due to high levels of anti-chaperonin 60 antibodies or may suggest that the receptor for chaperonin 60 is polymorphic (Khan and Henderson, unpublished). Kinetic studies of the binding of human chaperonin

60 to murine macrophages revealed a saturable binding site which was competed for by human chaperonin 60 but not by HSP90, HSP70 or α_2 macroglobulin [50]. It has been claimed that human HSP70 and human chaperonin 60 share the same receptor and can be bound by CD14-/TLR4-negative cells [51]. It is of interest in this context that human HSP70 is reported to bind to CD14/TLR4 [52] whereas the HSP70 protein of *M. tuberculosis* binds to CD40 [53]. As has been discussed, various chaperonin 60 proteins have been reported to bind to TLR4 [38, 39] or to TLR2 and TLR4 [40]. Human chaperonin 60 has also been reported to bind to high-density lipoprotein [54], the HIV transmembrane glycoprotein gp41 [55] and the integrin $\alpha_3\beta_1$ [56].

It may be that chaperonin 60 proteins behave in a similar manner to LPS molecules. There is sufficient variation in their structure that they may have evolved to bind to different cell surface "receptors". Perhaps one of the most curious facets of chaperonin 60 is the fact that the human protein binds to TLR receptors. This is unexpected, as it would identify human chaperonin 60 released from cells as a PAMP and could therefore confuse a system designed to recognise invading pathogens. This is particularly puzzling as in the next section the possibility that chaperonin 60 may be a hormone will be considered. If the release of chaperonin 60 in mammals is part of a homeostatic mechanism it may be that bacteria have evolved to evade such a recognition system by manipulating the non-folding structure/sequence of their chaperonin 60 proteins. It is now realised that some bacteria, for example *Clostridium difficile*, have evolved to utilise their chaperonin 60 proteins as adhesins for binding to human cells [57]. On the other side of the coin, *Staphylococcus aureus* uses surface-exposed human chaperonin 60 in combination with integrins to invade host cells [58].

Chaperonin 60 as an endocrine hormone

One of the major dogmas in molecular chaperone biology is that molecular chaperones are intracellular proteins that cannot be released from cells. There is, however, mounting evidence (reviewed in [59]) that bacteria and eukaryotic cells can release chaperonin 60. The mechanism of release has not been defined but then neither has the mechanism of release of a number of key cell-signalling proteins such as interleukin (IL)-1 [59]. In the past few years evidence has begun to accumulate to suggest that not only is human chaperonin 60 released from cells but that it is released in sufficient amounts to be found in the circulation. In a recent, report the author and colleagues examined plasma obtained from healthy British civil servants. The majority of these individuals had measurable levels of immunoreactive human chaperonin 60 in their blood [60]. This confirms and expands on the literature on chaperonin 60 levels in blood, which has concentrated on serum levels of human chaperonin 60 in normal individuals or those with cardiovascular disease (reviewed in

[61]). The source of the circulating chaperonin 60 in normal individuals is not known. What was interesting in the British civil service study was the correlation between circulating levels of human chaperonin 60 and the psychological profile of the individuals [60].

Conclusions

In 1993, chaperonin 60 was a very highly conserved intracellular protein evolved to aid the folding and refolding of proteins. In 2003, when this chapter was written, chaperonin 60 and other molecular chaperones are now viewed as moonlighting proteins with a variety of functions. In the context of *Homo sapiens*, chaperonin 60 proteins from a variety of microbial sources can interact with myeloid cells and alter their functions. What is particularly striking is that even with the enormous degree of sequence conservation, chaperonin 60 proteins from different sources can have profoundly different effects. In other words chaperonin 60 proteins can take on many different guises like a chameleon. The most striking illustration of this is the recent report of the insect neurotoxin produced by the antlion or doodlebug. This insect paralyses its prey using a salivary neurotoxin. The neurotoxin is produced by a symbiotic bacteria resident in the saliva, *Enterobacter aerogenes*. It turns out that the neurotoxin is the chaperonin 60 of this bacterium. Moreover, the chaperonin 60 protein of this bacterium is almost identical to GroEL. The most fascinating detail of this story is that single residue changes in GroEL (e.g., N101T) can turn this chaperonin from one which has no insect neurotoxic activity to one with potent activity [62]. These mutations occur on the outside of the molecule and it is hypothesised that chaperonin 60 is a topologically-constrained protein with two distinct surfaces. In the heptameric and tetradecameric structures the inner surface functions to fold proteins. In contrast, the outer surface of the protein may contain the cell-cell signalling motifs enabling this protein to transfer information from one cell to another. What is the nature of the information transferred by chaperonin 60. An obvious hypothesis is that extracellular chaperonin 60 transfers information about the existence of stress in a cells' environment. This process has been termed stress broadcasting [59]. It is postulated that this signalling may act as part of a continuum between psychological stress, physiological stress and cellular stress. The next few years will see rapid advances in our understanding of the signalling role of molecular chaperones and their participation in the generation of homeostasis.

Acknowledgement
The financial help of the Arthritis Research Campaign (Programme Grant HO600) is gratefully acknowledged.

References

1 Hemmingsen SM, Woolford C, van der Vies SM, Tilly K, Dennis DT, Georgopoulos CP, Hendrix RW, Ellis RJ (1988) Homologous plant and bacterial proteins chaperone oligomeric protein assembly. *Nature* 333: 330–334

2 Saibil H (2000) Molecular chaperones: containers and surfaces for folding, stabilising or unfolding proteins. *Curr Opin Struct Biol* 10: 251–258

3 Thirumalai D, Lorimer GH (2001) Chaperonin-mediated protein folding. *Annu Rev Biophys Biomol Struct* 30: 245–269

4 Brocchieri L, Karlin S (2000) Conservation among Hsp60 sequences in relation to structure, function and evolution. *Protein Sci* 9: 476–486

5 Jeffery CJ (1999) Moonlighting proteins. *TIBS* 24: 8–11

6 Haga A, Ninaka Y, Raz A (2000) Phosphohexose isomerase/autocrine motility factor/neuroleukin/maturation factor is a multifunctional phosphoprotein. *Biochim Biophys Acta* 1480: 235–244

7 Bermudez LE, Petrofsky M, Shelton K (1996) Epidermal growth factor-binding protein in *Mycobacterium avium* and *Mycobacterium tuberculosis*: a possible role in the mechanism of infection. *Infect Immun* 61: 830–835

8 Winram SB, Lottenberg R (1996) The plasmin-binding protein Plr of group A streptococci is identified as glyceraldehyde-3-phosphate dehydrogenase. *Microbiology* 142: 2311–2320

9 Zhang X, Beuron F, Freemont PS (2002) Machinery of protein folding and unfolding. *Curr Opin Struct Biol* 12: 231–238

10 Coates AR, Shinnick TM, Ellis RJ (1993) Chaperonin nomenclature. *Mol Microbiol* 8: 787

11 Friedland JS, Shattock R, Remick DG, Griffin GE (1993) Mycobacterial 65 kD heat shock protein induces release of proinflammatory cytokines from human monocytic cells. *Clin Exp Immunol* 91: 58–62

12 Peetermans WE, Raats CJ, Langermans JA, van Furth R (1994) Mycobacterial heat-shock protein 65 induces proinflammatory cytokines but does not activate human mononuclear phagocytes. *Scand J Immunol* 39: 613–617

13 Retzlaff C, Yamamoto Y, Hoffman PS, Friedman H, Klein TW (1994) Bacterial heat shock proteins directly induce cytokine mRNA and interleukin-1 secretion in macrophage cultures. *Infect Immun* 62: 5689–5693

14 Kirby AC, Meghji S, Nair SP, White P, Reddi K, Nishihara T, Nakashima K, Willis AC, Sim R, Wilson M, Henderson B (1995) The potent bone resorbing mediator of *Actinobacillus actinomycetemcomitans* is homologous to the molecular chaperone GroEL. *J Clin Invest* 96: 1185–1194

15 Gouhlen F, Hafezi A, Uitto V-J, Hinode D, Nakamura R, Grenier D, Mayrand D (1998) Subcellular localisation and cytotoxic activity of the GroEL-like protein isolated from *Actinobacillus actinomycetemcomitans. Infect Immun* 66:5307–5313

16 Paju S, Goulhen F, Asikainen S, Grenier D, Mayrand D, Uitto V (2000) Localization of

heat shock proteins in clinical *Actinobacillus actinomycetemcomitans* strains and their effects on epithelial cell proliferation. *FEMS Microbiol Lett* 182: 231–235

17 Wong PM, Chugn SW, Sultzer BM (2000) Genes, receptors, signals and responses to lipopolysaccharide endotoxin. *Scand J Immunol* 51: 123–127

18 Horwitz MC, Xi Y, Wilson K, Kacena MA (2001) Control of osteoclastogenesis and bone resorption by members of the TNF family of receptors and ligands. *Cyt Growth Fact Revs* 12: 9–18

19 Reddi K, Meghji S, Nair SP, Arnett TR, Miller AD, Preuss M, Wilson M, Henderson B, Hill P (1998) The *Escherichia coli* chaperonin 60 (groEL) is a potent stimulator of osteoclast formation. *J Bone Miner Res* 13: 1260–1266

20 Nair SP, Meghji S, Poole S, Miller AD, Henderson B (1999) Molecular chaperones stimulate bone resorption. *Calcif Tissue Intl* 64: 214–218

21 Boachie-Adjei O, Squillante RG (1996) Tuberculosis of the spine. *Orthop Clin North Am* 27: 95–103

22 Kong TH, Coates ARM, Butcher PD, Hickman CJ, Shinnick TM (1993) *Mycobacterium tuberculosis* expresses two chaperonin-60 homologs. *Proc Natl Acad Sci USA* 90: 2608–2612

23 Meghji S, White PA, Nair SP, Reddi K, Heron K, Henderson B, Zaliani A, Fossati G, Mascagni P, Hunt JF et al (1997) *Mycobacterium tuberculosis* chaperonin 10 stimulates bone resorption: A potential contributory factor in Pott's disease. *J Exp Med* 186: 1241–1246

24 Meghji S, Lillicrap M, Maguire M, Gaston JSH, Henderson B. Human chaperonin 60 is a bone resorbing agonist. *Bone; in press*

25 Margulis L (1993) *Symbiosis in cell evolution.* 2nd Ed, WH Freeman, New York

26 Lewthwaite JC, Coates ARM, Tormay P, Singh M, Mascagni P, Poole S, Roberts M, Sharp L, Henderson B (2001) *Mycobacterium tuberculosis* chaperonin 60.1 is a more potent cytokine stimulator than chaperonin 60.2 (hsp 65) and contains a CD14-binding domain. *Infect Immun* 69: 7349–7355

27 Billingham MEJ (1995) Adjuvant arthritis: The first model. In: B Henderson, JCW Edwards, ER Pettipher (eds): *Mechanisms and models in rheumatoid arthritis.* Academic Press, London, 389–409

28 Cobelens PM, Heijnen CJ, Nieuwenhuis ES, Kramer PGP, van der Zee R, van Eden W, Kavelaars A (2000) Treatment of adjuvant arthritis by oral administration of mycobacterial has 65 during disease. *Arthritis Rheum* 43: 2694–2702

29 Winrow VR, Coates ARM, Tormay P, Henderson B, Singh M, Blake DR, Morris CJ (2002) Chaperonin 60.1 prevents bone destruction in Wistar rats with adjuvant-induced arthritis. *Rheumatology* 41 (Abstr Suppl 1): 47

30 Campagnuolo G, Bolon B, Feige U (2002) Kinetics of bone protection by recombinant osteoprotegerin therapy in Lewis rats with adjuvant arthritis. *Arthritis Rheum* 46: 1926–1936

31 Han Z, Boyle DL, Chang L, Bennett B, Karin M, Yang L, Manning AM, Firestein GS

(2001) c-Jun N-terminal kinase is required for metalloproteinase expression and joint destruction in inflammatory arthritis. *J Clin Invest* 108: 73–81

32 Kong Y-Y, Feige U, Sarosi I, Bolon B, Tafuri A, Morony S, Capparelli C, Li J, Elliott R, McCabe S et al (1999) Activated T cells regulate bone loss and joint destruction in adjuvant arthritis through osteoprotegerin ligand. *Nature* 402: 304–308

33 Price N, Kelly SM, Wood S, auf de Mauer A (1991) The aromatic amino acid content of the bacterial chaperone protein groEl (Cpn60): evidence for the presence of a single tryptophan. *Febs Lett* 292: 9–12

34 Houry WA, Frishman D, Eckerskorn C, Lottspeich F, Hartl FU (1999) Identification of *in vivo* substrates of the chaperonin GroEL. *Nature* 402: 147–154

35 Maguire M, Coates ARM, Henderson B (2002) Cloning expression and purification of three chaperonin 60 homologues. *J Chromatography B Analyt Technol Biomed Life Sci* 786: 117–125

36 Tabona P, Reddi K, Khan S, Nair SP, Crean StJ, Meghji S, Wilson M, Preuss M, Miller AD, Poole S et al (1998) Homogeneous *Escherichia coli* chaperonin 60 induces IL-1 and IL-6 gene expression in human monocytes by a mechanism independent of protein conformation. *J Immunol* 161: 1414–1421

37 Kol A, Lichtman AH, Finberg RW, Libby P, Kurt-Jones EA (2000) Cutting edge: heat shock protein (HSP) 60 activates the innate immune response: CD14 is an essential receptor for HSP60 activation of mononuclear cells. *J Immunol* 164: 13–17

38 Ohashi K, Burkart V, Flohe S, Kolb H (2000) Cutting edge: heat shock protein 60 is a putative endogenous ligand of the toll-like receptor-4 complex. *J Immunol* 164: 558–561

39 Bulut Y, Faure E, Thomas L, Karahashi H, Michelsen KS, Equils O, Morrison SG, Morrison RP, Arditi M (2002) Chlamydial heat shock protein 60 activates macrophages and endothelial cells through toll-like receptor 4 and MD2 in a MyD88-dependent pathway. *J Immunol* 168: 1435–1440

40 Vabulas RM, Ahmad-Nejad P, da Costa C, Miethke T, Kirschning CJ, Hacker H, Wagner H (2001) Endocytosed HSP60s use toll-like receptor 2 (TLR2) and TLR4 to activate the toll/interleukin-1 receptor signaling pathway in innate immune cells. *J Biol Chem* 276: 31332–31329

41 Wallington EJ, Lund PA (1994) *Rhizobium leguminosarum* contains multiple chaperonin (Cpn60) genes. *Microbiology* 140: 113–122

42 Erbse A, Yifrach O, Jones S, Lund PA (1999) Chaperone activity of a chimeric GroEL protein that can exist in a single of double ring form. *J Biol Chem* 274: 20351–20357

43 Lewthwaite JC, George R, Lund PA, Poole S, Tormay P, Sharp L, Coates ARM, Henderson B (2002) *Rhizobium leguminosarum* chaperonin 60.3, but not chaperonin 60.1, induces cytokine production by human monocytes: activity is dependent on interaction with cell surface CD14. *Cell Stress & Chaperones* 7: 130–136

44 Henderson B, Poole S, Wilson M (1998) *Bacterial/cytokine interactions in health and disease.* Portland Press, London

45 Girardin SE, Sansonetti PJ, Philpott DJ (2002) Intracellular vs extracellular recognition of pathogens – common concepts in mammals and flies. *Trends Microbiol* 10: 193–199

46 Bouchon A, Facchetti F, Weigand MA, Colonna M (2001) TREM-1 amplifies inflammation and is a crucial mediator of septic shock. *Nature* 410: 1103–1107

47 Amar S, Oyaisu K, Li L, Van Dyke T (2001) Moesin: a potential LPS receptor on human monocytes. *J Endotoxin Res* 7: 281–286

48 Triantafilou M, Triantafilou K (2002) Lipopolysaccharide recognition: CD14, TLRs and the LPS-activation cluster. *Trends Immunol* 23: 301–314

49 Yoshimura A, Kaneko T, Kato Y, Golenbock DT, Hara Y (2002) Lipopolysaccharides from periodontopathic bacteria *Porphyromonas gingivalis* and *Capnocytophaga ochracea* are antagonists for human toll-like receptor 4. *Infect Immun* 70: 218–225

50 Habich C, Baumgart K, Kolb H, Burkart V (2002) The receptor for heat shock protein 60 on macrophages is saturable, specific, and distinct from receptors for other heat shock proteins. *J Immunol* 168: 569–576

51 Lipsker D, Ziylan U, Spehner D, Proamer F, Bausinger H, Jeannin P, Salamero J, Bohbot A, Cazenave JP, Drillien R et al (2002) Heat shock proteins 70 and 60 share common receptors which are expressed on human monocyte-derived but not epidermal dendritic cells. *Eur J Immunol* 32: 322–332

52 Asea A, Kraeft SK, Kurt-Jones EA, Stevenson MA, Chen LB, Finberg RW, Koo GC, Calderwood SK (2000) HSP70 stimulates cytokine production through a CD14-dependant pathway, demonstrating its dual role as a chaperone and cytokine. *Nat Med* 6: 435–442

53 Wang Y, Kelly CG, Karttunen JT, Whittall T, Lehner PJ, Duncan L, MacAry P, Younson JS, Singh M, Oehlmann et al (2001) CD40 is a cellular receptor mediating mycobacterial heat shock protein 70 stimulation of CC-chemokines. *Immunity* 15: 971–983

54 Bocharov AV, Vishnyakova TG, Baranova IN, Remaley AT, Patterson AP, Eggerman TL (2000) Heat shock protein 60 is a high-affinity high-density lipoprotein binding protein. *Biochem Biophys Res Commun* 277: 228–235

55 Speth C, Prohaszka Z, Mair M, Stockl G, Zhu X, Jobstl B, Fust G, Dierich MP (1996) A 60 kD heat-shock protein-like molecule interacts with the HIV transmembrane glycoprotein gp41. *Mol Immunol* 36: 619–628

56 Barazi HO, Zhou L, Templeton NS, Krutzsch HC, Roberts DD (2002) Identification of heat shock protein 60 as a molecular mediator of $\alpha_3\beta_1$ integrin activation. *Cancer Res* 62: 1541–1548

57 Hennequin C, Porcheray F, Waligora-Dupriet A-J, Collignon A, Barc M-C, Bourlioux P, Karjalainen T (2001) GroEL (Hsp60) of *Clostridium difficile* is involved in cell adherence. *Microbiology* 147: 87–96

58 Dziewanowska K, Carson AR, Patti JM, Deobald CF, Bayles KW, Bohach GA (2000) Staphylococcal fibronectin binding protein interactions with heat shock protein 60 and integrins: Role in internalisation by epithelial cells. *Infect Immun* 68: 6321–6328

59 Maguire M, Coates ARM, Henderson B (2002) Chaperonin 60 unfolds its secrets of cellular communication. *Cell Stress & Chaperones* 7: 317–329

60 Lewthwaite J, Owen N, Coates ARM, Henderson B, Steptoe AD (2002) Circulating heat shock protein (Hsp)60 in the plasma of British civil servants: Relationship to physiological and psychosocial stress. *Circulation* 106: 196–201
61 Pockley AG (2002) Hear shock proteins, inflammation, and cardiovascular disease. *Circulation* 105: 1012–1017
62 Yoshida N, Oeda K, Watanabe E, Mikami T, Fukita Y, Nishimura K, Komai K, Matsuda K (2001) Chaperonin turned insect toxin. *Nature* 411: 44
63 Zhang Y, Doerfler M, Lee TC, Guillemin B, Rom WN (1993) Mechanisms of stimulation of interleukin-1 beta and tumor necrosis factor-alpha by *Mycobacterium tuberculosis* components. *J Clin Invest* 91: 2076–2083
64 Chopra U, Vohra H, Chhibber S, Ganguly NK, Sharma S (1997) TH1 pattern of cytokine secretion by splenic cells from pyelonephritic mice after in-vitro stimulation with hsp-65 of *Escherichia coli. J Med Microbiol* 46: 139–144
65 Kol A, Bourcier T, Lichtman AH, Libby P (1999) Chlamydial and human heat shock protein 60s activate human vascular endothelium, smooth muscle cells, and macrophages. *J Clin Invest* 103: 571–577
66 Billack B, Heck DE, Mariano TM, Gardner CR, Sur R, Laskin DL, Laskin JD (2002) Induction of cyclooxygenase-2 by heat shock protein 60 in macrophage and endothelial cells. *Am J Physiol Cell Physiol* 283: C1267–C1277
67 Sharma SA, Miller GG, Peek RA, Perez-Perez G, Blaser MJ (1997) T-cell, antibody, and cytokine responses to homologs of the 60-kilodalton heat shock protein in *Helicobacter pylori* infection. *Clin Diagn Lab Immunol* 4: 440–446
68 Galdiero M, de l'Ero GC, Marcatili A (1997) Cytokine and adhesion molecule expression in human monocytes and endothelial cells stimulated with bacterial heat shock proteins. *Infect Immun* 65: 699–707
69 Verdegaal ME, Zegveld ST, van Furth R (1996) Heat shock protein 65 induces CD62e, CD106, and CD54 on cultured human endothelial cells and increases their adhesiveness for monocytes and granulocytes. *J Immunol* 157: 369–376
70 Marcatili A, Cipollaro DL, Galdiero M, Folgore A, Petrillo G (1997) TNF-alpha, IL-1 alpha, IL-6 and ICAM-1 expression in human keratinocytes stimulated *in vitro* with *Escherichia coli* heat-shock proteins. *Microbiology* 143: 45–53
71 Hinode D, Yoshioka M, Tanabe S, Miki O, Masuda K, Nakamura R (1998) The GroEL-like protein from *Campylobacter rectus*: immunological characterization and interleukin-6 and-8 induction in human gingival fibroblast. *FEMS Microbiol Lett* 167: 1–6
72 Yamaguchi H, Osaki T, Kurihara N, Taguchi H, Kamiya S (1999) Reactivity of monoclonal antibody to HSP60 homologue of *Helicobacter pylori* with human gastric epithelial cells and induction of IL-8 from these cells by purified *H. pylori* HSP60. *J Gastroenterol* 34 (Suppl 11): 1–5

Heat shock protein receptors, functions and their effect on monocytes and dendritic cells

Thomas Lehner[1], Yufei Wang[1] and Charles Kelly[2]

Departments of [1]Immunobiology and [2]Oral Medicine and Pathology, Guy's, King's and St. Thomas' Medical Schools, King's College, London SE1 9RT, UK

Introduction

Molecular chaperons of the heat shock protein (HSP) family are conserved proteins that modulate intracellular protein folding. By binding to unfolded or folding intermediate polypeptides, chaperons prevent misfolding and aggregation, and promote folding and translocation [1]. Human and microbial HSP70 consist of an N-terminal ATPase fragment of 44 kD and a C-terminal peptide-binding fragment of 28.5 kD. HSP70 is a highly conserved protein among various species and there is about 60% homology between human and mycobacterial and 70% between *E. coli* (Dnak) and mycobacterial HSP70. However, it should be noted that a homology plot between human and *E. coli* HSP70 shows differences along the molecule between 20 and 80% [2]. This variation must be taken into account when considering HSP70 epitopes. Mammalian HSP70 is found in all nucleated cells as a constitutive protein in the cytosol, but can be induced by a variety of stress stimuli (e.g., heat, toxic elements, infection) to give rise to the inducible HSP70. Another type of HSP70 (BiP) is found in the endoplasmic reticulum. Microbial HSP70 is found in bacteria, fungi, parasites and viruses. The gut associated lymphoid tissue is activated not only by LPS which are found exclusively in Gram-negative bacteria, but also HSP70 found in both Gram-positive and negative bacteria. Hence, T- and B-cells in Peyer's patches of the gut are activated by HSP70 and LPS of gut bacteria. Some of the properties of HSPs are summarised in Table 1, but this review will emphasize the novel findings of HSP70.

Structure of HSP70

HSP70 comprise an N-terminal domain (amino acid residues 1–384 *E. coli* DnaK) with ATP-ase activity, linked by a short conserved sequence to the C-terminal domain (amino acid residues 394–638, *E. coli* DnaK) which binds substrate. Although the overall three-dimensional structure of HSP70 is not known, the struc-

Heat Shock Proteins and Inflammation, edited by Willem van Eden

Table 1 - Major properties of HSPs

HSPs exert chaperone functions, by binding unfolded polypeptides to prevent misfolding and aggregation.
They bind peptides with a hydrophobic motif by non-covalent linkage.
HSP70 and HSP90 deliver exogenous antigen into the MHC Class I pathway, playing an important role in cross-priming free or antigen released from apoptotic cells.
HSPs stimulate production of the CC chemokines, CCL3, CCL4 and CCL5.
HSPs stimulate production of cytokines, especially TNF-α, IL-12 and NO.
HSPs stimulate maturation of DC to a similar extent of that of CD40L.
They exert robust adjuvant function when linked to antigens and they are effective when administered both systemically and by the mucosal route.
Generation of IL-12 and TNF-α may induce Th1 polarization of the adjuvant function of HSP70.
Tumour or virus-specific peptides, non-covalently bound to HSP70 or HSP96 exert a protective effect against the specific tumour or virus.

tures of the two domains from various members of the family have been solved separately.

The crystal structures of the ATP-ase domains of bovine HSC70 (heat shock constitutive protein) and human HSP70 have been determined at resolutions of 2.2 Å and 1.84 Å, respectively [3, 4]. The domain consists of two approximately equal sized lobes with a deep cleft between them. ATP binds at the base of the cleft. Two crystal forms of the human ATP-ase fragment were obtained that differed by a shift (1–2 Å) in one of the sub-domains. This shift may be important in ATP-binding and ADP-release and indicates some degree of flexibility in this domain.

The crystal structure of the substrate-binding domain of DnaK from *E. coli* with bound substrate (a 7-residue peptide with the sequence NRLLLTG) has been determined at 2 Å resolution [5], and consists of a β-sandwich sub-domain followed by an α-helical sub-domain. The β-sandwich sub-domain is formed by two stacked anti-parallel four stranded β-sheets. The upper sheet, forms the substrate binding site with loops L1,2 and L3,4 (between β-strands 1 and 2 and between β-strands 3

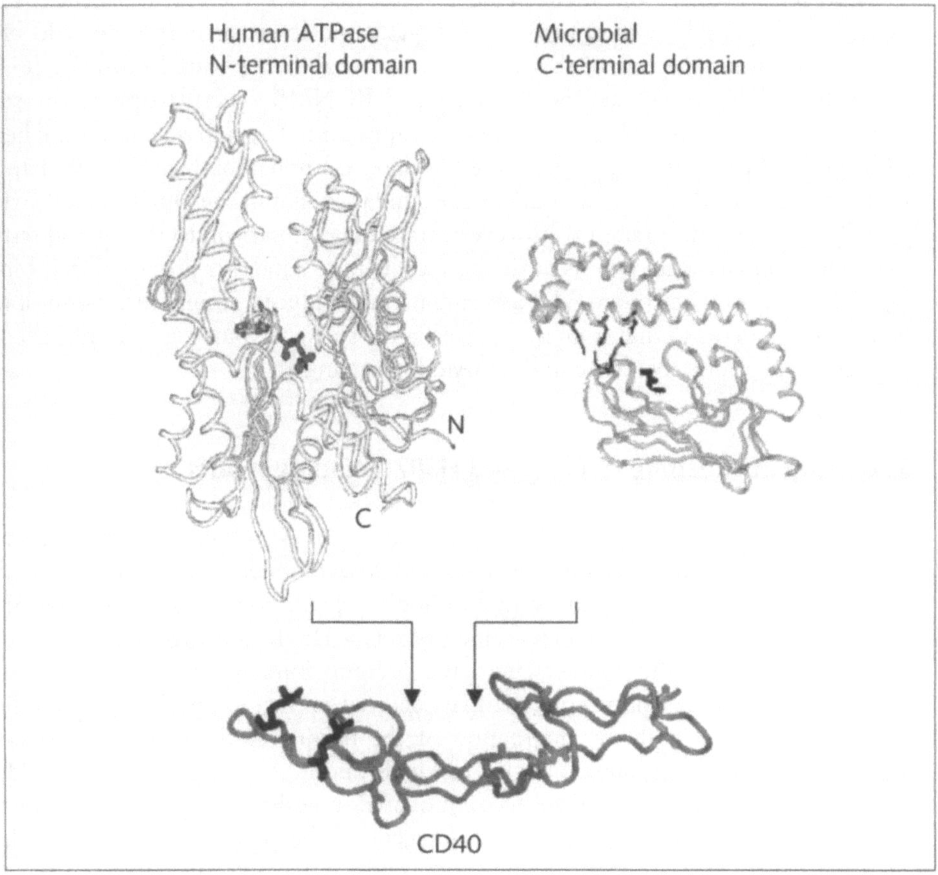

Figure 1
Structure of the N-terminal ATPase domain of HSP70 [5a], the C terminal peptide binding domain of DnaK [5] and a molecular model of CD40 [5b].

and 4, respectively), forming the sides of a channel that is the primary site of interaction, with substrate. The outer loop, L4,5 stabilises L1,2 by hydrogen bonds and hydrophobic interaction while L5,6 forms hydrogen bonds that stabilise L3,4. The α-helical sub-domain comprises five helices with the first and second helices (αA and αB) forming hydrophobic side chain contacts with the β-sandwich. Helix αB extends over the entire substrate binding site and may stabilize substrate binding by interaction with all four loops that form this site (see below) but does not interact directly with substrate. The C-terminal half of helix αB also forms an anti-parallel bundle of three helices with αC and αD while helix αE lies across one end of the bundle (Fig. 1).

Mutagenised forms of DnaK (from *E. coli*) that lack the entire α-helical sub-domain bind peptide with approximately 5–10 fold lower affinity than wild type DnaK, confirming the role of this sub-domain in stabilizing bound substrate [6–8]. The solution structures have been determined by NMR spectroscopy of the substrate binding domain of DnaK truncated at approximately halfway along the helix αB [9] and of the homologous truncated domain from mammalian HSC70 [10]. In both proteins, the β-sandwich domains are similar to that of the DnaK crystal structure. The helical sub-domain is however rotated in comparison to the crystal structure. The flexible hinge point is at the base of helix αA and it is proposed that binding of ATP by the N-terminal ATP-ase domain may induce a similar alteration of the helical sub-domain so that it does not overlie the substrate binding site. In this conformation, peptide substrate could be readily exchanged.

Specificity of substrate binding and HSP70 binding motifs

The peptide substrate (NRLLLTG) complexed to DnaK, adopts an extended conformation in which main-chain atoms form hydrogen bonds with DnaK while side-chain contacts are predominantly hydrophobic [5]. The central residue (Leu 4) is buried in a relatively large hydrophobic pocket of DnaK and together with Leu 3 contributes most of the contacts with the protein. Binding is almost completely determined by a five residue core (RLLLT) centred on Leu 4. Specificity of binding is determined principally by interaction of the residue at the centre of the core sequence with the hydrophobic pocket of DnaK and it was suggested that Ile, Met, Thr, Ser or possibly Phe could be accommodated in addition to Leu. Hydrophobic residues are preferred at the positions adjacent to the central residues and while there are fewer constraints on residues at the ends of the motif, negative charges are generally excluded. The NMR spectroscopic analyses of the truncated DnaK and mammalian Hsc70 substrate binding domains (discussed above) also indicated that Leu residues occupied the hydrophobic pocket of the substrate binding sites.

The observations that HSP70 substrates are predominantly hydrophobic correlates with the chaperone function of HSP70 family members in stabilizing partially unfolded polypeptides and are generally in agreement with studies of substrate specificity performed with a variety of synthetic peptides and peptide libraries. Affinity panning of a phage display library with DnaK originally identified NRLLLTG as a substrate and demonstrated a motif enriched in hydrophobic residues that could accommodate terminal basic or polar residues [11].

Some differences in substrate specificity between members of the HSP70 family have however been described which may reflect differences in function. Thus investigation of the substrate specificity of the endoplasmic reticulum-associated chaperone BiP by bulk sequencing of bound synthetic peptides from a random library [12] identified a heptapeptide motif that was enriched in aliphatic residues. Similarly,

Table 2 - Four major receptors interacting with HSP and stimulating cytokine and chemokine production and DC maturation

Receptor	HSP	Function	Refs.
CD14	Human HSP60, HSP70	Stimulation of TNF-α, IL-12 and IL-6	[16, 17]
CD91	Human HSP70, HSP90, gp96 and calreticulin	Stimulation of TNF-α, IL-12 and IL-1β	[69]
CD40	Microbial and human HSP70	Stimulation of chemokines, TNF-α, IL-12 and DC maturation	[20, 21, 26]
TLRs	Human HSP60, HSP70 and HSP90	Stimulation of TNF-α, IL-12 and DC maturation	[22, 23, 47–49]

affinity panning of phage libraries, displaying peptides (eight or 12 residues) [13], identified a heptapeptide motif in which alternate positions were occupied by large hydrophobic or aromatic residues but charged residues, particularly Lys were excluded. In contrast, HSC73, a member of the HSP70 family that targets proteins to lysosomes for degradation, recognises sequences related to KFERQ [14]. A systematic comparison of the substrate specificities of DnaK, BiP and HSC70 confirmed the importance of an anchoring large hydrophobic residue at position four of a heptapeptide motif and demonstrated differences in specificity associated with positions two and six [15].

HSP receptors and co-receptors

Search for HSP receptors over the last few years has led to the discovery of several receptors which are used by HSP (Tab. 2). CD14 [16, 17], CD91 [18, 19], CD40 [20, 21] and Toll-like receptors [22, 23] interact with HSP. It appears that different member of HSP family, although they have no similarities in protein sequences, can use the same receptor. On the other hand, the same members of HSP derived from different species, despite their highly conserved sequence use different receptors. This diversity of receptor usage by HSP may have important biological implications. Interactions between HSPs and their receptors may elicit two different but related functions; non-specific stimulation of antigen presenting cells to generate the production of chemokines [24] and cytokines [25, 26] and to mediate internalization of HSP-peptide complexes by endocytosis and translocation of HSP into the cell [27, 28].

CD14 is expressed on the cell surface of monocytes and macrophages and to lesser extent on other myeloid cells, such as neutrophils. CD14 molecule is a receptor binding LPS and LPS-binding protein [29]. Human HSP60, but not the homologue derived from *E. coli* and mycobacterial HSP65, also uses CD14 to activate the innate immune system [16]. Another report suggested that CD14 may also be used by human HSP70 to stimulate monocytes to produce TNF-α [17]. However, CD14 is a glycosylphosphatidylinositol (GPI)-anchored protein lacking a transmembrane domain, so it is unlikely that CD14 can directly lead to cellular activation. Indeed, CD14 is closely associated with TLRs in intracellular signalling [30]. Transfection of CD14 into a human cell line (astrocytoma cells, U373) or CHO (Chinese hamster ovarian cells) also demonstrated that CD14 is necessary, but not sufficient for cellular responsiveness to HSP60 [16]. These findings suggest that CD14 may be a component of a multi-receptor complex for HSP as suggested for LPS [31].

LPS contamination of HSP preparations

HSPs stimulate innate immune cells to produce inflammatory cytokines, including TNF-α, IL-1β, IL12, GM-CSF, nitric oxide and chemokines, such as MIP-1α, MIP-1β, RANTES, MCP-1 and MCP-2. These activities of HSP are similar to those of LPS which can contaminate HSP preparations, and it is essential to exclude LPS. Currently it is difficult to prepare LPS-free HSP preparations, especially those expressed in *E. coli*. LPS activity is abrogated or greatly reduced by polymixin B treatment [17, 20], whereas HSP stimulation is reduced by heat denaturation [17, 32]. Other reagents, such as RSLP derived from *Rhodopseudomonas spheroides* and lipid IVa also inhibit LPS activity and have no effect on HSP stimulation [17, 32]. The intracellular calcium chelator BAPTA-AM differentiates between LPS and HSP70 functions [17, 20]. The C3H/HeJ and C57BL/10ScCr inbred mouse strains are homozygous for a mutant *LPS* allele (*LPS*$^{d/d}$) which confers hyporesponsiveness to LPS challenge [33] and provide a model to study immunological functions of HSP. Genetic analysis revealed that C3H/HeJ mice have a point mutation within the coding region of the TLR4 gene, whereas C57BL/ 10ScCr mice exhibit a deletion of TLR4 [34]. In human studies using microbial HSP70, inhibition with antibodies to CD40 but not to CD14 should readily discriminate between HSP and LPS [20, 26].

Most of the LPS contamination (95–99%) in HSP preparations can be removed by using polymixin B immobilized on agarose affinity column. Polymixin B is a cationic cyclopeptide which can neutralize the biological activity of LPS by binding to its lipid A portion. It is noteworthy that HSP contains a hydrophobic domain which can interact with the affinity column and result in protein loss. This may be enhanced by HSP binding to the LPS on the column [35, 36]. Our experience with this method indicates that 60–70% of proteins can be recovered after one treatment with polymixin and there is further protein loss if a second treatment is required. The residual LPS concentration in recovered HSP preparation is usually less than 5

U per mg protein, as determined by the Limulus amebocyte lysate assay [26]. Although the significance of this low level of LPS in HSP preparation depends on the function under investigation, such low levels of LPS do not stimulate production of cytokines, CC-chemokines by cultured monocytes or maturation of DC [20, 26].

CD91 was identified as a receptor for α_2 macroglobulin (α_2M) which is a protease inhibitor that binds to microbial pathogens and mediates phagocytosis by monocytes [37]. Several members of the HSP family, including gp96, HSP90, HSP70 and calreticulin, despite being structurally distinct, interact with CD91 and are internalized by the receptor and the HSP-bound peptides. This interaction is critical to the induction of $CD8^+$ cytotoxic T lymphocytes by HSP mediated cross-priming mechanisms [18, 19]. HSP binds gp96 directly to CD91, which can be inhibited either by antibodies to CD91 or the α_2M ligand [18]. The CD91 mediated endocytosis of HSP-peptide complexes can direct the peptide to the proteosome-dependent transporter which is associated with the MHC Class 1 pathway. However, it is not clear whether CD91 engagement by HSP activates antigen presenting cells to produce cytokines, co-stimulatory molecules or maturation of dendritic cells.

CD40 is a 40–50 kDa glycoprotein and is a member of the tumor necrosis factor receptor superfamily. CD40 is primarily expressed on B-cells, monocytes and dendritic cells, but can also be found on epithelial cells and $CD8^+$ T-cells [38, 39]. The natural ligand for CD40 is CD40 ligand (CD154) expressed on activated T-cells. Interaction between CD40 and CD154 is essential for T-cell-dependent antigen activation of B-cells, activation of monocytes, maturation of dendritic cells and T-cells help for $CD8^+$ T-cell cytotoxic function [40].

We reported that CD40 is a receptor for microbial HSP70 [20] and this was later confirmed and extended to human HSP70 [21]. We have postulated the hypothesis that the adjuvanticity of HSP70 and HSP65 is accounted for by stimulating production of the CC-chemokines CCL3, CCL4 and CCL5 which attract the entire repertoire of immunological cells [24, 41]. As both major co-stimulatory pathways, CD80/86-CD28 and CD40-CD40L, stimulate these CC-chemokines [42–44], we explored the possibility that HSPs might interact with those co-stimulatory molecules [20]. Whereas antibodies to CD80 or CD86 had no effect, those to CD40 blocked HSP70 stimulation of CC-chemokine production. Further in depth investigations demonstrated that HSP70 stimulated the production of CC-chemokines only if HEK293 cells (human embryonic kidney cell lines) were transfected with human CD40, but not with control molecules. Immuno-precipitation studies revealed that HSP70 physically associates with cell membrane CD40 when incubated with CD40 expressing cells and using surface plasmon resonance showed that HSP70 can directly bind to CD40 molecules [20]. HSP70-peptide complexes binding CD40 may deliver the peptide into MHC Class 1 and 2 pathways and this process is dependent on the ADP-loaded state of HSP70 [21]. Utilization of the co-stimulation molecule CD40 as a receptor for HSP70 raises the paradigm that HSP70 may function at an interface between innate and adaptive immunity.

Toll-like receptors (TLRs), also known as pattern-recognition receptors are expressed by the innate immune system and recognize specific pathogen-associated molecular patterns (PAMPs) expressed on microbial components [45]. So far about ten TLRs have been described within the TLR family and each receptor seems to recognize different microbial pathogenic elements. TLRs are expressed mainly in the cell types that are involved in the first line of defence, such as dendritic cells, monocytes, neutrophils, epithelial and endothelial cells. Activation of TLRs leads to production of inflammatory cytokines, chemokines, nitric oxide, complement proteins, enzymes (such as cyclo-oxygenase-2, COX-2), adhesion molecules and immune receptors [46]. These innate immune responses are essential for elimination of pathogens and regulation of adaptive immunity.

TLRs are favourite candidates for HSP, as these are highly conserved among microbial organisms and may serve as PAMPs to activate the innate immune system by interacting with pattern recognition receptors. However, only human HSP60 and HSP70, but not microbial HSP have so far been found to stimulate TLRs [23, 47]. Bone marrow-derived macrophages from the mouse strain C3H/HeJ, carrying a mutant TLR4 is non-responsive to HSP60 [47]. A recent study suggests that both TLR2 and 4 mediate HSP60 activation of Toll/interleukin-1 receptor signalling pathway, to produce TNF-α [23]. Human HSP70 and HSPgp96 are also able to activate the TLR 2 and 4/IL-1 receptor signal pathways and stimulate production of TNF-α similar to HSP60 [48, 49].

It is, however, not clear whether TLRs act as receptors for HSP or are involved in signalling cellular activation. Some studies suggest that cell surface TLR4 is essential for human HSP60 activation of monocytes. Other studies indicate that endocytosis of HSP60 or HSP70 is a prerequisite for activation of the intracellular TLR2 and TLR4 signalling pathways and the endocytosis process is independent of TLR2 and TLR4 [48, 49]. At present it is not clear how HSP60 and HSP70 stimulate TLR2 and TLR4. There is no evidence to indicate that there is a direct interaction between human HSP and TLRs. The finding that human HSP can activate TLRs suggests that these not only serve as receptors for PAMPs derived from pathogenic microbes but also recognize endogenous ligands. Endogenous HSP are present predominantly in the cell cytoplasm and can be induced and released in pathological conditions, such as injury, necrosis and stress. Induction of local inflammatory responses by TLRs and endogenous HSP interaction in the milieu around damaged tissues is probably helpful for tissue repair and wound healing, but it can also play an important part in the chronic inflammatory diseases.

Signalling pathways

The findings that CD40 and TLRs are cellular receptors for HSP70 suggests that intracellular signalling following engagement of HSP70 with receptor will be medi-

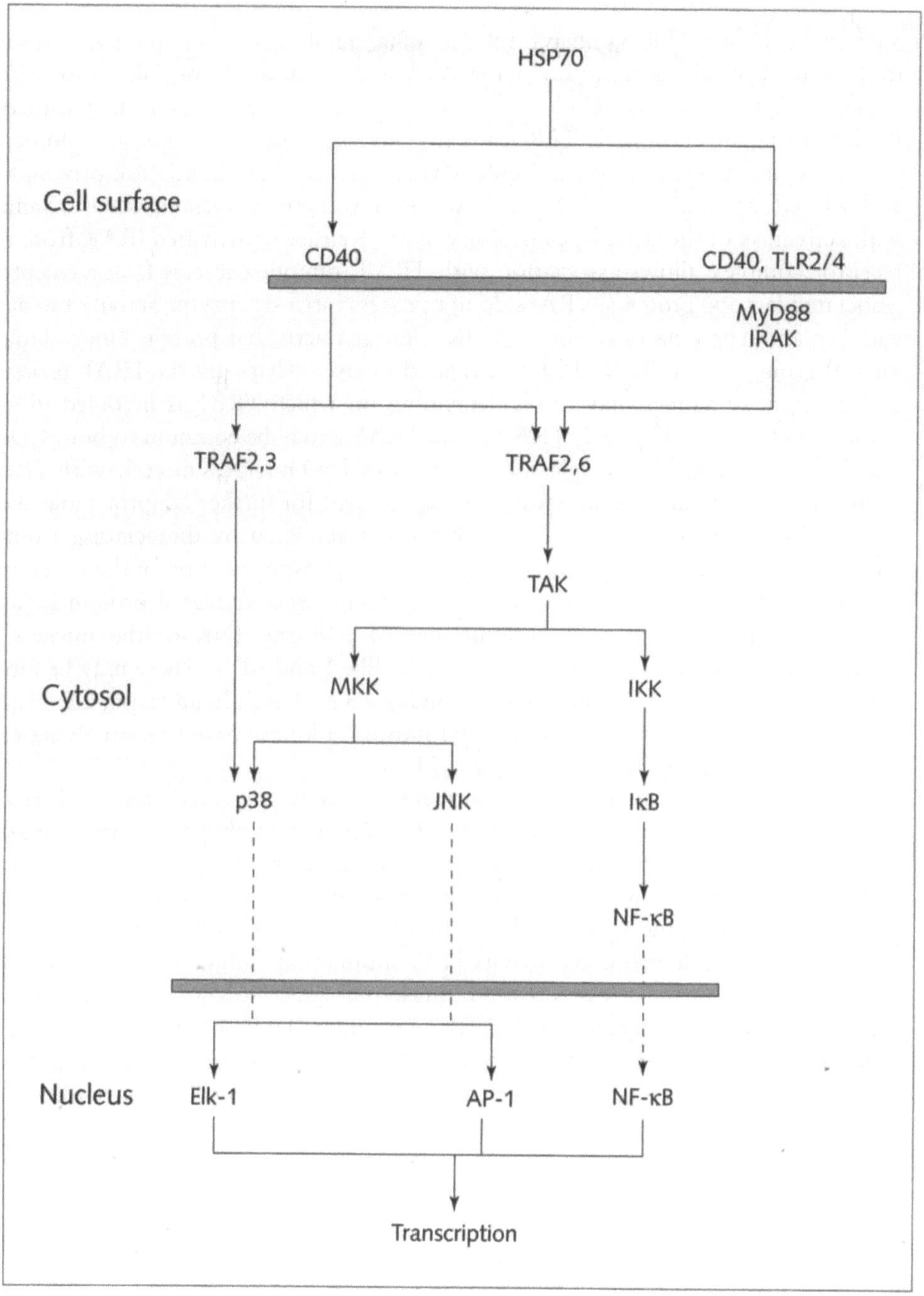

Figure 2
Signalling pathway of HSP70.

ated by CD40 and TLR pathways. Intracellular signalling by these pathways results in the activation of transcription factors such as NF-κB, AP-1 and Elk-1 [50–53] as shown in Figure 2 which includes an outline of major components of the pathways. Binding of ligand to dimeric TLR leads to association of the cytoplasmic domains with the cytoplasmic adaptor molecule MyD88 (myeloid differentiation protein 88) and serine/threonine kinases of the IRAK (IL-1 receptor associated kinase) family with activation of the latter by phosphorylation. Release of activated IRAK from the receptor complex allows association with TRAF (tumour necrosis factor receptor-associated factors) proteins. TRAFs do not possess intrinsic enzyme activity but activate Ser and Thr kinases of the MAP3K (mitogen activated protein kinase kinase kinase) group such as TAK (TGF-β-activated kinase). There are six TRAF proteins and different outcomes may result depending on which TRAF is involved in signalling. As shown in Figure 2, TRAF 6 and TRAF 2 may be common to both CD40 and TLR signalling. The cytoplasmic domain of CD40 interacts directly with TRAF proteins leading to activation without a requirement for further adaptor molecules.

In subsequent downstream events, NF-κB is activated by dissociating from a complex with the inhibitory protein IκB following phosphorylation of the latter and the kinases p38 and JNK (both members of the mitogen activated protein kinase, MAPK, family) are activated. Recruitment of p38 and JNK to the nucleus is required for activation of transcription factors Elk-1 and AP-1. There may be alternative pathways for activation of p38 involving TRAF2 or 3. In addition, CD40 and TLR may also signal by the phosphatidyl inositol 3 kinase pathway, involving calcium mobilisation (not shown in Fig. 2) [54–56].

A few studies have investigated intracellular signalling mediated by HSP70. Treatment of monocytes with mammalian HSP70 induces phosphorylation of IκB, the cytoplasmic inhibitor of the NF-κB [17]. The adaptor protein MyD88 (myeloid differentiation protein 88) undergoes relocalisation on stimulation of cells with mammalian HSP70 and is required for stimulation of IL-12 and TNF [49] and for stimulation of NF-κB promoter activity [17]. Stimulation with HSP70 also results in phosphorylation of the stress activated kinase p38 [21]. However, the finding that the intracellular calcium chelator BAPTA-AM inhibits HSP70 but not LPS stimulation of chemokine production [20] and HSP70-stimulated but not LPS-stimulated phosphorylation of IκB [17] indicates divergence from LPS-mediated signalling.

Interphase between innate and adaptive immunity

Both human and microbial HSP70 bind the CD40 receptor, but surprisingly a human N terminal domain of HSP70 binds one site, whereas a microbial C terminal domain binds another site of the CD40 molecule [20, 21, 26]. The B7 (CD80/CD86) and CD28 co-stimulatory interaction plays a central role in providing the second signal necessary for the adaptive immune function between HLA-

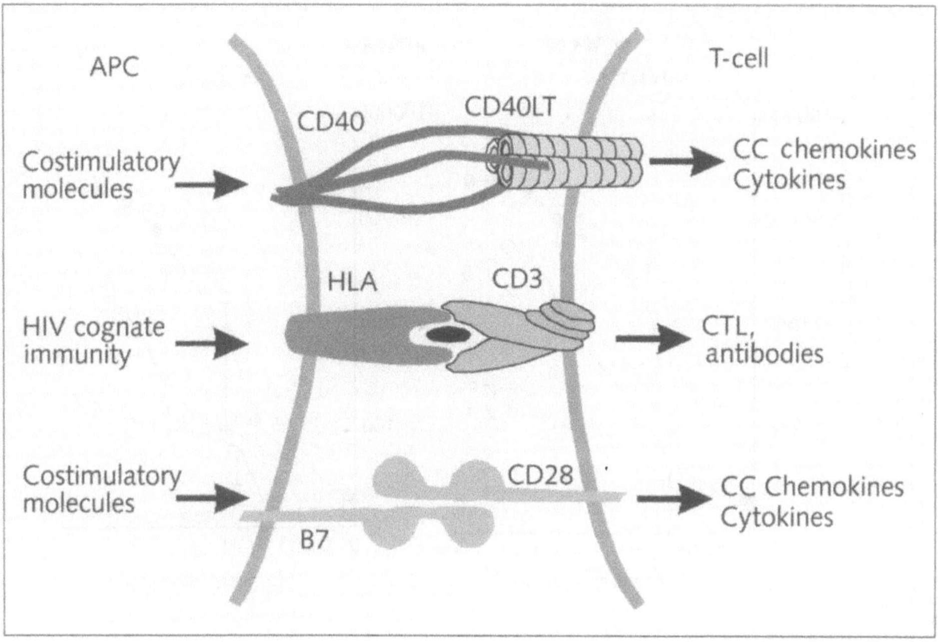

Figure 3
Interphase between innate and cognate immunity via *costimulatory molecules by HLA-peptide-TCR interaction.*

peptide and TCR [57]. The HLA-bound peptide and co-stimulatory molecule CD40 on an antigen presenting cell interacts with the corresponding TCR and CD40L on T-cells, respectively to elicit an effective immune response (Fig. 3). It is of interest to note that ligation of CD40 by CD40L (or HSP70) and that of CD80/86 by CD28 elicits a number of CC chemokines [42–44] which suggest a non-cognate immune response responsible for attracting the immunological repertoire of cells (monocytes, immature DC, T- and B-cells). The interaction between CD40 and CD40L also elicits production of some cytokines (such as IL-12 and TNF-α) and may induce Th1 polarization of the immune response. Thus, the innate immune chemokines and cytokines that are produced as a result of the interaction between HSP70 and CD40 may drive the specific immune responses to a HSP70-bound antigen. HSP70 functions as a multi-purpose molecule in acting as a carrier of antigens, which attracts the immune cells to the HSP70-antigen site and elicits maturation of DC. There is also the possibility that the HSP70-bound antigen is processed by an APC and chaperoned by HSP70 to bind HLA for presentation to T-cells [58]. In addition to the two signals necessary for an immune response to be elicited, a third signal may be

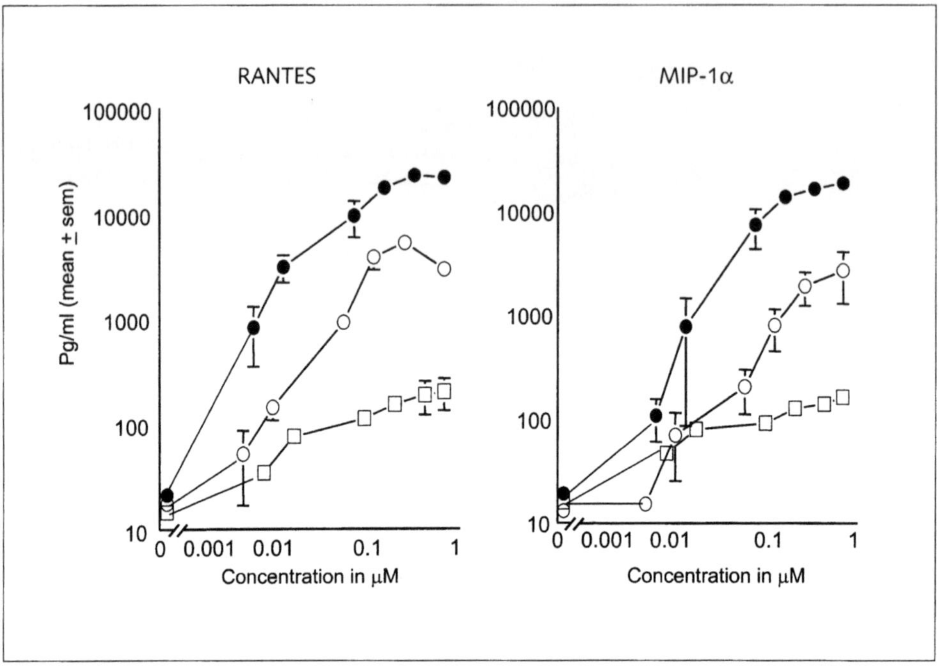

Figure 4
Effects of HSP70 (○), HSP70$_{1-358}$ (□) and HSP70$_{359-610}$ (•) on production of RANTES and MIP-1α by THP1 cells [26].

required [59] and this might be provided by interaction between HSP70 and the CD40 receptor.

Stimulation of chemokines

There is evidence that CC chemokines are generated by a variety of lymphoid, endothelial and epithelial cells [60, 61]. Microbial HSP70 and HSP65 stimulate monocytes and DC to generate CC chemokines *in vitro* (Fig. 4) [20, 24]. This innate response is greatly enhanced if mononuclear cells are used from HSP70 immunized animals. Indeed, stimulation of production of CC chemokines with HSP70 or HSP65 has been demonstrated in mononuclear cells of macaques immunized either by the systemic or mucosal route [41]. Furthermore, both TCRαβ and γδ T-cells produce CCL3, CCL4 and CCL5 (MIP-1α, MIP-1β and RANTES), when stimulated by microbial HSP70 [24, 41, 62]. Any increase in production of these CC chemokines

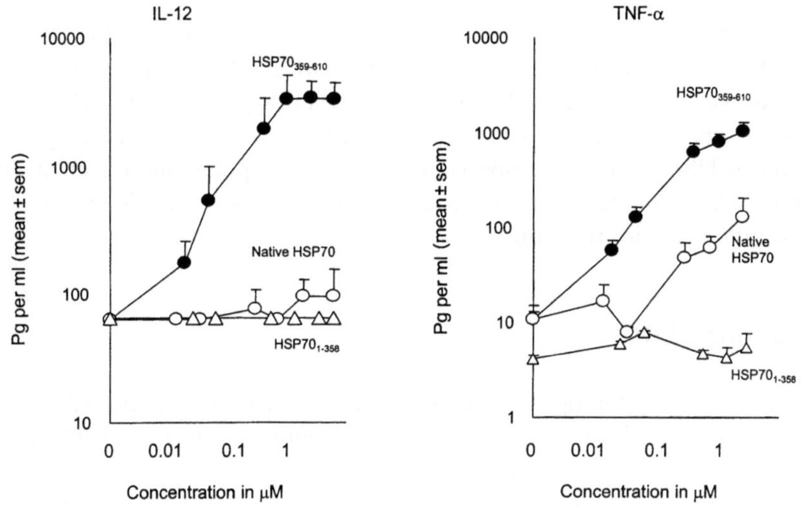

Figure 5
Effects of native HSP70, $HSP70_{1-358}$ and $HSP70_{359-610}$ on production of IL-12 and TNF-α by THP1 cells.

has been demonstrated on repeated administration of HSP70 in macaques [62], suggesting a memory-type of response, that is dependent on sensitised T-cells. Whilst DC are the most potent producers of CC chemokines [63], monocytes, CD4 and CD8 cells are all capable of generating chemokines [64].

Stimulation of cytokines

Stimulation of the human or murine monocyte-macrophage series of cells with mycobacterial HSP65 induces production of IL-1β and TNF-α [65–67]. Bacterial HSPs from *E. coli*, GroEL (60 kD) and DnaK (70 kD) stimulated human monocytes (and endothelial cells) to produce IL-6, TNF-α and GMCSF [68]. Mammalian HSP90, gp96 and HSP70 stimulate mouse peritoneal macrophages to produce small amounts of IL-1β, TNF-α and IL-12 [69]. Similarly, human HSP60 stimulates human or mouse macrophages to produce TNF-α and gene expression of IL-12 and

IL-13 [70]. An investigation of mycobacterial HSP70 revealed that whilst the wild type of HSP70 stimulates minimal generation of IL-12, TNF-α or NO (nitric oxide) from human monocytes, the C-terminal portion of HSP70 (p359–610) stimulates about ten-fold higher concentration of IL-12 or TNF-α (Fig. 5) [26]. Similar results were obtained by stimulating immature DC with the C-terminal portion of HSP70 [26]. As we have evidence that HSP70 may undergo lysosomal digestion which breaks up the molecule to a functional C-terminal fragment, we suggest that this portion of HSP70 may be responsible for Th1-like polarizing effect of HSP70. In addition, the pro-inflammatory cytokines play an important role in innate immunity and inflammatory reactions.

Maturation of DC

DC are professional antigen presenting cells and display an extraordinary capacity to stimulate naive T-cells and initiate primary immune responses [71]. Recent studies suggest that DC can also play a critical role in the induction of peripheral immunological tolerance, although the mechanisms of this function remain largely unknown [72, 73]. DC are also an important component of the innate immune system and represent the first line of defence against microbial pathogens [74]. These diverse functions depend on different DC populations, lineages and differentiation stages of DC [75]. Immature DC are continuously produced from haematopoietic stem cells within the bone marrow. Two cytokines FLT-3 and GM-CSF are key growth factors for DC differentiation *in vivo* from the hematopoietic stem cells. Immature DC resides in epithelia and can terminally differentiate into mature DC by various stimuli, such as microbial pathogens, inflammatory cytokines, or other danger signals. Maturation of DC appears to be a critical process for DC function, particularly for priming naive T-cells [71].

DC can be generated *in vitro* by culturing PBMC-derived monocytes for 5–7 days in GM-CSF and IL-4 conditioned medium [76]. Under these culture conditions, DC express immature phenotypic markers: low levels of CD80, CD86 and MHC Class 2, but little or no CD83 and CCR7 maturation markers. Human HSP60, HSP70, HSP gp90 and HSP90 released from necrotic cells, or tumour cells or recombinant human HSP70 stimulate DC maturation by up-regulation of co-stimulatory molecules and produce IL-12 and TNF-α [69, 77, 78].

Microbial HSP70 applied to immature DC cultures for two days, induces dramatic changes in the cell-surface expression of DC maturation markers, similar to those elicited by the CD40 ligand. There is an increase in MHC Class 2 molecules, the co-stimulatory molecules CD80, CD86, as well as CD83 and CCR7 (Fig. 6) [26]. The Th1-polarizing cytokines IL-12 and TNF-α are also produced [26], but the C-terminal portion of microbial HSP70 (aa359-610) is much more potent than the full length of HSP70 in stimulating these cytokines and in DC maturation. In

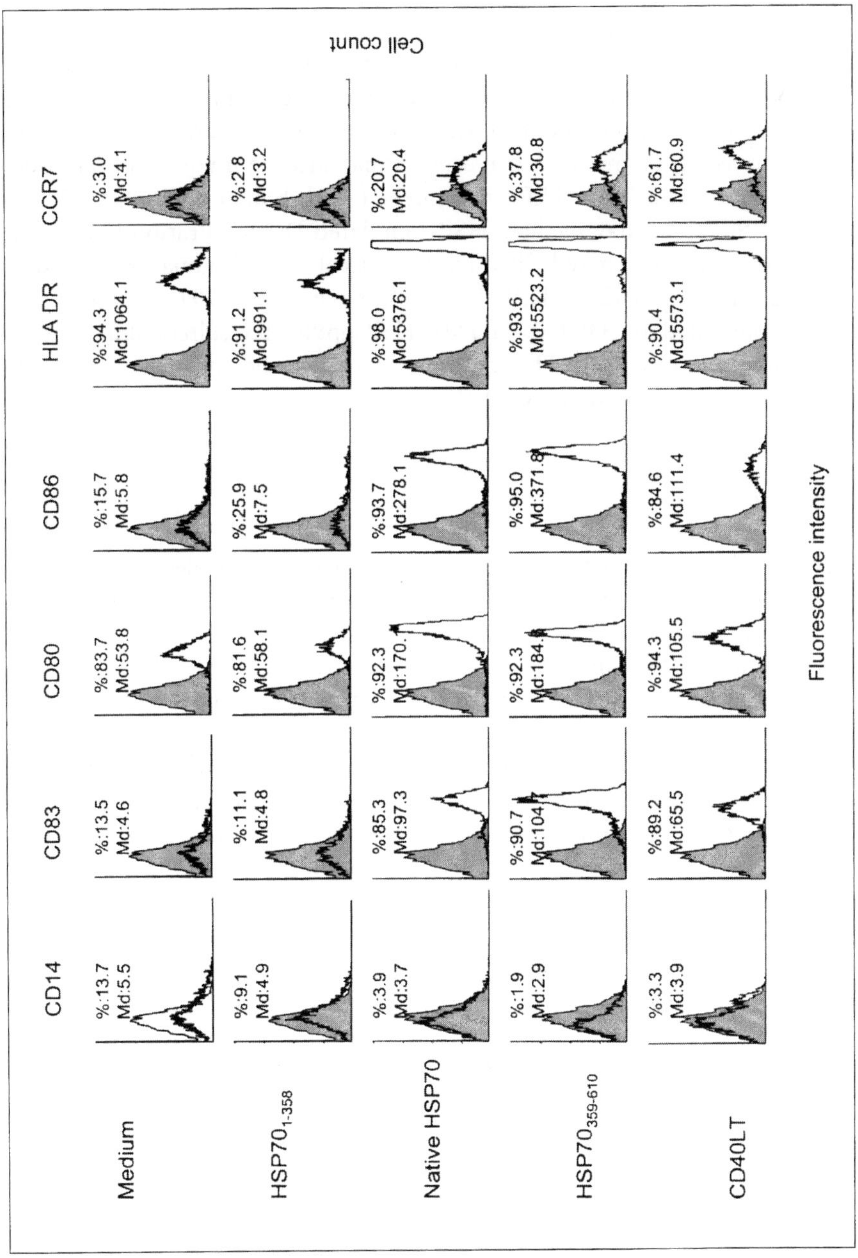

Figure 6
The effect of native HSP70, C- and N-terminal fragments and CD40LT on DC maturation.

contrast, the N-terminal ATP-ase domain of microbial HSP70 fails to stimulate expression of any of these DC phenotypes or production of cytokines (Fig. 6). This remarkable finding indicates that the C-, but not N-terminal fragment of HSP70 stimulates the innate activity and this is consistent with the peptide binding function of the C-terminal domain that delivers peptides into the MHC Class 1 pathway and induces $CD8^+$ cytotoxic T lymphocytes.

The mechanism that underlines the process of maturation from immature DC is not fully understood. Signalling through CD40 by engagement of the trimerized CD40 ligand is one of the best defined pathways driving DC maturation [51]. Other receptors, such as TLRs and TNF-α receptors may also be involved in the process of DC maturation. TLR2 and TLR4 can interact with human HSP60, HSP70 and HSPgp96 and therefore may mediate HSP stimulated DC maturation [23, 48, 49]. By the same token, CD40 may be involved in microbial and human HSP70 stimulation of DC maturation [20, 21]. It is not clear whether CD91 can mediate DC maturation stimulated by HSP70, HSP90, HSP gp96 and calreticulin. However, CD91 mediated endocytosis of HSP may play a role in the endocytosis-dependent pathway of HSP activation of TLRs [18].

Adjuvanticity

Microbial HSP70 and HSP65 have been used as carrier molecules or adjuvants to enhance systemic immune responses when covalently linked to synthetic peptides [79–81]. Indeed, these and HSP96 can be fused, covalently linked or loaded with peptides to elicit specific immunity to tumours or viruses [82–85]. The adjuvanticity of microbial HSP70 and HSP65 have been demonstrated not only on systemic immunization but also by atraumatic mucosal administration to non-human primates [24]. Both systemic and mucosal adjuvanticity is dependent on stimulating the production of 3 CC chemokines – CCL3, 4 and 5 (or RANTES, MIP-1α and MIP-1β). CCL5 is a potent chemo-attractant of monocytes, CD4 cells and activated CD8 cells [86-89]. CCL3 and CCL4 attract $CD4^+$ T- and B-cells [90] and the three chemokines attract immature DC [91]. Monocytes and DC take up antigen which is processed and presented on the cell surface. DC undergo maturation and with monocytes migrate to the regional lymph nodes, and present the processed antigen to T- and B-cells which elicit cellular and antibody responses.

The presence of HSP in most microorganisms [92, 93] and their function of generating CC chemokines, raises the hypothesis that the well-recognized immunogenicity of whole organisms, as compared with a subunit antigen, might be mediated by the CC chemokines generated by HSP [24]. This is consistent with the principle that the innate immune system may drive adaptive immunity [94, 95]. HSP in micro-organisms may function as a natural adjuvant generating CC chemokines, somewhat akin to microbial agents stimulating the complement pathway to generate C3d which has potent adjuvant activity [96]. This concept is also consistent with the "danger hypothesis" of infection [97], with the HSP alerting the innate system to secrete CC chemokines and to mobilize the immune repertoire of cells and generate specific immune responses against the invading organism.

Recent investigations have demonstrated that the chemokine stimulating function of HSP70 is dependent on interaction with the CD40 molecule expressed on the cell-surface of monocytes and DC [20, 21]. Whilst the wild type of HSP70 stimulates production of small amounts of IL-12, TNF-α and NO, this is greatly enhanced with the C-terminal fragment of p336–610) [26]. Because IL-12 is one of the most potent cytokines inducing Type 1 polarization [98], these findings may have important implications in using $HSP70_{359-610}$ fragment as a Th1-polarizing adjuvant. Indeed, $HSP70_{359-610}$-linked peptide elicited higher serum IgG2a and IgG3 sub-classes of antibodies than the wild type HSP70-bound peptide, consistent with a Th1-polarizing response [26]. Furthermore, the Th2 type of cytokine (IL-4) was not produced in immunized macaques. Thus, $HSP70_{359-610}$ might be used as a microbial adjuvant that attracts the entire immunological repertoire of cells by virtue of stimulating the production of CC chemokines and eliciting a Th1 response by generating IL-12.

Acknowledgements

This work was supported by the European Commission Biomed Grant (QLK2-CT-1999-01321) and the Guy's and St. Thomas' Charitable Foundation.

References

1 Hartl FU (1996) Molecular chaperones in cellular protein folding. *Nature* 381: 571

2 Hunt C, Morimoto RI (1985) Conserved features of eukaryotic HSP70 genes revealed by comparison with the nucleotide sequence of human HSP70. *Proc Natl Acad Sci USA* 82: 6455–6459

3 Flaherty KM, Deluca-Flaherty C, McKay DB (1990) Three-dimensional structure of the ATPase fragment of a 70K heat-shock cognate protein. *Nature* 346: 623–628

4 Osipiuk J, Walsh MA, Freeman BC, Morimoto RI, Joachimiak A (1999) Structure of a new crystal form of human hsp70 ATPase domain. *Acta Cryst* D55: 1105–1107

5 Zhu X, Zhao X, Burkholder WF, Gragerov A, Ogata CM, Gottesman ME, Hendrickson WA (1996) Structural analysis of substrate binding by the molecular chaperone DnaK. *Science* 272: 1606–1614

5a Harrison CJ, Hayer-Hartl M, Di Liberto M, Hartl F, Kuriyan J (1997) Crystal structure of the nucleotide exchange factor GrpE bound to the ATPase domain of the molecular chaperone DnaK. *Science* 276: 431–435

5b Bajorath J (1998) Detailed comparison of two molecular models of the human CD40

ligand with an x-ray structure and critical assessment of model-based mutagenesis and residue mapping studies. *J Biol Chem* 273: 24603–24609
6 Mayer MP, Schröder H, Rüdiger S, Paal K, Laufen T, Bukau B (2000) Multistep mechanism of substrate binding determines chaperone activity of Hsp70. *Nature Structural Biology* 7: 586–593
7 Pellecchia M, Montgomery DL, Stevens SY, Kooi CWV, Feng H-P, Giersach LM, Terlecky SR, Chiang H-L, Olson TS, Dice JF (1992) Protein and peptide binding and stimulation of *in vitro* lysosomal proteolysis by the 73 kDa heat shock cognate protein. *J Biol Chem* 267: 9202–9209
8 Buczynski G, Slepenkov SV, Sehorn MG, Witts SN (2001) Characterization of a lidless form of the molecular chaperone DnaK. *J Biol Chem* 276: 27231–27236
9 Wang H, Kurochkin AV, Pang Y, Hu W, Flynn GC, Zuiderweg ERP (1998) NMR solution structure of the 21kDa chaperone protein DnaK substrate binding domain: a preview of chaperone-protein interaction. *Biochemistry* 37: 7929–7940
10 Morschauser RC, Hu W, Wang H, Pang Y, Flynn GC, Zuiderweg ERP (1999) High resolution solution structure of the 18kDa substrate-binding domain of the mammalian chaperone protein Hsc70. *J Mol Biol* 289: 1387–1403
11 Gragerov A, Zeng L, Zhao X, Burkholder W, Gottesman ME (2000) Specificity of DnaK-peptide Binding. *J Biol Chem* 275: 33329–33335
12 Flynn GC, Pohl J, Flocco MT, Rothman JE (1991) Peptide-binding specificity of the molecular chaperone BiP. *Nature* 353: 726–30
13 Blond-Elguindi S, Cwirla SE, Dower WJ, Lipshutz RJ, Sprang SR, Sambrook JF, Gething MJ (1993) Affinity panning of a library of peptides displayed on bacteriophages reveals the binding specificity of BiP. *Cell* 75: 717–728
14 Terlecky SR, Chiang H-L, Olson TS, Dice JF (1992) Protein and peptide binding and stimulation of *in vitro* lysosomal proteolysis by the 73 kDa heat shock protein. *J Biol Chem* 267: 9202–9209
15 Fourie AM, Sambrook JF, Gething MJH (1994) Common and divergent peptide binding specificities of hsp70 molecular chaperones. *J Biol Chem* 269: 30470–30478
16 Kol A, Lichtman AH, Finberg RW, Libby P, Kurt-Jones EA (2000) Cutting edge: heat shock protein (HSP) 60 activates the innate immune response: CD14 is an essential receptor for HSP60 activation of mononuclear cells. *J Immunol* 164: 13017
17 Asea A, Kraeft S-K, Kurt-Jones EA, Stevenson MA, Chen LB, Finberg RW, Koo GC, Calderwood SK (2000) HSP70 stimulates cytokine production through a CD14-dependent pathway, demonstrating its dual role as a chaperone and cytokine. *Nat Med* 6: 435–442
18 Binder RJ, Han DK, Srivastava PK (2000) CD91: a receptor for heat shock protein gp96. *Nat Immunol* 1: 151–155
19 Basu S, Binder RJ, Ramalingam T, Srivastava PK (2001) CD91 is a common receptor for heat shock proteins gp96, hsp90, hsp70, and calreticulin. *Immunity* 14: 303–313
20 Wang Y, Kelly CG, Kartunen JT, Whittall T, Lehner PJ, Duncan L, MacAry P, Younson

JS, Singh M, Oehlmann W et al (2001) CD40 is a cellular receptor mediating mycobacterial heat shock protein 70 stimulation of CC chemokines. *Immunity* 15: 971–983

21 Becker T, Hartl F-U, Wieland F (2002) CD40, an extracellular receptor for binding and uptake of Hsp70-peptide complexes. *J Cell Biol* 158: 1277–1285

22 Ohashi K, Burkart V, Flohë S, Kolb H (2000) Cutting edge: Heat shock protein 60 is a putative endogenous ligand of the Toll-like receptor-4 complex. *J Immunol* 164: 558–561

23 Vabulas RM, Ahmad-Nejad P, Costa da C, Miethke T, Kirschning CJ, Hacker H, Wagner H (2001) Endocytosed HSP60s use Toll-like receptor 2 (TLR2) and TLR4 to activate the Toll/interleukin-1 receptor signaling pathway in innate immune cells. *J Biol Chem* 276: 31332–31339

24 Lehner T, Bergmeier LA, Wang Y, Tao L, Singh M, Spallek R, van der Zee R (2000) Heat shock proteins generate β-chemokines which function as innate adjuvants enhancing adaptive immunity. *Eur J Immunol* 30: 594–603

25 Retzlaff C, Yamamoto Y, Hoffman PS, Friedman H, Klein TW (1994) Bacterial heat shock proteins directly induce cytokine mRNA and interleukin-1 secretion in macrophage cultures. *Infect Immun* 62: 5689–5693

26 Wang Y, Kelly CG, Singh M, McGowan EG, Carrara A-S, Bergmeier LA, Lehner T (2002) Stimulation of Th1-polarizing cytokines, CC chemokines, maturation of dendritic cells, and adjuvant function by the peptide binding fragment of heat shock protein 70. *J Immunol* 169: 2422–2429

27 Castellino F, Boucher PE, Eichelberg K, Mayhew M, Rothman JE, Houghton AN, Germain RN (2000) Receptor-mediated uptake of antigen/heat shock protein complexes results in major histocompatibility complex class I antigen presentation *via* two distinct processing pathway. *J Exp Med* 191: 1957

28 Fujihara SM, Nadler SG (1999) Intranuclear targeted delivery of functional NF-κB by 70 kDa heat shock protein. *EMBO J* 18: 411

29 Wright SD, Ramos RT, Tobias PS, Ulevitch RJ, Mathison JC (1990) CD14, a receptor for complex of lipopolysaccharide (LPS) and LPS binding protein. *Science* 249: 1431–1433

30 Yang RB, Mark MR, Gurney AL, Godwski PJ (1999) Signalling events induced by lipopolysaccharide-activtaed toll-like receptor 2. *J Immunol* 163: 639–643

31 Da Silva Correia J, Soldau K, Christen U, Tobias PS, Ulevitch RJ (2001) Lipopolysaccharide is in close proximity to each of the proteins in its membrane receptor complex transfer from CD14 to TLR4 and MD-2. *J Biol Chem* 276: 21129–21135

32 Panjwani NN, Popova L, Srivastava PK (2002) Heat shock proteins gp96 and hsp70 activate the release of nitric oxide by APCs. *J Immunol* 168: 2997–3003

33 Coutinho A, Meo T (1978) Genetic basis for unresponsiveness to lipopolysaccharide in C57BL/10Cr mice. *Immunogenetics* 7: 17–24

34 Quereshi ST, Larivière L, Leveque G, Clermont S, Moore KJ, Gros P, Malo D (1999) Endotoxin-tolerant Mice Have Mutations in Toll-like receptor 4 (TLR4). *J Exp Med* 189: 615–625

35 Randow F, Seed B (2001) Endoplasmic reticulum chaperon gp96 is required for innate immunity but not cell viability. *Nat Cell Biol* 3: 891–896

36 Triantafilou K, Triantafilou M, Dedrick RL (2001) A CD14-independent LPS receptor cluster. *Nat Immunol* 2: 338–344

37 Armstrong PB, Quigley JP (1999) α2 macroglobulin: an evolutionary conserved arm of the innate immune system. *Dev Comp Immunol* 23: 375–390

38 Banchereau J, Bazan F, Blanchard D, Briere F, Galizzi JP, van Kooten C, Liu YJ, Rousset F, Saeland S (1994) The CD40 antigen and its ligand. *Annu Rev Immunol* 12: 881–922

39 Young LS, Eliopoulos AG, Gallagher NJ, Dawson CW (1998) CD40 and epithelial cells: across the great divide. *Immunol Today* 19: 502–506

40 Grewal IS, Flavell RA (1998) CD40 and CD154 in cell-mediated immunity. *Annu Rev Immunol* 16: 111

41 Lehner T, Mitchell E, Bergmeier L, Singh M, Spallek R, Cranage M, Hall G, Dennis M, Villinger F and Wang Y (2000) The role of γδ T cells in generating antiviral factors and β-chemokines in protection against mucosal simian immunodeficiency virus infection. *Eur J Immunol* 30: 2245–2256

42 Herold KC, Lu J, Rulifson I, Vezys V, Taub D, Grusby MJ, Bluestone JA (1997) Cutting Edge: Regulation of C-C chemokine production by murine T cells by CD28/B7 costimulation. *J Immunol* 159: 4150–4153

43 Kornbluth RS, Kee K, Richman DD (1998) CD40 ligand (CD154) stimulation of macrophages to produce HIV-1 suppressive beta-chemokines. *Proc Natl Acad Sci USA* 95: 5205–5210

44 McDyer JF, Dybul M, Goletz TJ, Kinter AL, Thomas EK, Berzofsky JA, Fauci AS, Seder RA (1999) Differential effects of CD40 ligand/trimer stimulation on the ability of dendritic cells to replicate and transmit HIV infection: evidence for CC-chemokine-dependent and –independent mechanisms. *J Immunol* 162: 3711–3717

45 Janeway CA, Medzhitov R (2002) Innate immune recognition. *Annu Rev Immunol* 20: 197–216

46 Medzhitov R (2001) Toll-like receptors and innate immunity. *Nat Rev Immunol* 1: 135–145

47 Ohashi K, Burkart V, Flohë S, Kolb H (2000) Cutting edge: Heat shock protein 60 is a putative endogenous ligand of the Toll-like receptor-4 complex. *J Immunol* 164: 558–561

48 Vabulas RM, Ahmad-Nejad P, Ghose S, Kirschning CJ, Issels RD and Wagner H (2002) HSP70 as endogenous stimulus of the Toll/interleukin-1 receptor signal pathway. *J Biol Chem* 277: 15107–15112

49 Vabulas RM, Bradedel S, Hilf N, Sigh-Jasuja H, Herter S, Ahmad-Nejad P, Kirschning CJ, Costa da C, Rammensee H-G, Wagner H et al (2002) The endoplasmic reticulum-resident heat shock protein Gp96 activates dendritic cells *via* the toll-like receptor 2/4 pathway. *J Biol Chem* 277: 20847–20853

50 Medzhitov R, Preston-Hurlburt P, Janeway CA (1997) A human homologue of the Drosophila Toll protein signals activation of adaptive immunity. *Nature* 388: 394–397

51 Lee, HH, Dempsey PW, Parks TP, Zhu X, Baltimore D, Cheng G (1999) Specificities of CD40 signaling: involvement of TRAF2 in CD40-induced NF-κB activation and intercellular adhesion molecule-1 up-regulation. *Proc Natl Acad Sci USA* 96: 1421–1426

52 Chung KC, Kim SM, Rhang S, lau LF, Gomes I, Ahn YS (2000) Expression of immediate early gene pip92 during anisomycin-induced cell death is mediated by the JNK- and p38-dependent activation of Elk 1. *Eur J Biochem* 267: 4676–4684

53 Mann J, Oakley F, Johnson PW and Mann DA (2002) CD40 induces interleukin-6 gene transcription in dendritic cells: regulation by TRAF2, AP-1, NF-κB, and CBF1. *J Biol Chem* 277: 17125–17138

54 Andjelic S, Hsia C, Suzuki H, Kadowaki T, Koyasu S, Liou H-C (2000) Phosphatidyl inositol 3-kinase and NF-κB/Rel are at the divergence of CD40-mediated proliferation and survival pathways. *J Immunol* 165: 3860–3867

55 Dadgostar H, Zarnegar B, Hoffmann A, Qin X-F, Truong U, Rao G, Baltimore D, Cheng G (2002) Cooperation of multiple signaling pathways in CD40-regulated gene expression in B lymphocytes. *Proc Natl Acad Sci USA* 99: 1497–1502

56 Arbibe L, Mira JP, Teusch N, Kline L, Guha M, Mackman N, Godowski PJ, Ulevitch RJ, Knaus UG (2000) Toll-like receptor 2-mediated NF-κB activation requires a Rac1-dependent pathway. *Nat Immunol* 1: 533–540

57 Bretcher PA (1999) A two-step, two-signal model for the primary activation of precursor helper T cells. *Proc Natl Acad Sci USA* 96: 185–190

58 Roth S, Willcox N, Rzepka R, Mayer MP, Melchers I (2002) Major differences in antigen-processing correlate with a single Arg71 ↔ Lys substitution in HLA-DR molecules predisposing to rheumatoid arthritis and with their selective interactions with 70 kDa heat shock protein chaperones. *J Immunol* 169: 3015–3020

59 Medzhitov RM, Janeway CA Jr (1997) Innate immunity: impact on the adaptive immune response. *Curr Opin Immunol* 9: 4–9

60 Baggiolini M (1998) Chemokines and leukocyte traffic. *Nature* 392: 565–568

61 Ward SG, Bacon K, Westwick J (1998) Chemokines and T lymphocytes: more than an attraction. *Immunity* 9: 1–11

62 Wang Y, Tao L, Mitchell E, Bergmeier L, Doyle C, Lehner T (1999) The effect of immunization on chemokines and CCR5 and CXCR4 coreceptor functions in SIV binding and chemotaxis. *Vaccine* 17: 1826–1836

63 Sallusto F, Palermo B, Lenig D, Miettinen M, Matikainen S, Julkunen I, Forster R, Burgstahler R, Lipp M, Lanzavecchia A (1999) Distinct patterns and kinetics of chemokine production regulate dendritic cell function. *Eur J Immunol* 29: 1617–1625

64 Babaahmady K, Bergmeier LA, Whittall T, Singh M, Wang Y, Lehner T (2002) A comparative investigation of CC chemokines and SIV suppressor factors generated by $CD8^+$ and $CD4^+$ T cells and $CD14^+$ monocytes. *J Immunol Methods* 264: 1–10

65 Zhang Y, Doefler M, Lee TC, Guillemin B, Rom WN (1993) Mechanisms of stimulation of interleukin-1β and tumor necrosis factor-α by mycobacterium tuberculosis components. *J Clin Invest* 91: 2076

66 Friedland JS, Shattock R, Remake DG, Griffin E (1993) Mycobacterial 65 kd heat shock

protein induces release of proinflammatory cytokines from human monocytic cells. *Clin Exp Immunol* 91: 58

67 Peeterman WE, Raats CJI, van Furth R, Langermans JAM (1995) Mycobacterial 65 kilodalton heat shock protein induces tumor necrosis factor α, interleukin 6, reactive nitrogen immediates and toxo-plasmastic activity in murine peritoneal macrophages. *Infect Immun* 63: 3454

68 Galdiero M, De L'Ero GC, Marcatili A (1997 Cytokine and adhesion molecule expression in human monocytes and endothelial cells stimulated with bacterial heat shock proteins. *Infect Immun* 65: 699–707

69 Basu S, Binder RJ, Suto R, Anderson KM, Srivastava PK (2000) Necrotic but not apoptotic cell death releases heat shock proteins, which deliver a partial maturation signal to dendritic cells and activate the NF-κB pathway. *International Immunol* 12: 1539–1546

70 Chen W, Syldath U, Bellmann K, Burkart V, Kolb H (1999) Human 60 kDa heat-shock protein: A danger signal to the innate immune system. *J Immunol* 162: 3212–3219

71 Mellman I, Steinman RM (2001) Dendritic cells: specialized and regulated antigen processing machines. *Cell* 106: 255–258

72 Steinman RM, Nussenzweig MC (2002) Inaugural Article: Avoiding horror autotoxicus: The importance of dendritic cells in peripheral T cell tolerance. *Proc Natl Acad Sci USA* 99: 351–358

73 Albert ML, Jegathesan M, Darnell RB (2001) Dendritic cell maturation is required for the cross-tolerization of $CD8^+$ T cells. *Nat Immunol* 2: 1010–1017

74 Moretta A (2002) Natural killer cells and dendritic cells: rendezvous in abused tissues. *Nat Rev Immunol* 2: 957–964

75 Liu Y-J (2001) Dendritic cell subsets and lineages, and their functions in innate and adative immunity. *Cell* 106: 259–262

76 Sallusto F, Lanzavecchia A (1999) Mobilizing dendritic cells for tolerance, priming, and chronic inflammation. *J Exp Med* 189: 611–614

77 Somerdan S, Larsson M, Fonteneau JF, Basu S, Srivastava PK, Bhardwaj N (2001) Primary tumor tissue lysates are enriched in heat shock proteins and induce the maturation of human dendritic cells. *J Immunol* 167: 4844–4852

78 Kuppner MC, Gastpar R, Gelwer S, Nässner E, Ochmann O, Scharner A, Issels RD (2001) The role of heat shock protein (HSP70) in dendritic cells maturation: HSP70 induces the maturation of immature dendritic cells but reduces DC differentiation from monocyte precursors. *Eur J Immunol* 31: 1602–1609

79 Lussow AR, Barrios C, van Embden J, van der Zee R, Verdini AS, Pessi A, Louis JA, Lambert P-H, Giudice GD (1991) Mycobacterial heat-shock proteins as carrier molecules. *Eur J Immunol* 21: 2297–2302

80 Barrios C, Lussow JA, van Embden J, van der Zee R, Rappouli R, Costantino P, Louis JA, Lambert P-H, Giudice GD (1992) Mycobacterial heat-shock proteins as carrier molecules. II The use of the 70 kDa mycobacterial carrier for conjugated vaccines can circumvent the need for adjuvants and Bacillus Calmette Guerin priming. *Eur J Immunol* 22: 1365–1372

81 Perraut R, Lussow AR, Gavoille S, Garraud O, Matile H, Tougne C, van Embden J, van der Zee R, Lambert P-H, Gysin J, Giudice GD (1993) Successful primate immunization with peptide conjugated to purified protein derivative or mycobacterial heat shock proteins in the absence of adjuvants. *Clin Exp Immunol* 93: 382–386
82 Suzue K, Young RA (1996) Adjuvant-free hsp70 fusion protein system elicits humoral and cellular immune responses to HIV-1 p24. *J Immunol* 156: 873–879
83 Udono H, Srivastava PK (1993) Heat shock protein 70-associated peptides elicit specific cancer immunity. *J Exp Med* 178: 1391–1396
84 Nieland TJF, Tan MCA, Monnet-van-Muigen M, Koning F, Kruisbeek AM, van Bleek B M (1996) Isolation of an immunodominant viral peptide that is endogenously bound to the stress protein GP96/GRP94. *Proc Natl Acad Sci USA* 93: 6135–6139
85 Ciupitu A-MT, Petersson M, O'Donnell CL, Williams K, Jindal S, Kiessling R, Welsh RM (1998) Immunization with a lymphocytic choriomeningitis virus peptide mixed with heat shock protein 70 results in protective antiviral immunity and specific cytotoxic T lymphocytes. *J Exp Med* 187: 685–691
86 Schall TJ, Bacon K, Toy KJ, Goedall DV (1990) Selective attraction of monocytes and T lymphocytes into the peripheral tissues of mice with severe combined immune deficiency. *Eur J Immunol* 24: 1823–1827
87 Murphy WJ, Taub DD, Anver M, Conlon K, Oppenheim JJ, Kelvin DJ, Longo DL (1994) Human RANTES induces the migration of human T lymphocytes into the peripheral tissues of mice with severe combined immune deficiency. *Eur J Immunol* 24: 1823–1827
88 Meurer R, Van Riper G, Feeney W, Cunningham P, Hora D Jr, Springer MS, MacIntrye DE, Rosen H (1993) Formation of eosinophilic and monocytic intradermal inflammatory sites in the dog by injection of human RANTES but not human monocyte chemoattractant protein 1, human macrophage inflammatory protein 1 alpha, or human interleukin 8. *J Exp Med* 178: 1913–1921
89 Kim JJ, Nottingham LK, Sin JI, Tsai A, Morrison L, Oh J, Dang K, Hu Y, Kazahaya K, Bennett M et al (1998) CD8 positive influence antigen-specific immune responses through the expression of chemokines. *J Clin Invest* 102: 1112–1124
90 Schall TJ, Bacon K, Camp RD, Kaspari JW, Goeddel DV (1993) Human macrophage inflammatory protein alpha (MIP-1 alpha) and MIP-1 beta chemokines attract distinct populations of lymphocytes. *J Exp Med* 177: 1821–1826
91 Dieu MC, Vanbervliet B, Vicari A, Bridon JM, Oldham E, Ait-Yahia S, Briere F, Zlotnik A, Lebecque S, Caux C (1998) Selective recruitment of immature and mature dendritic cells by distinct chemokines expressed in different anatomic sites. *J Exp Med* 188: 373–386
92 Thole JE, van Schooten WC, Keulen WJ, Hermans PW, Janson AA, de Vries RR, Kolk AH, van Embden JD (1988) Use of recombinant antigens expressed in Escherichia coli K-12 to map B-cell and T-cell epitopes on the immunodominant 65 kilodalton protein of *Mycobacterium bovis* BCG. *Infect Immun* 56: 1633–1640

93 Ivanyi J, Sharp K, Jackett P, Bothamley G (1988) Immunological study of defined constituents of mycobacteria. *Springer Semin Immunopathol* 10: 279–300
94 Janeway CA Jr (1989) Approaching the asymtote? Evolution and revolution in immunology. *Cold Spring Harbor Symp Quant Biol* 54: 1–13
95 Fearon DT, Locksley RM (1996) The instructive role of innate immunity in the acquired immune response. *Science* 272: 50–53
96 Dempsey PW, Allison MED, Akkaraju S, Goodnow CC, Fearon DT (1996) C3d of complement as a molecular adjuvant: bridging innate and acquired immunity. *Science* 271: 348–350
97 Matzinger P (1994) Tolerance, danger and the extended family. *Annu Rev Immunol* 12: 991–1045
98 Trichieri G (1994) Interleukin-12: a cytokine produced by antigen presenting cells with immunoregulatory functions in the generation of T-helper cells type 1 and cytotoxic lymphocytes. *Blood* 84: 4008

Heat shock protein expression in transplanted kidney

Klemens Trieb

Department of Orthopedics, University of Vienna, Währingergürtel 18-20, A-1090 Vienna, Austria

Kidney transplantation

Clinical development

Kidney transplantation is the standard treatment for end-stage renal failure and improves quality of life. The first animal kidney transplantation was done by Emerich Ullmann in 1902 in Vienna and in the USA in 1905 by Alexis Carrel. The problem of graft rejection was recognized later and in 1923 the differences between autografts and homografts were defined. The first transplantation, so far without function, was reported in 1936 by the Russian Voronoy [1, 2]. In 1951, David Hume in Boston transplanted a series of cadavar kidney allografts in the thigh of the patients of whom one functioned for almost half a year, whereas all others were rejected within the first weeks. This was done without specific drug support, because immunosuppressive therapy in transplantation was not developed that time. Then in 1954 the first successful kidney transplantation between identical teens was performed, which was done then in several cases. Despite this success the problem of allograft rejection remained due to the lack of immunosuppression. Total body irradiation was the only immunosuppression possible at this time. The introduction of irradiation of the graft failed because of unacceptable complications [1, 2].

Immunosuppressive therapy

In the early 1960s, azathioprine, a less toxic derivate of the anti-cancer agent 6-mercaptopurine, was introduced. In combination with steroids, this standard immunosuppressive therapy enabled a breakthrough of kidney transplantation. The introduction of Cyclosporine A, a fungal metabolite and calcineurin inhibitor, in the early 1980s led to the next breakthrough in kidney transplantation. For a powerful induction therapy anti-lymphocyte globulin or a pan-T lymphocyte monoclonal antibody (OKT 3) reacting with the CD3 complex are available for combination therapy.

Heat Shock Proteins and Inflammation, edited by Willem van Eden

New immunosuppressive drugs have been developed to improve rejection therapy. FK506 (tacrolimus) is a macrolide antibiotic and inhibits T-cell activation and cytokine transcription [3]. New drugs like mycophenolate mofetil and rapamycin are about to come to market. Rapamycin (sirolimus) is another macrolide antibiotics derived from *Streptomyces hygroscopicus* and mycophenolate mofetil is a potent inhibitor of inosine monophosphatase dehydrogenase, which in turn down-regulates adhesion molecules and inhibits lymphocyte proliferation. To achieve a more specific immunosuppression, monoclonal antibodies for instance against CD25 (the interleukin (IL)-2 receptor), LFA-1, CD4, CD28 (inhibition of binding with the B7 molecule) or adhesion molecules are developed. Clinical trials are ongoing and early results suggest the introduction of new potent immunosuppressive agents in routine use.

Complications related to immunosuppressive therapy are a higher susceptibility to opportunistic infections (fungal or viral) and an increased risk for developing lymphomas, squamous cell carcinomas or Kaposi sarcoma. They may also contribute to allograft fibrosis by induction of pro-fibrotic gene expression [3, 4].

Graft survival or failure

Although the one year survival rates increased in the last years due to a decrease in acute allograft rejection, chronic allograft failure still remains as a main problem limiting long-term survival [5–7]. The mechanisms underlying long-term allograft failure are still not fully understood. The one year allograft survival increased to over 80% for cadaveric and about 95% for living related donors due the introduction of potent immunosuppressants, such as cyclosporin and tacrolimus, and the consequent treatment of rejection episodes. However, at this time no effect on the long-term survival can be seen, because chronic allograft failure did not decrease with an unchanged half-life of renal allografts. The prevention or immediate treatment of early acute rejection episodes is one of the most important factors for long-term survival [8–12]. Patients with more than one rejection episode have a significant decreased graft survival. Other risk factors for failure are the histological grade (vascular rejection is more deleterious), glomerulo-nephritis of the graft, repeated rejection episodes and a delay of the first rejection. Patients with an early rejection episode have a better prognosis than patients with a later onset [13–17]. The successful treatment of the first rejection is the best indicator for a favourable survival prognosis, patients resistent to steroids have a worser outcome. Other factors like plasma creatinine, donor age, reperfusion time, non-compliance, blood pressure, high cholesterine or metabolic disorders are of prognostic importance, too [18].

The cause of renal allograft failure is still not fully understood, although both immunological and non-immunological factors play a role. Deterioration of renal function by chronic graft nephropathy results in an increased serum creatinine level, proteinuria and increased diastolic blood pressure. Unfortunately, these clinical findings are not specific to transplant failure, because they might be found in non-failed kidneys too. Chronic graft nephropathy is characterized by interstitial fibrosis, intimal hyperplasia, glomerulosclerosis, tubular atrophy and cellular infiltration [4]. The mononuclear cellular infiltrate consists mainly of T-lymphocytes and macrophages. The end-stage consists of an accumulation of extracellular proteins leading to fibrosis and organ failure [4, 19, 20].

Graft rejection and histocompatibility

Graft rejection depends on the recognition of the grafted tissue as foreign by the host. The antigens responsible for this recognition are the (major) histocompatibility complex antigens (MHC or human leukocyte antigen HLA). The effect of transplantation is determined by the degree of matching of the MHC complex and of the blood group. A degree of incompatibility will always be present, except for identical twins. This mismatch will lead to recognition of the allograft as foreign and result in a reaction of the immune system of the host against the graft (rejection). A matching for MHC Class I antigens has a modest effect on graft acceptance. Additional matching for MHC Class II has a pronounced effect, because CD4 T lymphocytes are activated by Class II. CD4 T-cells play a central role in the induction of cellular and humoral immunity [17]. As a mismatch is always the case, immunosuppressive therapy is necessary to prevent graft rejection and prolong graft survival. An additional response is mediated by the minor histocompatibility complex antigens, which are non-self peptides derived from polymorphic cellular proteins bound to graft MHC I molecules, responsible for structural integrity and expression on the cell surface of MHC I molecules [21].

Immunology of graft rejection

Graft rejection is a complex immunological process of cellular and humoral contribution. T lymphocyte mediated reactions are based on two mechanisms [19]; a direct recognition of allogeneic MHC molecules on antigen presenting cells of the graft. Dendritic cells seem to play a crucial role in this way. CD4 T lymphocytes of the recipient (helper cells) are activated by MHC Class II molecules of the donor. In turn they proliferate, secret cytokines and play a central role. At the same time, precursor CD8 T lymphocytes (killer cells) are activated by MHC Class I molecules of the donor. They proliferate and differentiate to mature killer cells. In the second (indirect) way of lymphocyte activation, antigens from the donor are presented by the recipient's own antigen-presenting cells in the same way as physiologic processing and presentation of foreign antigens [20, 21].

Classification of graft rejection

On the basis of the morphology, rejection episodes are classified as hyperacute, acute and chronic rejections.

Hyperacute rejection occurs within minutes and end in a destruction of the endothelial by antibodies with thrombosis in the glomeruli capillaries, fibrin deposit in arteries and allograft necrosis following infarction. Immunoglobulin and complement are found together with neutrophils.

Acute rejection is a process-mediated cellular and humoral and occurs within days after transplantation or months or years later. It is accompanied with elevated serum creatinine and clinical signs of renal failure. The kidney is infiltrated with activated (CD25 positive) CD 4 and 8 T lymphocytes. Infiltration of capillaries with CD 8 and antibody mediated endothelial cell destruction (endothelitis) results in oedema and necrosis.

Chronic rejection is characterized by vascular changes, interstitial fibrosis and loss of renal parenchyma. Vessels consist of an end-stage artheritis with intimal fibrosis mainly in the arteries. The interstitium is infiltrated with mononuclear cells containing plasma cells and eosinophils. Patients with chronic rejection present after four to six months with a progressive rise in serum creatinine [22–24].

Injury to the transplanted kidney

Injury to the allograft kidney is unavailable during transplantation by the temporary discontinuation of renal blood supply, occurring during organ retrieval and storage. This includes warm and cold ischaemia, ischaemia reperfusion injury, nephrotoxic drugs and clinical changes mentioned above. The kidney might respond to this injury phase by a temporary repair mechanism with an influx of mononuclear cells and fibroblasts. On the other hand the allograft responds to the injury by expression of genes that are necessary for organ and cell survival. Among them the heat shock proteins are remarkably important [4, 25–27].

Heat shock proteins and kidney allografts

Heat shock proteins and immunity

Heat shock proteins are physiologically essential; highly conserved proteins executing a variety of vital intracellular chaperoning functions including protein processing. In case of harmful events they are up-regulated to protect cells from death by protecting DNA and proteins from degradation. This phenomenon was seen first in *Drosophila* after elevation of temperature [28]. The mechanism of increased synthesis of HSPs is uniform and independent from the stimuli. The HSPs consist of constitutive and inducible forms, and they are sub-divided into families according to their molecular weight. For instance, HSP60 is a mitochondrial HSP, HSP73 is the cytoplasmic/nuclear constitutive member of the 70-kDa family and HSP72 the stress-inducible nuclear HSP70 member [29]. Beside these functions, HSPs have been shown to be strongly involved in immunologically mediated reactions; they can act as antigens and stimulate immune competent cells. In autoimmunity they can stimulate the immune system *via* "molecular mimicry". In murine models HSPs elicit cancer immunity and additionally, T lymphocytes specific for HSPs with cytotoxic potential were described and isolated from human osteosarcomas [30–33]. Mice develop T-cell lymphomas when they express the human HSP70 gene, whereas transfection of tumours with HSP65 *in vivo* resulted in the rejection of the tumour by development of an immune response, and tumour cells transfected with HSPs *in vitro* lost in tumourgenicity as compared to un-transfected cells [34, 35]. Over-expression of HSP70 and HSP27 by gene transfection results in protection of many apoptotic stimuli as heat shock, anticancer drugs, radiation or tumour necrosis factors [33]. In sarcoma cell lines, HSP72 has been shown to be selectively expressed on their cell surface thereby overcoming protection and acting as the target for natural killer cells. HSPs are frequently over-expressed in different human cancers and correlation of expression with prognosis and resistance or response to chemotherapy is not uniform in all tumours. It has been reported that HSP27 over-expression in human osteosarcomas is related to poor prognosis [36]. The author's group has shown that HSP72 is *de novo* expressed in human osteosarcoma. This HSP72 *de novo* expression correlates with a good response to neo-adjuvant chemotherapy. Expression of HSPs in kidneys has been investigated in normals and in several diseases including allografts with correlation to severity of injury. It has been shown that HSP expression is modulated *in vitro* by hypoxia, cytokines, glucocorticoids and cyclosporin [36–39].

Expression of HSPs in normal kidney

Studies on the expression of HSPs in the normal kidney are so far scanty and focus on the rat and human kidney. Mainly the expression of the HSP70 family has been investigated by different methods. The expression of HSP73 was demonstrable in the human kidney in all compartments, i.e., also including the smooth muscle layer and the endothelium of blood vessels in the human normal kidney [40]. This is in accordance with studies in the rat [41], whereas HSP72 was reported negative for glomerular staining in one investigation [42], in another study HSP72/73 was demonstrated in the normal human kidney predominantly in distal tubules and collecting ducts similar to other findings [43]. Different methodologies used, like Northern blotting, Western blotting and immunohistochemistry might explain such differences. For immunohistochemistry, antigen retrieval may be relevant. HSP60, HSP72 and HSP73 are constitutively expressed in normal kidney tissue in a distinct compartmental pattern and cellular distribution. HSP60 was predominantly present in the distal convoluted tubules, the loop of Henle, and in collecting ducts, and was weakly expressed in the epithelial lining of proximal tubules and in epithelial cells of the glomeruli. Observations in human kidneys are scanty and somewhat controversial. Thus, in the latter study HSP60 could be detected in normal kidneys by immunohistochemistry as well as by Western blot. This is somewhat in contrast to findings of another group [44], which, however, based on m-RNA detection by Northern blot. Another group identified HSP60 as a membrane fraction of the mammalian kidney, but did not specify further localisation [45]. For HSP47, a collagen specific stress protein, a lack of expression in normal human control and one hour post-transplantation kidneys was observed [46]. In a rat model, the constitutive expression of HSPs 32 and 90, but not 72 is investigated by Western blot analysis without further immunohistochemical specification [47]. HSP90 has been localised to the cytoplasm of distal tubular and glomerular cells in the rat, reports about localization in human kidney are lacking.

Expression of HSPs in the transplanted kidney

Studies of T lymphocyte reactivity isolated from human and rat heart biopsies gave the first evidence for an involvement of HSP in transplant immunology because T lymphocytes were reactive with HSP65 and 70. Based on these studies, investigation of HSPs in the transplanted kidney were stimulated. An increased expression of HSP60 and HDJ-2 m-RNA was observed in acute or chronic rejected kidneys, but not in post-translational kidneys without rejection (Fig. 1; [44]). The immunohistochemical expression of HSP47 in myofibroblasts and tubular epithelial cells correlates with the degree of interstitial renal allograft fibrosis [46]. One third of rejected allografts were positive for HSP72 (Fig. 2) in a study performed on nine rejections [42]. In contrast, five of eight chronic rejected kidneys were positive for HSP72 in the glomerular and tubular compartment, whereas the vascular was negative in all cases [40]. Remarkable is the *de novo* expression of HSP60 in the vascular compartment during rejection (Fig. 3), which might indicate an additional mechanism of the involvement of HSP60 in the vascular compartment [40].

To date, the role of increased HSP expression and the somehow controversial data on HSP expression during rejection episodes are not completely clear.

HSP expression can be increased during transplantation by reperfusion, ischaemia and even by surgery. HSPs are assumed to protect cell from damage in this stage and to increase transplant survival. If this up-regulation is prolonged by persisting stimuli, this protecting effect might change later on. As precursor T lymphocytes specific for HSPs can be isolated from normal individuals, the ongoing

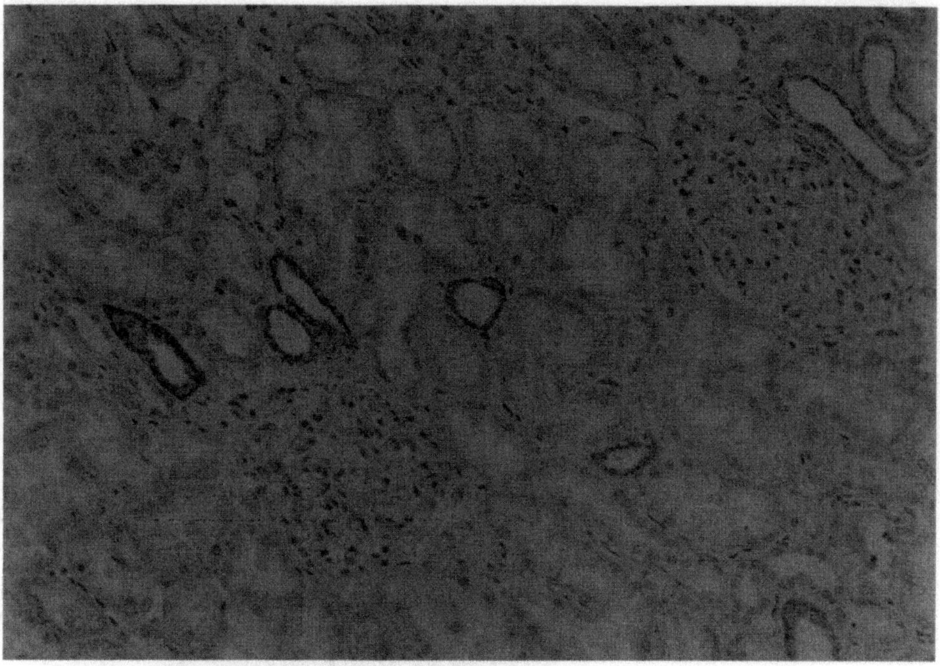

Figure 1
Expression of HSP60 in the renal cortex of normal human kidney predominantly in distal tubules (× 200).

increased expression of HSPs might lead to a release of HSPs from dying cells. As HSP72 seems to be exclusively expressed in tubules and glomeruli without any expression in the vascular compartment [40], this distinct distribution might indicate that HSP72 is immunologically more relevant as an antigen released by tubular cell damage and thus pathogenetically linked with the tubulo-interstitial type of acute rejection. HSP72 and the other released HSPs are in turn internalized and processed by mononuclear antigen presenting cells already present in the graft. This leads to a specific presentation of HSPs *via* MHC-molecules to specific T lymphocytes and not to a polyclonal activation. The thus activated T lymphocytes recognize the still over-expressed HSPs in the transplanted kidney cells, which results in aggravation and longer duration of the rejection process. This theory is supported by the fact that T-cells reactive for HSP were recently isolated from rejected human kidney allografts [48] and from rejected cardiac allografts [49, 50]. These T-lymphocytes could be stimulated with recombinant human HSP72 in combination with antigen-presenting cells suggesting a role of HSP72 either as a transplantation antigen or as an unspecific trigger of T-cell activation associated with severe inflamma-

Figure 2
Expression of HSP72 in normal human kidney with staining in glomeruli (podocytes/epithelia) and the distal tubules (×200).

tion. On the other hand, hyperthermia induced HSP expression correlated with an improved renal rat isograft survival in a seven day study [47].

The distinct expression of HSPs in rejected kidneys, together with corresponding *in vitro* findings of HSP-specific T-cells, suggest a participation as immunogens in graft rejection. Whether heat shock proteins thereby act as primary inducers of rejection or as secondary stimulators of an immune response is subject of future investigations. If a temporary induction of HSPs in the allograft by hyperthermia leads to new procedures and prolonged graft survival remains open.

HSPs and resistance

A cell exposed to high temperature (45 °C) dies by apoptosis. If this cell is exposed to a sub-lethal temperature (for instance 41.5 °C for 30–60 minutes) and exposed after a recovery time to lethal temperature, the cell will survive. This is called ther-

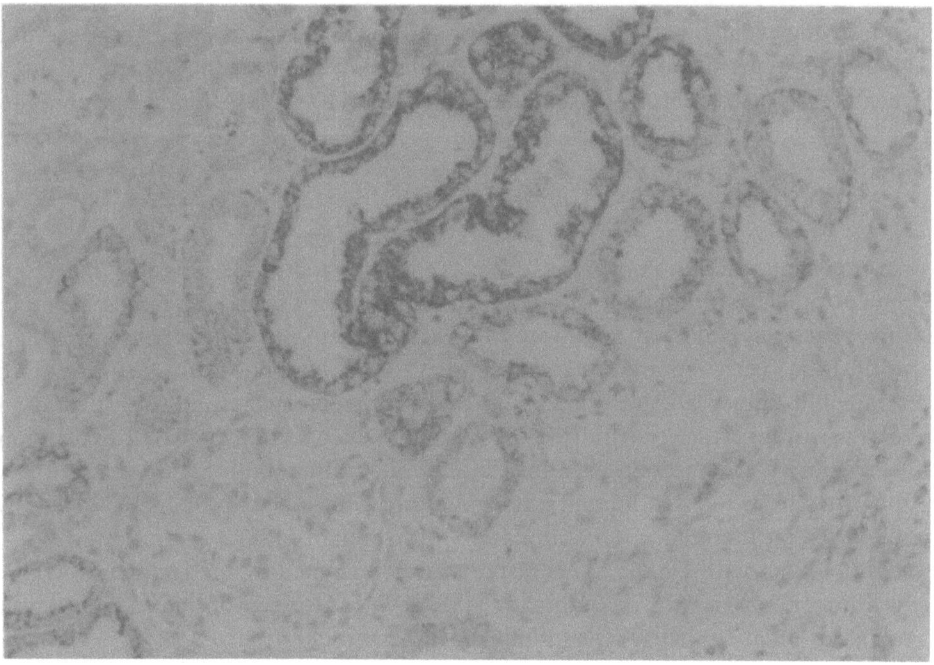

Figure 3
Expression of HSP60 in a rejected human allograft with cytoplasmic immunoreactivity in the tubular epithelium, glomerulas and blood vessels (×200).

motolerance and has been firstly investigated in T lymphoctes. Later on this phenomen came true for other forms of preconditioning by different stressors. It is accepted that (thermo-) tolerance is mainly related to the induction of heat shock proteins by the first stressor, including "chemical preconditioning". The most important role is mediated by an increased expression of HSP72, which protects cells from death. HSP72 in renal tubular cells has been induced by heat, ischemia, nephrotoxic agents, hyperosmosis and hormones [39, 50–56]. Porcine tubular cells stably transfected with the HSP72 gene (not influencing HSP27, 60, 73 and 90 expressions) were treated with hydrogen peroxide or cisplatin. Transfected cells were significantly more resistant to treatments than controls [41]. In ischaemia/reperfusion injury might cause delayed graft function in the first period after transplantation. Cell death in ischaemia/reperfusion is caused by different mechanisms including necrosis and apoptosis and may predict allograft failure [57–59] and several methods have been investigated to injury. Pre-treatment with FK506-inducing HS72 expression has been shown to be reno-protective and improving renal function and histology [60, 61]. The heat shock response during ischaemia/reperfusion seems not to be regulated by the heat shock factor 1, because heat shock factor 1 m-RNA remained constitutively expressed in a rat model [62].

Heat shock proteins and T lymphocytes

In transplant immunology, HSPs have been suggested to serve as new target molecules for the stimulation of allograft infiltrating T-cells, as T-cells reactive to mycobacterial HSP65 and 70 were isolated from heterotrophic rat cardiac allografts [50]. It has been reported in recent studies that T lymphocytes infiltrating rejected heart and kidney allografts recognize HSPs, thereby suggesting a participation of HSPs in transplant rejection [48–50], accompanied by an increased expression of HSPs during the rejection of human kidney allografts [40, 44]. HSP-reactive T lymphocytes from rejected human kidney allografts have been propagated. These lymphocytes proliferated upon stimulation with human HSP72 (about five-fold), but did not react with HSP65. This response was restricted by the recipient's MHC. T-cell lines reactive with recombinant HSP72 were also responsive to the respective allograft renal epithelial cells (about four fold proliferation).

Antibodies to heat shock proteins in sera

Although T lymphocytes play a major role in the development of both cellular and inflammatory immune responses leading to rejection of an allograft, the humoral part of the immune system is also of critical importance. HSPs are immunogenic, but their specific participation in the rejection process is still not completely clear, although an active role in the pathogenesis of rejection seems be possible. An association of antibodies to HSP60 and high anti-heart antibodies in cardiac transplantation has been reported [63]. In kidney allografts, so far anti-HSP60 and 70 antibody titers were evaluated by enzyme linked immunosorbent assay, whether they are of diagnostic, predicative, or prognostic value and were compared to age-matched healthy controls [64, 65]. Antibodies to HSP70 were found in 30% of the patients and to HSP60 in 50% before transplantation. This did not differ from post-transplant or normal levels. An ongoing reversible or irreversible rejection episode was not accompanied by a change in anti-HSP60 or 70 antibody titres. No correlation was found either to severity of rejection or an immunologically uncomplicated course or with the severity of immunosuppression. These antibodies might represent a part of naturally occurring antibodies. They may play a role in autoimmunity *via* cross-sharing of epitopes [31], due to high conservation on pathogens and on human HSPs. But in transplant rejection, there is so far no evidence of a primary involvement of antibodies against HSPs.

References

1 Morris PJ (1994) *Kidney transplantation: principles and practice.* 4th Ed. WB Saunders, Philadelphia

2 Murray JE (1992) Human organ transplantation: Background and consequences. *Science* 256: 1411–1416

3 Vincenti F (2000) *Tacrolimus kidney transplant study group. Tacrolimus vs. cyclosporine in kidney transplantation: five year results of the US multicenter randomized comparative study.* Rom: Transplantation Society; abstract

4 Waller JR, Nicholson ML (2001) Molecular mechanisms of renal allograft fibrosis. *British J Surg* 88: 1429–1441

5 Morris PJ, Johnson RJ, Fuggle SV, Belger MA, Briggs JD (1999) Analysis of factors that affect outcome of primary cadaveric renal transplantations in the UK. *Lancet* 354: 1147–1152

6 Dennis MJS, Foster MC, Ryan JJ, Burden RP, Morgan AG, Blamey RW (1989) The increasing importance of chronic rejection as a cause of renal allograft failure. *Transpl Int* 2: 214–217

7 Bergmann L, Roper L, Bow LM, Hull D, Bartus SA, Schweizer RT (1992) Late graft loss in cadaveric renal transplantation. *Transplant Proc* 24: 2718–2719

8 Matas AJ, Gillingham KJ, Payne WD, Najararian JS (1994) The impact of an acute rejection episode on long-term renal allograft survival. *Transplantation* 57: 857–859

9 Kasiske BL, Heim-Duthoy KL, Tortorice KL, Rao KV (1991) The variable nature of chronic declines in renal allograft function. *Transplantation* 51: 330–334

10 Lindholm A, Ohlmann S, Albrechtsen D, Tuvfeson G, Persson H, Persson NH (1993) The impact of acute rejection episodes on long-term graft function and outcome in 1347 primary renal transplants treated by 3 cyclosporine regimens. *Transplantation* 56: 307–315

11 Dickenmann MJ, Nickeleit V, Tsinalis D, Gürke L, Mihatsch MJ, Thiel G (2002) Why do kidney grafts fail? A long-term single-center experience. *Transpl Int* 15: 508–514

12 Vanrenterghem YFC (1995) Acute rejection and renal allograft outcome. *Nephrol Dial Transpl* 10 (Suppl 1): 29–31

13 Wissing KM, Abramovicz D, Broeders N, Vereestraeten P (2000) Hypercholesterolemia is associated with increased kidney graft loss caused by chronic rejection in male patients with previous acute rejection. *Transplantation* 70: 464

14 Matas AJ, Gillingham KJ, Sutherland DER (1993) Half-life and risk factors for kidney transplant outcome – importance of death with function. *Transplantation* 55: 757–761

15 Basadonna GP, Matas AJ, Gillingham KJ, Payne WD, Dunn DL, Sutherland DE, Gores PF, Gruessner RW, Najararian JS (1993) Early vs. late acute renal acute renal allograft rejection: impact on chronic rejection. *Transplantation* 55: 993–995

16 Foss A, Leivestad T, Fauchald P, Bentdal O, Husberg B, Lien B, Pfeffer P, Oyen O, Scholz T, Gjertsen H et al (2000) *Episodes of acute rejection in kidney transplantation have a major impact on long term graft survival.* Rom: Transplantation Society: abstract

17 Madden RL, Mulhern JG, Benedetto BJ, O'Shea MH, Germain MJ, Braden GL, O'Shaugnessy J, Lipkowitz GS (2000) Completely reversed acute rejection is not a significant risk factor for the development of chronic rejection in renal allograft recipients. *Transpl Int* 13: 344–350
18 Kouwenhoven EA, Ijzermans JNM, De Bruin RWF (2000) Etiology and pathophysiology of chronic transplant dysfunction. *Transpl Int* 13: 385–401
19 Mason DW, Morris PJ (1986) Effector mechanisms in allograft rejection. *Annu Rev Immunol* 4: 119–145
20 Hata Y, Ozawa M, Takemoto S, Cecka JM (1996) HLA matching. *Clin Transpl* 381–396
21 Zantvoort FA, D´Amaro J, Persijn GG, Cohhen B, Schreuder GM, van Rood JJ, Thorogood J (1996) The impact of HLA-A matching on long-term survival of renal allografts. *Transplantation* 61: 841–844
22 Solez K, Axelsen RA, Benediktsson H, Burdick JF, Cohen AH, Colvin RB, Croker BP, Droz D, Dunnhill MS, Halloran PF et al (1993) International standardization of criteria for the histologic diagnosis of renal allograft rejection: The Banff working classification of kidney transplant pathology. *Kidney Int* 44: 411–422
23 Colvin RB, Cohen AH, Saiontz C, Bonsib S, Buick M, Burke B, Carter S, Cavallo T, Haas M, Lindblad A et al (1997) Evaluation of pathological criteria for acute renal allograft rejection: reproducability, sensitivity, and clinical correlation. *J Am Soc Nephrol* 8: 1930–1941
24 Van Saase JLCM, Van der Woude FJ, Thorogood J, Hollander AA, Van Es LA, Weening JJ, Bockel JM, Bruijn JM (1995) The relation between acute vascular and interstitial renal allograft rejection and subsequent chronic rejection. *Transplantation* 59: 1280–1285
25 Peters TG, Shaver TR, Ames JE, Santiago-Delpin EA, Jones KW, Blanton JW (1995) Cold ischemia and outcome in 17,937 cadaveric kidney transplants. *Transplantation* 59: 191–196
26 Shoskes DA, Cecka M (2001) The Effect of preservation and recipient immune factors on delayed graft function and its sequelae in cadaveric renal transplantation. *Transplantation* Proceedings 33: 2967
27 Campistol JM, Sacks SH (2000) Mechanisms of nephrotoxicity. *Transplantation* 69 (Suppl 12): SS5–10
28 Ritossa F (1962) A new puffing pattern induced by temperature shock and DNP in *Drosophila. Experimentia* 18: 571–578
29 Trieb K, Kotz R (2001) Proteins expressed in osteosarcoma and serum levels as prognostic factors. *Int J Biochem Cell Biol* 33: 11–17
30 Jäättelä M (1999) Escaping cell death: survival proteins in cancer. *Exp Cell Res* 248: 30–43
31 Kaufmann S (1990) Heat shock proteins and the immune response. *Immunol Today* 11: 129–133

32 Suto R, Srivastava P (1995) A mechanism for specific immunogenicity of heat shock protein-chaperoned peptides. *Science* 269: 1585–1588
33 Multhoff G, Botzler C, Jennen L, Schmidt J, Ellwart J, Issels R (1997) Heat shock protein 72 on tumor cells. *J Immunol* 158: 4341–4350
34 Lukacs KV, Nakakes A, Atkins CJ, Lowrie DB, Colston MJ (1997) *In vivo* gene therapy of malignant tumours with heat shock protein 65 gene. *Gene Ther* 4: 346–350
35 Soo J, Park Y, Kim J, Shim E, Kim C, Jang J, Kim S, Lee W (1996) T cell lymphoma in transgenic mice expressing the human hsp 70 gene. *Biochem Biophys Res Comm* 218: 582–587
36 Trieb K, Lechleitner T, Lang S, Windhager R, Kotz R, Dirnhofer S (1998) Heat shock protein 72 expression in osteosarcomas correlates with good response to neoadjuvant chemotherapy. *Hum Pathol* 10: 1050–1055
37 Trieb K, Windhager R, Lang S, Kotz R, Dirnhofer S (1998) Evaluation of HLA-DR expression and T lymphocyte infiltrate in osteosarcoma. *Pathol Res Pract* 194: 679–684
38 Trieb K, Lang S Kotz R (2000) Heat shock protein reactive T lymphocytes in human osteosarcomas. *Ped Hemat Oncol* 17: 1–10
39 Yuan C, Bohen E, Musio F, Carome M (1996) Sublethal heat shock and cyclosporine exposure produce tolerance against subsequent cyclosporine toxicity. *Am J Physiol* 271F: 571–578
40 Trieb K, Dirnhofer S, Krumbock N, Blahovec H, Sgonc R, Margreiter R, Feichtinger H (2001) Heat shock protein expression in the transplanted human kidney. *Transpl Int* 14: 281–286
41 Komatsuda A, Wakui H, Oyama Y, Imai H, Miura A, Itoh H, Tashima Y (1999) Overexpression of the human 72 kDa heat shock protein in renal tubular cells confers resistance against oxidative injury and cisplatin toxicity. *Nephrol Dial Transplant* 14: 1385–1390
42 Dodd SM, Martin JE, Swash M, Mather K (1993) Expression of heat shock protein epitopes in renal disease. *Clin Nephrol* 39:239–244
43 Venkataseshan VS, Maarquet E (1996) Heat shock protein 72/73 in normal and diseased kidneys. *Nephron* 73: 442–449
44 Alevy YG, Brennan D, Durriya S, Howard T, Mohanakumar T (1996) Increased expression of the HDJ-2 heat shock protein in biopsies of human rejected kidney. *Transplantation* 61:963–967
45 Ross W, Bertrand W, Morrison A (1992) Identification of a processed protein related to the human chaperonins (hsp60) protein in mammalian kidney. *Biochem Biophys Res Comm* 185: 683–687
46 Abe K, Ozono Y, Miyazaki M, Koji T, Shioshita K, Furusu A, Tsukasaki S, Matsuya F, Hosokawa N, Harada T et al (2000) Interstitial expression of heat shock protein 47 and alpha-smooth muscle actin in renal allograft failure. *Nephrol Dial Transplant* 15: 529–535
47 Redaelli CA, Wagner M, Kulli C, Tian YH, Kubulus D, Mazzucchelli L, Wagner AC, Schilling MK (2001) Hyperthermia-induced HSP expression correlates with improved

rat renal isograft viability and survival in kidneys harvested from non-heart-beating donors. *Transpl Int* 14: 351–360

48 Trieb K, Grubeck-Loebenstein B, Eberl T, Margreiter R (1996) T cells from rejected human kidney allografts respond to heat shock protein 72. *Transplant Immunol* 4: 43–45

49 Moliterno R, Woan M, Bentlejewski C, Qian J, Zeevi A, Pham S, Griffith BP, Duquesnoy RJ (1995) Heat shock protein-induced T-lymphocyte propagation from endomyocardial biopsies in heart transplantation. *J Heart Lung Transplant* 14: 329–336

50 Moliterno R, Valdivia L, Pan F, Duquesnoy RJ (1995) Heat shock protein reactivity of lymphocytes isolated from heterotopic rat cardiac allografts. *Transplantation* 59: 598–604

51 Komatsuda A, Wakui H, Oyama Y, Imai H, Miura A, Itoh H, Tashima Y (1999) Overexpression of the human 72 kDa heat shock protein in renal tubular cells confers resistance against oxidative injury and cisplatin toxicity. *Nephrol Dial Transplant* 14: 1385–1390

52 Wakui H, Komatsuda A, Miura A (1995) Heat-shock proteins in animal models for acute renal failure. *Renal Fail* 17: 641–649

53 Cohen D, Wassermann J, Gullans S (1991) Immediate early gene and hsp70 expression in hyperosmotic stress in MDCK cells. *Am J Physiol* 261C: 594–601

54 Andrus L, Altus M, Pearson D, Grattan M, Nagamine Y (1988) Hsp 70 mRNA accumulates in LLC-PK1 pig kidney cells treated with calcitonin but not with 8-bromo-cyclic AMP. *J Biol Chem* 263: 6183–6187

55 Brown M, Upender R, Hightower L, Renfro J (1992) Thermoprotection of a functional epithelium: heat stress on transepithelial transport by flounder renal tubule in primary monolayer culture. *Proc Natl Acad Sci USA* 89: 3246–3250

56 Xu Q, Ganju L, Fawcett T, Holbrook N (1996) Vasopressin induced heat shock protein expression in renal tubular cells. *Lab Invest* 74: 178–187

57 Van Es A,Hermans J, van Bockel J, Persijn G, van Hooff J, Graeff J (1983) Effect of warm ischemia time and HLA (A and B) matching on renal cadaveric graft survival and rejection episodes. *Transplantation* 36: 255–258

58 Cole E, Naimark D, Aprile M, Wade J, Cattran D, Pei Y, Fenton S, Robinette M, Zaltsman J, Bear R et al (1995) An analysis of predictors of long-term cadaveric renal allograft survival. *Clin Transplant* 9: 282–288

59 Troppmann C, Gillingham K, Benedetti E, Almond P, Gruessner R, Najarian J, Matas A (1995) Delayed graft function, acute rejection, and outcome after cadaver renal transplantation: a multivarante analysis. *Transplantation* 59: 962–968

60 Yang C, Ahn H, Han H, Kim W, Shin M, Kim S, Park J, Kim Y, Moon I, Bang B (2001) Pharmacological preconditioning with low-dose cyclosporine or FK506 reduces subsequent ischemia/reperfusion injury in rat kidney. *Transplantation* 72: 1753–1759

61 Sakr M, Zetti G, McClain C, Gavaler J, Nalesnik M, Todo S, Starzl, van Thiel D (1992) Protective effect of FK506 pretreatment against renal ischemia/reperfusion injury in rats. *Transplantation* 53: 987–991

62 Akcetin Z, Pregla R, Darmer D, Bromme H, Holtz J (2000) During ischemia-reperfusion in rat kidneys, heat shock response is not regulated by expressional changes of heat shock factor 1. *Transpl Int* 13: 297–302

63 Latif N, Yacoub MH, Dunn MJ (1997) Association of pre-transplant anti-heart antibodies against human heat shock protein 60 with clinical course following cardiac transplantation. *Transplant Proc* 29: 1039–40

64 Trieb K, Gerth R, Windhager R, Grohs J, Holzer G, Berger P, Kotz R (2000) Serum antibodies against the heat shock protein 60 are elevated in patients with osteosarcoma. *Immunobiology* 201: 368–376

65 Trieb K, Gerth R, Holzer G, Grohs J, Berger P, Kotz R (2000) Antibodies to heat shock protein 90 in osteosarcoma patients correlate with response to neoadjuvant chemotherapy. *Br J Cancer* 82: 85–87

Mycobacterial heat shock proteins and the bovine immune system

Ad P. Koets

Department of Farm Animal Health and Division of Immunology, Department of Infectious Diseases and Immunology, Faculty of Veterinary Medicine, Utrecht University, PO Box 80.165, 3508 TD Utrecht, The Netherlands

Introduction

Mycobacterial infections constitute a major thread to cattle populations worldwide. The major mycobacterial infections are tuberculosis, caused by infection with *M. bovis* (MB), and paratuberculosis, caused by infection with *M. avium* ssp. *paratuberculosis* (MAP). Evidence of (bovine) tuberculosis goes back to before the domestication of cattle (8000–4000 BC), however the battle against mycobacteria was significantly boosted by the discovery of the tubercle bacillus in 1882 by Robert Koch [1, 2]. Bovine tuberculosis can reside in many different organs, although the pulmonary form is usually considered to be the "classical" form, the latter being restricted to the lung and its draining lymph nodes. The description of a chronic granulomatous infection of the small intestine in a cow, by Johne and Frottingham in 1895, was the first report on paratuberculosis; although at the time they considered it to be an unusual case of bovine tuberculosis [3]. Apart from the fact that MB and MAP have different tissue trophisms, they share many other characteristics. Both cause slow developing diseases, with a long asymptomatic period during which disease is spread between individuals, eventually causing a wasting syndrome in animals progressing to the clinical stage of the disease, months or more likely years after infection. Both diagnosis, especially in the early asymptomatic stages of the disease, as well as protective vaccination are notoriously difficult. Both diseases also represent a threat to human health as bovine tuberculosis is a zoonosis and, although still controversial, MAP has been implicated in the etiology of human Crohn's Disease [1, 4].

Another major point of interest lies in the physiology of the bovine immune system in which γδ T-cells are the major T-cell population in the circulation during the first 6–12 months of life [5, 6]. In comparison with αβ T-cells many questions on the functionality of γδ T-cell populations, during infection and inflammation, are still unanswered. Heat shock proteins (HSP) have been shown to be immunodomi-

Heat Shock Proteins and Inflammation, edited by Willem van Eden

nant antigens for cells in the B- and αβ T-cell compartments, and they also have been shown to be important antigens for the γδ T-cells [7, 8].

This chapter provides an overview of interactions between mycobacteria and the bovine immune system with special focus on reactivity towards mycobacterial HSPs in paratuberculosis.

Bovine paratuberculosis

Young calves, generally in the first six months of life, are most susceptible to acquire the disease. Bacteria are ingested *via* contaminated food sources most notably milk and environmental contamination [9]. MAP is taken up by M-cells in the mucosa of the Peyers patches of the small intestine. Following expulsion at the basal side of the M-cell the bacteria are taken up by macrophages underlying the domes of the Peyers patches [10]. The bacteria persist inside the macrophages, surviving its microbiocidal mechanisms among others by altering the normal phagolysosomal maturation pathways [11, 12]. Dying macrophages release the bacteria which in turn are taken up by other macrophages in the vicinity. The macrophages also appear to be restricted in their capacity to migrate and accumulate at the lesional site, thus forming a granulomatous lesion. These lesions are poorly organised and usually characterized as lepromatous type lesions in analogy to the classification of leprosy lesions [13].

Clinically the disease has a prolonged asymptomatic phase in which neither immune responses nor excretion of bacteria in the faeces can be measured. This period may take as long as two years before the first signs of infections become apparent [13]. Most notably these signs are IFN-γ production in response to challenge with mycobacterial antigen preparations such as purified protein derivative (PPD), or similar indications of activation of cell mediated immunity, and (intermittent) shedding of bacteria in the faeces. During the years to follow, bacterial excretion will increase, immunological responses become more apparent as also anti-mycobacterial antibody response becomes detectable. Cows can remain in this stage of the disease for years. A limited number of animals rapidly progress to the clinical stage of the disease characterized by incurable diarrhoea due to the granulomatous lesions in the intestinal wall. Ultimately these animals die of emaciation caused by the protein loosing enteropathy. In the animals that develop clinical disease, antigen-specific immune response wane while excretion of the pathogen increases exponentially, thus contributing to spread of the disease [14].

In conjunction with other mycobacterial diseases, e.g., tuberculosis, antibodies are considered to contribute little to protective immunity against those intracellular bacteria. Protective immunity against paratuberculosis is considered to be essentially cell mediated. As such, protection depends on the interaction between T-cells and infected antigen presenting cells [15].

Immunology of bovine paratuberculosis

Despite their high degree of conservation HSPs are essentially very immunogenic proteins. In mycobacterial diseases the HSP have been found to induce both T- and B-cell activity [16]. As reviewed by Matzinger [17] the HSP recognition by immune cells could deliver an efficient danger signal to activate immune responses. They are challenging antigens for the immune system as they may provide a universal signal for infection, based on epitopes unique for, or shared, by pathogens. Furthermore, expression of self-HSP derived epitopes by infected cells may also facilitate immune recognition by shared mechanisms [18]. The sharing of epitopes between host and pathogen may potentially lead to autoimmunity (reviewed in [19, 20]). However, pre-immunisation with HSP often leads to (partial) protection against, rather than induction of, autoimmunity in a number of experimental disease models (reviewed in [19]), probably *via* induction of regulatory self-HSP-reactive T-cells. Our research focuses on the unravelling of immune responses to MAP HSP60 and HSP70, as these constitute an evolutionary and functionally important strategy to deal with mycobacterial infections.

Heat shock proteins and B lymphocytes

The serological response to crude mycobacterial antigens (mainly purified protein derivate (PPD) type antigens) during paratuberculosis has been a subject in many studies with the primary aim to investigate the possibilities to improve diagnosis of this disease [21–27]. The results of serological studies have been generalized to argue in favour of an increased antibody response in later stages of the disease. It has been hypothesized that the decrease in CMI and the increased antibody responses reflects a switch in immune reactivity from Type 1 to Type 2 responses (reviewed in [13, 14]); based on the T helper cell dichotomy first described by Mosmann and co-workers [28]. Although this dichotomy is not as clear-cut in outbred species as it is in different murine strains, studies regarding bovine Type 1 and Type 2 immune responses have confirmed the crucial role of IL-4 and IFN-γ as driving cytokines [29], like observed in mice. Furthermore, as a functional classification, a destination can be made between IFN-γ dependent (Th1) antibody isotypes and IL-4 dependant Th2 related isotypes [30]. Likewise for cattle, it has been shown that IgG1 and IgA as opposed to IgG2 and IgM isotypes can be classified as Type 2 associated isotypes and Type 1, respectively [29, 31–33].

To further study the claims that high antibody titres signal a switch from a protective Type 1 to a non-protective Type 2 immune response, we studied isotype specific antibody responses to mycobacterial antigens. Significantly elevated HSP65 specific IgG1 was measured in sera of vaccinated and shedding animals when compared to controls. More HSP65 specific IgG1 was detected in vaccinated animals as compared to non-shedders. In shedders and animals with clinical signs less HSP65 specific IgG2 was detected as compared to vaccinated animals. Decreased HSP65 specific IgG2 and IgG1 responses were observed in animals with clinical disease when compared with shedders. In comparison to other antigens antibody responses to HSP70 in general were relatively low. Although shedders had higher responses compared to the other groups, no significant differences were found in HSP70 specific IgG1 between the groups. HSP70 specific IgG2 was significantly lower in the clinical diseased animals as compared to shedders and vaccinated animals.

Previous studies have reported a decrease in cell-mediated immunity during progression of paratuberculosis and concomitant increase in antibody responses during the disease. We were the first to show that that observation depends highly on the antigens and isotypes used to study the disease. We were able to show the "classical" pattern only for PPDP antigens and the IgG1 isotype. For other antigens and isotypes the response pattern is different and indicates that the progression from the asymptomatic stage to the clinical stage is not uniformly associated with an increased Type 2 response. Additionally, the observation that total IgG levels decreased during the clinical stage of paratuberculosis indicated that there appears to be no generalized increase in antibody responses. We concluded that, with exception of PPDP specific IgG1, the change in Type 1 and Type 2 responses is characterized by a loss of Type 1 reactivity without an increase, or even with a concomitant decrease, in Type 2 reactivity [34].

Heat shock proteins and CD4$^+$ T-cells

In cross-sectional studies lymphoproliferative responses to MAP HSP70 and HSP60 were evaluated, and in addition also to purified protein derivate (PPD) preparations from *M. avium* ssp. *paratuberculosis*, *M. avium* and *M. bovis*. As lymphoproliferation in adult cattle is almost exclusively attributable to the activation of CD4$^+$ T-cells these studies provided insight into the basic relations between MAP HSPs and the bovine helper T-cell compartment [35].

Responses to PPD and HSP70 were higher in the vaccinated animals and in asymptomatic animals that shed the organism in their faeces. Compared with these animals, responses were lower in cows with clinical signs of paratuberculosis. Observations with short-term CD4$^+$ T-cell lines raised to PPD-P and to HSP70 indicated that the similarity between those two antigens was not due to the presence of HSP70 in PPD-P. Mycobacterial HSP60 induced less prominent responses compared with HSP70 but showed a similar pattern with regard to the stages of disease. In conclusion our study indicated that, as for PPD antigens the mycobacterial heat shock protein specific cell mediated immune responses decrease when comparing the asymptomatic stage to the clinical stage in bovine paratuberculosis [35].

Subsequently, the hypothesis that a loss of CD4+ Th cell activity could be due to redistribution of T-cells to the lesional site was tested. Cows with known infection status were sacrificed following blood collection by vena puncture. Mesenteric lymph nodes draining sections of the small intestine as well as the drained sections of the intestine were collected for further studies. The lymphocytes collected from those tissues showed similar responses to mycobacterial antigen when compared to blood derived lymphocytes when compared according to disease status. Hence a similar loss of reactivity was observed in local lesional tissues of animals with clinical signs of disease. Further analysis by immunohistochemistry and flow cytometry revealed a loss of CD4+ Th cells from those lesional sites [36].

Collectively these data on CMI suggest that observed changes in antigen specific immune reactivity during progressive bovine paratuberculosis are more likely due to a loss of CD4+ Th cells rather then redistribution of Th cells.

Heat shock proteins and CD8+ T-cells

Experiments with different strains of knock-out mice as well as studies on genetic defect in humans have indicated a pivotal role for the cytokines IFN-γ and IL-12 in mycobacterial diseases. Second to those cytokine mediated signalling systems the helper T-cell compartment is thought to be very important in resistance [37–40]. The role of the CD8+ T-cells has been subject of debate, however cytotoxic T-cells do contribute to resistance to mycobacterial infections [41–43]. Data on the induction of CD8+ T-cells during mycobacterial infections in cattle either using crude or recombinant antigens is scant, and for HSP reactive CD8+ T-cells no data is available. A limited number of studies, using murine models of mycobacterial infection, have indicated that mycobacterial HSP60 contains at least one CTL epitope [44].

In the case of mycobacterial HSP70 no specific CD8+ CTLs have been described; however this molecule has gained recent attention as it was shown to possess so-called cross-priming capabilities. In short HSP70 belongs to a group of molecules that are able to translocate protein epitopes, complexed to them or linked as a fusion protein, into the cellular MHC Class I presentation pathway when administered extracellularly. This contrary to unaltered protein which when offered extracellularly is predominantly processed and presented *via* the MHC Class II pathway (reviewed in [45]). This process is receptor mediated and to date several receptors have been described to be involved such as the α2 macroglobuline receptor [46, 47] and LOX [48]. Several studies have indicated that these properties can be used to generate CTLs that recognise tumor antigens, viral antigens, as well as intracellular pathogen related antigens in the context of MHC Class I. Such protein based, adjuvants free, vaccine systems may prove to have substantial benefits over using e.g., attenuated pathogens to induce CTLs [49]. Other interactions between HSP70 and receptors on APCs have been described which point to the fact that heat shock proteins, originating from pathogens and/or cellular damage, that are able to come in contact with APCs provide ample signalling to these cells to induce their activation [50]. We have recently demonstrated that similar interactions between MAP HSP70 and bovine APCs occur [51], thus fulfilling the primary criteria allowing further study of these interactions in cattle.

Heat shock proteins and γδ T-cells

The immune system of ruminants (cattle, sheep, and goats) has a number of characteristics unique among mammals. One of those features is the abundance of T lymphocytes with a γδ T-cell receptor present in blood of young animals. The γδ T-cells can represent up to 75% of blood T lymphocytes during the first 6–9 months of life [5, 6, 52]. The majority of these γδ T-cells express the workshop cluster 1 (WC1) molecule. Based on the expression of the WC1 molecule, which belongs to the scavenger receptor cysteine-rich domain family, the γδ T-cells can be divided into two sub-populations, $WC1^+$ and $WC1^-$. As no homologues of the WC1 have been described to be expressed on human or murine γδ T-cells this too can be considered a unique feature of ruminant γδ T-cells [53–55].

The recognition of antigen by γδ T-cells is incompletely understood, although several restriction elements have been described. While a minority of the γδ T-cells recognizes antigen in the context of MHC I or II, the majority of γδ T-cells see their antigen differently. Protein antigens may be recognized *via* non-classical MHC I (MHC Ib) molecules [56], lipid based antigens by presentation in CD1 molecules [57]. Ruminant γδ T-cells express a very diverse repertoire of T-cell receptors which is in sharp contrast to the limited VγVδ segment use in mice and humans. Hence the ruminant γδ T-cells may have a better developed capacity to recognize diverse ligands as compared to human and murine γδ T-cells [52, 58].

With regard to the antigenic specificity of the regulatory T-cells, peptides derived from (self) heatshock proteins (HSP) are interesting ligands that could be presented by APC, but also in some species, among which cattle, by activated αβ and γδ T-cells in the context of MHC Class II. In addition, both eukaryotic and prokaryotic HSPs are protein antigens that have frequently been observed to induce activation of γδ T-cells [59–61].

In the course of intracellular mycobacterial infections two, apparently opposite, functions of the γδ T-cell population have been described. Evidence has been presented that the γδ T-cell population contributes to clearance of infection by the secretion of IFN-γ *via* a positive feedback loop sustained by activated antigen specific $CD4^+$ T-cells and IL-12, possibly enhancing the conditions for a micro-environment that favours cell mediated immune responses in early stages of infection [62]. However, in low-dose infection models for tuberculosis, it has been shown that γδ T-cells may predominantly have an immunoregulatory (anti-inflammatory) function which involves regulation of granuloma formation [63]. Additionally, negative feedback loops *via* IL-10 have been described which appear to down-regulate Th1 and NK-based IFN-γ production, and also indicate an immunoregulatory function of γδ T-cells controlling Th1 responses [64, 65]. Studies using murine gene knock-out (KO) models e.g., γδ KO show that γδ KO mice are not severely hampered in their anti-mycobacterial capabilities, however the γδ T-cells appear to have an anti-inflammatory role aimed at reducing inflammation driven tissue damage (reviewed in [8, 40]). Similar results have been obtained with regard to the role of γδ T-cells in models for bovine mycobacterial diseases such as tuberculosis (*M. bovis*) and paratuberculosis (*M. paratuberculosis*). Bovine $WC1^+$ γδ T-cells in *M. bovis* infection in the xenogeneic SCID-bo system may have a pivotal role in the early stages of infection, and it is involved in the resulting architecture of the lesions, possibly by recruitment of cells to the lesions [66]. Tanaka et al., in a Balb/c mouse model for paratuberculosis, have shown that in the γδ KO Balb/c mice there is more organisation in the granulomas in the advanced stages of infection as compared to the wild type mice. The results indicate that the γδ T-cell population does not restrict the growth of the mycobacteria but allows for formation of epitheloid granulomata as are observed in progressive bovine paratuberculosis [67]. Recent studies on cattle experimentally infected with *M. bovis* show that there is early activation of $WC1^+$ γδ T-cells. Purified $WC1^+$ γδ T-cells have a strong proliferative response to *M. bovis* antigen *in vitro*, with relative low production of IFN-γ [68]. Our own studies on the immunopathogenesis of bovine paratuberculosis also show different kinetics of γδ T-cells both in early stages of infections and during the final progressive stages of the disease [36]. However, the role of γδ T-cells in protective immunity remains unclear, and regulatory functions may prevail over those inducing protective immunity.

Conclusions

In the course of bovine paratuberculosis dramatic changes in the immune reactivity to the mycobacterial pathogen occur that are related to the outcome of the infection. This review has focussed on some of the major players. Current evidence points to an antigen specific mechanism that causes the demise of $CD4^+$ helper T-cells that recognize mycobacterial antigens. As a consequence of the progressive loss of these $CD4^+$ T-cells the infection progresses beyond control leading to massive intestinal pathology eventually leading to the death of the cow. Major questions at this point relate to the mechanism of this loss of $CD4^+$ T-cells. The direct interaction between infected macrophages and dendritic cells on the one hand and $CD4^+$ T-cells on the other hand may lead to the induction of apoptosis in the $CD4^+$ T-cell population. This could occur *via* signalling mechanisms through cell membrane bound molecules (e.g., Fas-FasL interactions) or *via* cytokines [69–71]. Another possibility may be a regulatory function of γδ T-cells, which may start down-regulating the chronic activation of the CD4+ T-cells during these chronic infections, in a misdirected attempt to limit tissue damage [8]. On the point of intervention and protection more insight in the role of the CD8+ T-cells during these intracellular infections warrants attention. The role of NK cells, functioning as a bridge between innate and specific immunity, probably together with γδ T-cells, has not been discussed, as currently there are no tools to study these cells in ruminants.

Further studies on these complex mycobacterial diseases in a natural host are indispensable to further understand the pathogenesis as well as aid the rational design of eradication strategies both on the level of individual animals as well as the population, with potential benefits for the battle against mycobacterial diseases threatening human health as well.

References

1 Collins CH (2000) The bovine tubercle bacillus. *Br J Biomed Sci* 57: 234–240

2 Waddington K (2001) The science of cows: tuberculosis, research and the state in the United Kingdom, 1890–1914. *Hist Sci* 39: 355–381

3 Johne HA, Frothingham L (1895) Ein eigenthümlicher Fall von Tuberculose beim Rind. *Deutsche Zeitschrift f. Thiermed. u. vergl. Pathologie* XXV: 438–454

4 Chiodini RJ, Rossiter CA (1996) Paratuberculosis: a potential zoonosis? *Vet Clin North Am Food Anim Pract* 12: 457–467

5 Davis WC, Brown WC, Hamilton MJ, Wyatt CR, Orden JA, Khalid AM, Naessens J (1996) Analysis of monoclonal antibodies specific for the gamma delta TcR. *Vet Immunol Immunopathol* 52: 275–283

6 Wyatt CR, Brackett EJ, Perryman LE, Davis WC (1996) Identification of gamma delta T lymphocyte subsets that populate calf ileal mucosa after birth. *Vet Immunol Immunopathol* 52: 91–103

7 Boismenu R, Havran WL (1997) An innate view of gamma delta T cells. *Curr Opin Immunol* 9: 57–63

8 Born W, Cady C, Jones-Carson J, Mukasa A, Lahn M, O'Brien R (1999) Immunoregulatory functions of gamma delta T cells. *Adv Immunol* 71: 77–144

9 Sweeney RW (1996) Transmission of paratuberculosis. *Vet Clin North Am Food Anim Pract* 12: 305–312

10 Momotani E, Whipple DL, Thiermann AB, Cheville NF (1988) Role of M cells and macrophages in the entrance of *Mycobacterium paratuberculosis* into domes of ileal Peyer's patches in calves. *Vet Pathol* 25: 131–137

11 Pieters J (2001) Entry and survival of pathogenic mycobacteria in macrophages. *Microbes Infect* 3: 249–255

12 Kuehnel MP, Goethe R, Habermann A, Mueller E, Rohde M, Griffiths G, Valentin-Weigand P (2001) Characterization of the intracellular survival of *Mycobacterium*

avium ssp. *paratuberculosis*: phagosomal pH and fusogenicity in J774 macrophages compared with other mycobacteria. *Cell Microbiol* 3: 551–566
13 Clarke CJ (1997) The pathology and pathogenesis of paratuberculosis in ruminants and other species. *J Comp Pathol* 116: 217–261
14 Chiodini RJ (1996) Immunology: resistance to paratuberculosis. *Vet Clin North Am Food Anim Pract* 12: 313–343
15 Kaufmann SH (2001) How can immunology contribute to the control of tuberculosis? *Nat Rev Immunol* 1: 20–30
16 Ivanyi J, Norton P, Matsuzaki G (1996) Immune responses to stress proteins in mycobacterial infections. In: W van Eden, D Young (eds): *Stress proteins in medicine.* Marcel Dekker Inc, New York, 265–286
17 Matzinger P (1994) Tolerance, danger, and the extended family. *Annu Rev Immunol* 12: 991–1045
18 Kaufmann SH (1990) Heat shock proteins and the immune response. *Immunol Today* 11: 129–136
19 van Eden W, van der Zee R, Paul A, Prakken B, Wendling U, Anderton S, Wauben M (1998) Do heat shock proteins control the balance of T-cell regulation in inflammatory diseases? *Immunol Today* 19: 303–307
20 Rook GA, Stanford JL (1998) Give us this day our daily germs. *Immunol Today* 19: 113–116
21 Abbas B, Riemann HP (1988) IgG, IgM and IgA in the serum of cattle naturally infected with *Mycobacterium paratuberculosis. Comp Immunol Microbiol Infect Dis* 11: 171–175
22 Spangler L, Bech-Nielsen S, Heider L (1988) A study of sub-clinical paratuberculosis in three central Ohio dairy herds: fecal culture, serologic testing and milk production. *Acta Vet Scandinavica* S 84: 148–150
23 Vannuffel P, Gilot P, Limbourg B, Naerhuyzen B, Dieterich C, Coene M, Machtelinckx L, Cocito C (1994) Development of species-specific enzyme-linked immunosorbent assay for diagnosis of Johne's disease in cattle. *J Clin Microbiol* 32: 1211–1216
24 Sockett DC, Conrad TA, Thomas CB, Collins MT (1992) Evaluation of four serological tests for bovine paratuberculosis. *J Clin Microbiol* 30: 1134–1139
25 Collins MT, Angulo A, Buergelt CD, Hennager SG, Hietala SK, Jacobson RH, Whipple DL, Whitlock RH (1993) Reproducibility of a commercial enzyme-linked immunosorbent assay for bovine paratuberculosis among eight laboratories. *J Vet Diagn Invest* 5: 52–55
26 Hilbink F, West DM, de Lisle GW, Kittelberger R, Hosie BD, Hutton J, Cooke MM, Penrose M (1994) Comparison of a complement fixation test, a gel diffusion test and two absorbed and unabsorbed ELISA's for the diagnosis of paratuberculosis in sheep. *Vet Microbiol* 41: 107–116
27 Jark U, Ringena I, Franz B, Gerlach GF, Beyerbach M, Franz B (1997) Development of an ELISA technique for serodiagnosis of bovine paratuberculosis. *Vet Microbiol* 57: 189–198

28 Mosmann TR, Cherwinski H, Bond MW, Giedlin MA, Coffman RL (1986) Two types of murine helper T cell clone. I. Definition according to profiles of lymphokine activities and secreted proteins. *J Immunol* 136: 2348–2357
29 Brown WC, Estes DM (1997) Type I and type II responses in cattle and their regulation. In: MS Horzinek, VECJ Chijns (eds): *Cytokines in veterinary medicine*. CAB International, Wallingford, UK, 15–33
30 Abbas A, Murphy K, Sher A (1996) Functional diversity of helper T lymphocytes. *Nature* 383: 787–793
31 Estes DM, Closser NM, Allen GK (1994) IFN-g stimulates IgG2 production from bovine B cells co-stimulated with anti-m and mitogen. *Cell Immunol* 154: 287–295
32 Estes DM, Hirano A, Heussler VT, Dobbelaere DA, Brown WC (1995) Expression and biological activities of bovine interleukin 4: effects of recombinant bovine interleukin 4 on T cell proliferation and B cell differentiation and proliferation *in vitro*. *Cell Immunol* 163: 268–279
33 Brown WC, McElwain TF, Palmer GH, Chantler SE, Estes DM (1999) Bovine CD4(+) T-lymphocyte clones specific for rhoptry-associated protein 1 of *Babesia bigemina* stimulate enhanced immunoglobulin G1 (IgG1) and IgG2 synthesis. *Infect Immun* 67: 155–164
34 Koets AP, Rutten VP, de Boer M, Bakker D, Valentin-Weigand P, van Eden W (2001) Differential changes in heat shock protein-, lipoarabinomannan-, and purified protein derivative-specific immunoglobulin G1 and G2 isotype responses during bovine *Mycobacterium avium* subsp. *paratuberculosis* infection. *Infect Immun* 69: 1492–1498
35 Koets AP, Rutten VP, Hoek A, Bakker D, van Zijderveld F, Muller KE, van Eden W (1999) Heat-shock protein-specific T-cell responses in various stages of bovine paratuberculosis. *Vet Immunol Immunopathol* 70: 105–115
36 Koets A, Rutten V, Hoek A, van Mil F, Muller K, Bakker D, Gruys E, van Eden W (2002) Progressive bovine paratuberculosis is associated with local loss of CD4(+) T cells, increased frequency of gamma delta T cells, and related changes in T-cell function. *Infect Immun* 70: 3856–3864
37 Ottenhoff TH, Kumararatne D, Casanova JL (1998) Novel human immunodeficiencies reveal the essential role of type-I cytokines in immunity to intracellular bacteria. *Immunol Today* 19: 491–494
38 Thompson-Snipes L, Skamene E, Radzioch D (1998) Acquired resistance but not innate resistance to *Mycobacterium bovis* bacillus Calmette-Guerin is compromised by interleukin-12 ablation. *Infect Immun* 66: 5268–5274
39 Flesch IE, Hess JH, Huang S, Aguet M, Rothe J, Bluethmann H, Kaufmann SH (1995) Early interleukin 12 production by macrophages in response to mycobacterial infection depends on interferon gamma and tumor necrosis factor alpha. *J Exp Med* 181: 1615-1621
40 Schaible UE, Collins HL, Kaufmann SH (1999) Confrontation between intracellular bacteria and the immune system. *Adv Immunol* 71: 267–377
41 Cho S, Mehra V, Thoma-Uszynski S, Stenger S, Serbina N, Mazzaccaro RJ, Flynn JL,

Barnes PF, Southwood S, Celis E et al (2000) Antimicrobial activity of MHC class I-restricted CD8+ T cells in human tuberculosis. *Proc Natl Acad Sci USA* 97: 12210–12215.

42 Stenger S, Modlin RL (1998) Cytotoxic T cell responses to intracellular pathogens. *Curr Opin Immunol* 10: 471–477

43 Milon G, Louis J (1993) CD8+ T cells and immunity to intracellular pathogens. *Parasitology Today* 9: 196–197

44 Zugel U, Kaufmann SH (1997) Activation of CD8 T cells with specificity for mycobacterial heat shock protein 60 in *Mycobacterium bovis* bacillus Calmette-Guerin-vaccinated mice. *Infect Immun* 65: 3947–3950

45 Li Z, Menoret A, Srivastava P (2002) Roles of heat-shock proteins in antigen presentation and cross-presentation. *Curr Opin Immunol* 14: 45–51

46 Basu S, Binder RJ, Ramalingam T, Srivastava PK (2001) CD91 is a common receptor for heat shock proteins gp96, hsp90, hsp70, and calreticulin. *Immunity* 14: 303–313

47 Binder RJ, Han DK, Srivastava PK (2000) CD91: a receptor for heat shock protein gp96. *Nat Immunol* 1: 151–155

48 Delneste Y, Magistrelli G, Gauchat J, Haeuw J, Aubry J, Nakamura K, Kawakami-Honda N, Goetsch L, Sawamura T, Bonnefoy J et al (2002) Involvement of LOX-1 in dendritic cell-mediated antigen cross-presentation. *Immunity* 17: 353

49 Srivastava PK, Amato RJ (2001) Heat shock proteins: the "Swiss Army Knife" vaccines against cancers and infectious agents. *Vaccine* 19: 2590–2597

50 Wallin RP, Lundqvist A, More SH, von Bonin A, Kiessling R, Ljunggren HG (2002) Heat-shock proteins as activators of the innate immune system. *Trends Immunol* 23: 130–135

51 Langelaar M, Koets A, Muller K, van Eden W, Noordhuizen J, Howard C, Hope J, Rutten V (2002) *Mycobacterium paratuberculosis* heat shock protein 70 as a tool in control of paratuberculosis. *Vet Immunol Immunopathol* 87: 239–244

52 Hein WR, Mackay CR (1991) Prominance of $\gamma\delta$ T cells in the ruminant immune system. *Immunol Today* 12: 30–34

53 Takamatsu HH, Kirkham PA, Parkhouse RM (1997) A gamma delta T cell specific surface receptor (WC1) signaling G0/G1 cell cycle arrest. *Eur J Immunol* 27: 105–110

54 Wijngaard PL, MacHugh ND, Metzelaar MJ, Romberg S, Bensaid A, Pepin L, Davis WC, Clevers HC (1994) Members of the novel WC1 gene family are differentially expressed on subsets of bovine CD4-CD8- gamma delta T lymphocytes. *J Immunol* 152: 3476–3482

55 Clevers H, MacHugh ND, Bensaid A, Dunlap S, Baldwin CL, Kaushal A, Iams K, Howard CJ, Morrison WI (1990) Identification of a bovine surface antigen uniquely expressed on $CD4^-CD8^-$ T cell receptor gamma/delta$^+$ T lymphocytes. *Eur J Immunol* 20: 809–817

56 Soloski MJ, Szperka ME, Davies A, Wooden SL (2000) Host immune response to intracellular bacteria: A role for MHC-linked class-Ib antigen-presenting molecules. *Proc Soc Exp Biol Med* 224: 231–239

57 Porcelli SA, Segelke BW, Sugita M, Wilson IA, Brenner MB (1998) The CD1 family of lipid antigen-presenting molecules. *Immunol Today* 19: 362–368
58 Hein WR, Dudler L (1997) TCR gamma delta[+] cells are prominent in normal bovine skin and express a diverse repertoire of antigen receptors. *Immunology* 91: 58–64
59 Zugel U, Kaufmann SH (1999) Role of heat shock proteins in protection from and pathogenesis of infectious diseases. *Clin Microbiol Rev* 12: 19–39
60 Born W, Hall L, Dallas A, Boymel J, Shinnick T, Young D, Brennan P, O'Brien R (1990) Recognition of a peptide antigen by heat shock-reactive gamma delta T lymphocytes. *Science* 249: 67–69
61 Pfeffer K, Schoel B, Gulle H, Kaufmann S, Wagner H (1990) Primary responses of human T cells to mycobacteria: a frequent set of γ/δ T cells are stimulated by protease-resistant ligands. *Eur J Immunol* 20: 1175–1179
62 Garcia VE, Sieling PA, Gong J, Barnes PF, Uyemura K, Tanaka Y, Bloom BR, Morita CT, Modlin RL (1997) Single-cell cytokine analysis of gamma delta T cell responses to non-peptide mycobacterial antigens. *J Immunol* 159: 1328–1335
63 D'Souza CD, Cooper AM, Frank AA, Mazzaccaro RJ, Bloom BR, Orme IM (1997) An anti-inflammatory role for gamma delta T lymphocytes in acquired immunity to *Mycobacterium tuberculosis. J Immunol* 158: 1217–1221
64 Hsieh B, Schrenzel MD, Mulvania T, Lepper HD, DiMolfetto-Landon L, Ferrick DA (1996) *In vivo* cytokine production in murine listeriosis. Evidence for immunoregulation by gamma delta[+] T cells. *J Immunol* 156: 232–237
65 Flesch IE, Kaufmann SH (1994) Role of macrophages and alpha beta T lymphocytes in early interleukin 10 production during *Listeria monocytogenes* infection. *Int Immunol* 6: 463–468
66 Smith RA, Kreeger JM, Alvarez AJ, Goin JC, Davis WC, Whipple DL, Estes DM (1999) Role of CD8[+] and WC-1[+] gamma/delta T cells in resistance to *Mycobacterium bovis* infection in the SCID-bo mouse. *J Leukoc Biol* 65: 28–34
67 Tanaka S, Itohara S, Sato M, Taniguchi T, Yokomizo Y (2000) Reduced formation of granulomata in gamma(delta) T cell knockout BALB/c mice inoculated with *Mycobacterium avium* subsp. *paratuberculosis. Vet Pathol* 37: 415–421
68 Smyth AJ, Welsh MD, Girvin RM, Pollock JM (2001) *In vitro* responsiveness of gammadelta T cells from *Mycobacterium bovis*-infected cattle to mycobacterial antigens: predominant involvement of WC1(+) cells. *Infect Immun* 69: 89–96
69 Mustafa T, Bjune TG, Jonsson R, Pando RH, Nilsen R (2001) Increased expression of fas ligand in human tuberculosis and leprosy lesions: a potential novel mechanism of immune evasion in mycobacterial infection. *Scand J Immunol* 54: 630–639
70 Dockrell DH (2001) Apoptotic cell death in the pathogenesis of infectious diseases. *J Infect* 42: 227–234
71 Gao LY, Kwaik YA (2000) The modulation of host cell apoptosis by intracellular bacterial pathogens. *Trends Microbiol* 8: 306–313

Microbial infection generates pro-inflammatory autoimmunity against the small heat shock protein alpha B-crystallin and provides the fuel for the development of multiple sclerosis

Johannes M. van Noort

Division of Immunological and Infectious Diseases, TNO Prevention and Health, P.O. Box 2215, 2301 CE Leiden, The Netherlands

Introduction

The idea that microbial infections play a role in the development of multiple sclerosis (MS) is as old as the notion itself that MS represents a distinct neurodegenerative disease. Epidemiological data on MS, the appearance of characteristic oligoclonal populations of IgG in the cerebrospinal fluid, elevated serum Ig levels to several viruses, associations between MS relapses and infections and the frequent presence of viral infections in MS brains are all consistent with the idea that infections, in particular with Epstein-Barr virus (EBV) somehow play a role [1–5]. Also animal models of viral infections in the central nervous system (CNS) confirm the ability of such infections to trigger chronic demyelinating disease. Frustratingly however, EBV and other pathogens so far implicated in the pathogenesis of MS are all rather ubiquitous and so far, they have never been found to be specific for MS. Current data therefore point to a relationship between microbial infection and MS that is more complex than a simple "one agent – one disease" paradigm.

The apparent absence of an MS-specific infectious agent is paralleled by the notion that autoimmune T-cell responses to myelin components can also form the basis for chronic inflammatory demyelinating disease (reviewed in [6]). Animal models have shown that in appropriate genetic backgrounds, many myelin-associated proteins or peptides can trigger pathogenic experimental autoimmune responses. Presentation of such myelin-derived antigens in active MS lesions is abundant, as is apparent from neuropathological studies [7]. Myelin-associated antigens have therefore ample opportunity to contribute to the inflammatory process should they be able to activate T-cells. This autoimmune view on MS, however, raises the question how myelin-specific autoimmune responses develop in humans in the absence of the strong artificial adjuvants that are commonly used in experimental models.

Heat Shock Proteins and Inflammation, edited by Willem van Eden

This question seems to take us back to microbial infections since this is one of the major driving forces behind the shaping and maintenance of an adult human T-cell repertoire.

Several hypotheses have previously been put forward to explain how microbial infections could activate myelin-reactive T-cells. The focus on helper T-cells of the immune system in these hypotheses is inspired by the generally accepted notion that *via* secretion of interferon (IFN)-γ, helper T-cells control the activity of microglia and macrophages that are the prime effector cells of demyelination in MS. The two currently most popular ideas are generally referred to as either molecular mimicry or epitope spreading.

The concept of molecular mimicry proposes that structural similarities between microbial antigens and myelin-associated proteins can lead to functional cross-reactivity and can turn antimicrobial responses into autoimmune ones [3, 8]. Several examples have been described of individual T-cell clones that display cross-reactivity between myelin peptides and microbial sequences. That this would be a surprising phenomenon, however, is based on the implicit assumption that T-cell receptors (TCR) should be very specific and unlikely to react to any other peptide than their original target epitope. However, data have accumulated over the past few years to show that each TCR is, in fact, capable of reacting with hundreds of thousands, if not millions of different peptides [9–13]. These data prompt a re-evaluation of the concept of molecular mimicry in suggesting that molecular mimicry at the level of single TCR is much more frequent than once assumed, and probably of little relevance at the level of polyclonal T-cell responses.

A second popular hypothesis to explain how microbial infection could activate myelin-specific immune responses is bystander activation or epitope spreading (reviewed in [14]). This hypothesis proposes that infectious tissue damage could lead to such a powerful pro-inflammatory micro-environment that novel T-cell specificities are generated to local tissue antigens. Specific immune responses could thus spread from the initial infectious trigger to local self antigens and these secondary autoimmune responses could then become pathogenic in their own right [15–17]. In murine models of autoimmunity, newly emerging and regressing systemic autoimmune T-cell responses to myelin antigens have been convincingly documented shortly after the first episode of acute disease. However, abnormal systemic anti-myelin T-cell responses that define determinant spreading in animal models have never been found in MS patients, despite impressive efforts to reveal them (reviewed in [6]). All evidence that has been accumulated in over two decades of research only show that T-cell reactivities against myelin antigens in MS patients are indistinguishable from what is found in healthy subjects [18]. In conclusion, no evidence is available for either molecular mimicry or epitope spreading to occur in MS at levels that may explain the disease. Another problem is that none of the mechanisms can provide an explanation for the fact that MS only occurs in humans. Clearly, there is still room for alternatives.

Alpha B-crystallin as a dominant autoimmune target in myelin

In a set of experiments that turned out to be pivotal for our studies, we examined human T-cell responses to the complete collection of myelin-associated proteins in MS brains [19]. Following high-resolution HPLC fractionation of all myelin proteins derived from MS brains, a single protein was found that was much more potent in activating cultured myelin-primed human T-cells than any other myelin component. This protein was identified as the small HSP alpha B-crystallin. Figure 1 provides a representative example of the T-cell response profile against fractionated myelin proteins from MS brains. Like in all other cases reported, these T-cell response profiles were essentially the same in MS patients as compared to healthy control subjects. Using collections of myelin proteins derived from control brains, the dominant T-cell response to alpha B-crystallin did not occur. This difference corresponded to a locally very high expression of alpha B-crystallin inside oligodendrocytes (and thus in myelin) in areas of early inflammation in MS brains and very low levels of expression in normal brains [19]. Recent studies have confirmed this difference by showing *via* large-scale sequencing of random cDNA libraries that alpha B-crystallin is the single most strongly up-regulated gene product in areas of active inflammation in MS [20].

The small HSP alpha B-crystallin shares the classical characteristics of other HSP, i.e., stress-inducibility, strong sequence homologies among different mammalian species and molecular chaperone properties. Yet, alpha B-crystallin is quite different from other HSP in its pattern of expression (reviewed in [21, 22]). Alpha B-crystallin has a restricted tissue distribution in humans and it certainly does not become expressed in virtually all cells, as appears to be the rule for many other HSP including members of the HSP90, HSP70, HSP60 and HSP27 families of HSP. In the human body, significant expression of alpha B-crystallin can only be found intracellularly in components of the eye, cardiac muscle and skeletal muscles and in astrocytes and oligodendrocytes in the CNS but not in healthy lymphoid cells and organs. When expressed, alpha B-crystallin is generally tightly associated with cytoskeletal elements, where it plays a major role in conferring resistance against e.g., TNF and oxidative stress-induced damage. Another important distinction between alpha B-crystallin and many other HSP is the fact that alpha B-crystallin has no prokaryotic homolog. These features of alpha B-crystallin are key in allowing it to play a special immunological role in humans.

Since alpha B-crystallin is an HSP, a relevant issue is the question is whether or not the very high expression of this protein in oligodendrocytes/myelin from MS brains is only an epiphenomenon, and secondary to the inflammatory process. In two detailed studies, we conclusively showed that in fact, high expression of alpha B-crystallin in oligodendrocytes in MS brains and its uptake by phagocytosing macrophages is the earliest marker known to date for inflammatory demyelination lesions in MS [23, 24] and that its expression rapidly fades at later stages of inflam-

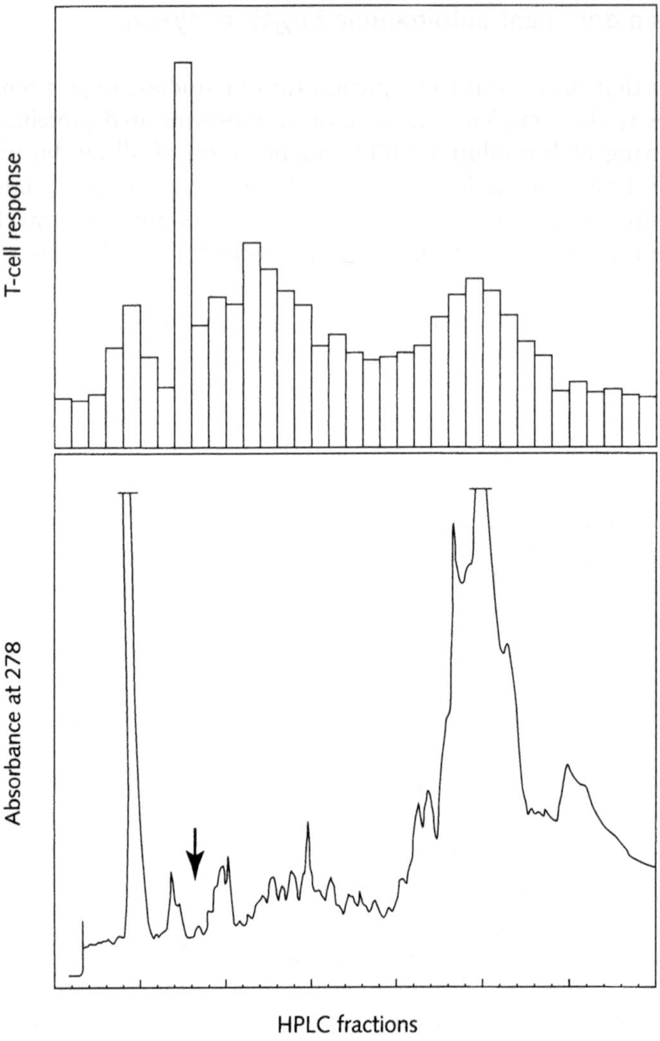

Figure 1
Alpha B-crystallin is a dominant immunogen in myelin from MS brains.
In the bottom part of Figure 1, a reversed-phase HPLC profile is shown of the complete collection of proteins extracted from pooled CNS myelin from MS patients. In the upper part, proliferative responses are shown of cultured myelin-primed T-cells from an MS patient to each of the separate protein fractions, as representative example for 24 donors tested. The result reveals that despite representing an only minor fraction relative to the total amount of protein, alpha B-crystallin (indicated by an arrow) triggers a stronger T-cell response than any other myelin protein. For further details, see [19].

mation: Quite the opposite therefore from being a secondary phenomenon. These studies also revealed that alpha B-crystallin is indeed functionally presented to T-cells *via* Class II MHC molecules in early MS lesions, accompanied by key co-stimulatory molecules such as CD40 and B7-1.

Autoimmune T-cell and antibody responses to alpha B-crystallin are unique to humans, and they are absent from other mammals

At the early stage of our studies, extensive efforts were devoted to induce experimental autoimmune encephalomyelitis (EAE) in rodents with alpha B-crystallin, to provide for a system for experimental manipulation of autoimmune responses against the HSP and to develop strategies for intervention. The consistent outcome of these efforts, however, was the finding that in any other mammal except for humans, a state of complete tolerance exists for self alpha B-crystallin. Not even with powerful adjuvants such as complete Freund's adjuvant could any detectable level of helper T-cell or antibody reactivity be experimentally induced against self alpha B-crystallin in various strains of rats or mice [25–27]. In association with this profound state of tolerance, readily detectable constitutive expression of alpha B-crystallin was found in primary (thymus) as well as secondary (spleen, lymph nodes and peripheral blood lymphocytes) lymphoid tissues and cells at the level of both mRNA and protein [28]. The same was found for rabbits, cattle, sheep and several primates including marmosets, aotus and rhesus monkeys.

Only in humans, alpha B-crystallin has remained undetectable in lymphoid cells by current methodologies and in concordance with this absence, the adult human immune repertoire is characterized by autoimmune reactivity against alpha B-crystallin at levels far beyond the background levels that are found for comparable other self proteins. This is most clearly demonstrated by results from recent studies on human antibody reactivity against alpha B-crystallin (unpublished results). In these studies, we compared auto-antibody reactivity in humans against alpha B-crystallin with other closely related protein targets, viz. other crystallins. Most of these share key physico-chemical properties with alpha B-crystallin, they are of similar size, and they all form the building blocks of specialized structures in the eye such as the eye lens. In particular alpha A crystallin shares about 50% sequence homology with alpha B-crystallin and is an almost perfect comparative subject of study. By comparing auto-antibody reactivities to HPLC-fractionated eye lens crystallins using western blotting, it was revealed that normal human serum antibodies contain a remarkable specificity towards alpha B-crystallin, but not to any of the other crystallins (Fig. 2C). Very similar response profiles and levels of reactivity were documented for antibodies in the serum of MS patients (Fig. 2D). This unique auto-antibody reactivity, dominated by IgG responses, parallels the previously documented helper T-cells reactivity against alpha B-crystallin.

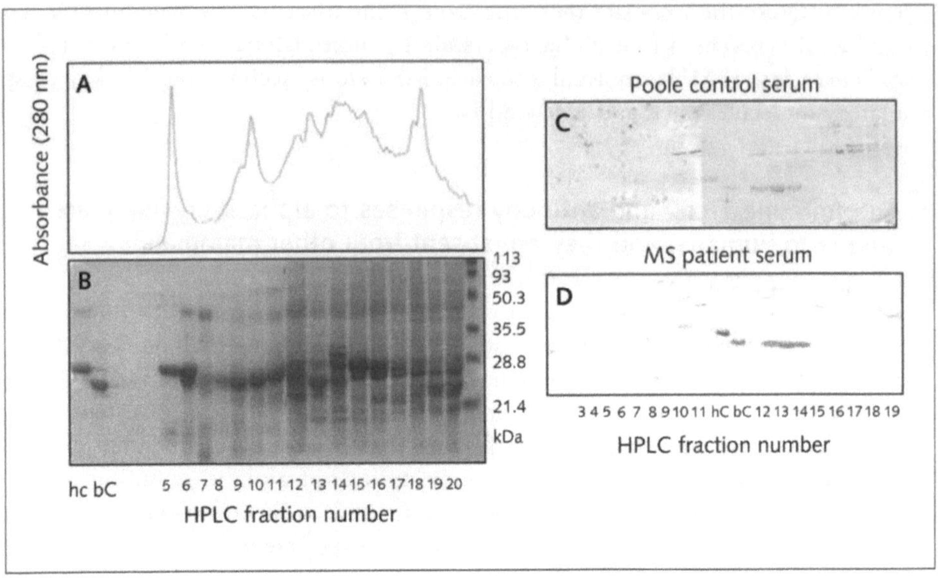

Figure 2
Alpha B-crystallin is a prime target for human serum antibodies.
Total protein extracts were prepared from pooled human eye lenses and fractionated by reversed phase HPLC (panel A). The protein fractions collected by HPLC were subjected to SDS-PAGE (panel B, note that Figure 2B is a composite of two separate gels). Figures 2A and B together reveal high-resolution fractionation of all eye lens proteins confining each individual protein to maximally three consecutive HPLC fractions. As references, purified recombinant human alpha B-crystallin supplied with a His-tag (hC) and purified bovine eye lens alpha B-crystallin (bC) were included.
Human serum IgG antibody recognition of eye lens proteins was evaluated by western blotting using the HPLC-fractionated material and SDS-PAGE conditions as illustrated in panels A and B. In panel 2C, the results are illustrated for pooled control serum (a composite of 70 different serum samples from healthy subjects) and in panel 2D for a clinically definite MS patient as a representative example for a total of 12 MS patient sera tested. Note marked IgG recognition of alpha B-crystallin in fractions 12 to 14 in both cases and negligible responses to other crystallins.

All available data indicate that the T-cell response to alpha B-crystallin in humans is dominated by secretion of the pro-inflammatory mediator IFN-γ and an almost complete lack of the regulatory cytokines IL-4 and IL-10 [19, 24, 29]. This has been best demonstrated by studies in which intracellular accumulation of cytokines inside helper T-cells in response to recombinant human alpha B-crystallin was examined. By using this technique [30, 31], acute antigen-specific cytokine responses were measured in peripheral T-cells without pre-culturing steps. At the same time, expression of CD4 and the CD45RO memory marker on responsive T-cells were assessed. The data consistently revealed that the anti-alpha B-crystallin T-cell response in humans is primarily pro-inflammatory and involves almost exclusively $CD4^+$ $CD45RO^+$ memory T-cells. This feature is found for healthy controls and MS patients alike.

Viral infection triggers bystander autoimmunity to alpha B-crystallin

As clarified above, a dramatic difference exists between antibody and T-cell reactivity to self alpha B-crystallin in humans as compared to other mammals. In a normal adult human immune repertoire, powerful pro-inflammatory responses exist while in rodents for example, such responses cannot even be induced with complete Freund's adjuvant. Importantly, the fact that in normal adult subjects we are dealing with an IgG and memory helper T-cell repertoire indicates that a naturally occurring event must have primed the human repertoire against alpha B-crystallin. Previous contact with alpha B-crystallin from exogenous sources could in principle explain this, but as already mentioned above, no alpha B-crystallin homologs exist in any bacterium or virus known to date. Also, several studies have indicated that the human response is rather specific for self alpha B-crystallin and frequently fails to cross-react even with highly homologous other mammalian alpha B-crystallins such as the bovine (98% identical) or murine (97% identical) homologs. The most likely source of the priming event is therefore the human body itself. What could it be?

A rather serendipitous event led us to examine the role of viral infection in this respect more closely. When using Epstein-Barr virus (EBV)-transformed cells as autologous antigen-presenting cells, we found unexpected spontaneous responses from specific T-cells lines against alpha B-crystallin even before antigen was added to the cultures. As it turned out, many EBV-infected and -transformed B-cell lines showed unexpectedly high expression of alpha B-crystallin. Follow-up studies revealed that also upon normal infection of B-cells with live EBV, *de novo* expression of alpha B-crystallin rapidly occurred. As the result of productive infection and lysis of B-cells, this could be shown to lead to MHC class II-restricted presentation of alpha B-crystallin to helper T-cells [28]. Also with other viruses such as human herpes virus-6, measles virus and influenza virus, some infection-induced expression could be detected in cultured lymphocytes. While EBV therefore may not be the only virus that produces this effect, it does so quite effectively and at least *in vitro* by far the strongest.

Although there is currently no direct evidence as yet, it seems likely that *de novo* expression of alpha B-crystallin in virus-infected human lymphocytes also occurs *in vivo*. This would provide a very plausible explanation for the presence of memory T-cells and IgG against self alpha B-crystallin in most humans. Priming against alpha B-crystallin would thus become the side effect of a natural viral infection such as with EBV. Natural EBV infection occurs in about 90% of the general population and, interestingly, in a full 100% of people with MS (reviewed in [5]).

How does a widely distributed self HSP play a role in organ-specific autoimmunity?

A final issue to be addressed in this review is the question how alpha B-crystallin could play a role in an organ-specific disease like MS. While alpha B-crystallin is certainly not ubiquitously expressed in humans, it is also not specifically expressed only by myelin-forming cells in the central nervous system. Also, as stressed several times in this review, the autoimmune T- and B-cell repertoire against this HSP is indistinguishable between healthy subjects and MS patients. Being neither organ specific in its expression, nor disease specific in its status as autoimmune target, how could the protein play a role as autoimmune target in MS?

The key consideration to understanding this issue is the notion that any antigen requires presentation by Class II MHC molecules as well as productive co-stimulation to be effective as T-cell activator. Cells and tissues that normally express alpha B-crystallin are non-lymphoid and they display no or very low levels of expression of Class II MHC molecules, and no detectable levels of co-stimulatory molecules. It is widely recognized that under such conditions, even non-self antigens remain silent since they are "invisible" to the immune system [32–34]. The visibility of alpha B-crystallin is even further diminished by the fact that the protein is tightly associated with insoluble intracellular cytoskeletal elements, preventing its release into the extracellular space that would allow access to the MHC Class II presentation pathway.

Thus, alpha B-crystallin that is routinely expressed in various human cells and organs will fail to trigger autoimmunity by being packed inside immunologically silent cells. In the CNS, however, a unique topological situation occurs that may change this. When alpha B-crystallin is produced by oligodendrocytes, it becomes part of the myelin sheaths that protect axons and thus, it effectively becomes part of the extracellular matrix. For reasons currently unknown, oligodendrocytes in MS brains sometimes produce at least ten-fold higher levels of alpha B-crystallin [20] and at the same time, the surrounding microglial cells as well as the blood-brain barrier endothelial cells become activated [35–37]. As this will lead to increased immune surveillance and influx of leukocytes, a dangerous mixture thus arises of antigen-presenting cells, T-cells, MHC molecules and co-stimulatory signals in a micro-environment where alpha B-crystallin is available at high levels in the extracellular space. It is this mixture that is more than likely to spark a pathogenic inflammatory reaction.

Concluding remarks

Above, we have clarified that in adult humans, a remarkably strong pro-inflammatory memory T-cell and antibody response exists to the small HSP alpha B-crystallin. *In vitro* data strongly suggest that primary infection of B-cells with EBV triggers *de novo* expression of alpha B-crystallin in the infected lymphoid tissue, and subsequent priming of human helper T-cells and antibody responses against alpha B-crystallin. Programmed for absence of alpha B-crystallin from healthy lymphoid tissues, human T- and B-cells will mistake the protein for another microbial antigen and they will treat it accordingly. Given the fact that viruses such as EBV are life-long persistent viruses that periodically become reactivated [38, 39], maintenance of this memory repertoire against alpha B-crystallin is certified. In itself, however, the alpha B-crystallin-reactive T-cell and antibody repertoire is not pathogenic since the target antigen remains invisible in healthy cells and tissues where it is expressed. Local pro-inflammatory factors in CNS white matter, however, including accumulation of extracellular alpha B-crystallin may in some cases and at some times lead to a spark for pathogenic inflammation, with the existing memory repertoire as the fuel for such inflammation to reach pathogenic levels. After all, the adult human immune repertoire is programmed to mistake alpha B-crystallin for a non-self microbial antigen.

The unusual tissue distribution of the small HSP alpha B-crystallin in humans therefore seems to have unusual consequences in turning it into a driving antigen in a disease that is also specific for humans: MS. Unfortunately, no animal model will ever be able to effectively mimic the immunology of alpha B-crystallin in humans and the final test for the above ideas will depend on experimental intervention in humans. The immunological and clinical response in MS patients to specific tolerance induction for alpha B-crystallin will be an interesting indicator for the validity of our ideas.

Acknowledgement
Our studies are supported by the Netherlands Foundation for the Support of MS Research.

References

1 Kurtzke JF and Hyllested K (1988) Validity of the epidemics of multiple sclerosis in the Faroe islands. *Neuroepidemiol* 7: 190–227
2 Gilden DH, Devlin ME, Burgoon MP and Owens GP (1996) The search for virus in multiple sclerosis brain. *Mult Scler* 2: 179–183
3 Gran B, Hemmer B, Vergelli M, McFarland HF,Martin R (1999) Molecular mimicry and

multiple sclerosis: degenerate T-cell recognition and the induction of autoimmunity. *Ann Neurol* 45: 559–567

4 Bujevac D, Flach HZ, Hop WC, Hijdra D, Laman J, Savelkoul HF, Van der Mechee FG, Van Doorn PA, Hintzen RQ (2002) Prospective study on the relationship between infections and multiple sclerosis exacerbations. *Brain* 125: 952–960

5 Munch M, Hvas J, Christensen T, Moller-Larsen A, Haahr S (1997) The implications of Epstein-Barr virus in multiple sclerosis – a review. *Acta Neurol Scand* 169: 59–64

6 Martin R, McFarland HF (1995) Immunological aspects of experimental allergic encephalomyelitis and multiple sclerosis. *Crit Rev Clin Lab Sci* 32: 121–182

7 Lassmann H, Raine CS, Antel J, Prineas JW (1998) Immunopathology of multiple sclerosis: report on an international meeting held at the Institute of Neurology of the University of Vienna. *J Neuroimmunol* 86: 213–217

8 Fujinami RS, Oldstone MB (1985) Amino acid homology between the encephalitogenic site of myelin basic protein and virus: mechanism for autoimmunity. *Science* 230: 1043–1045

9 Ausubel LJ, Kwan CK, Sette A, Kuchroo V, Hafler DA (1996) Complementary mutations in an antigenic peptide allow for cross-reactivity of autoreactive T-cell clones. *Proc Natl Acad Sci USA* 93: 15317–15322

10 Hemmer B, Vergelli M, Gran B, Ling N, Vonlon P, Pinilla C, Houghten R, McFarland HF, Martin R (1998) Predictable TCR antigen recognition based on peptide scans leads to identification of agonist ligands with no sequence homology. *J Immunol* 160: 3631–3636

11 Mason D (1998) A very high level of cross-reactivity is an essential feature of the T cell receptor. *Immunol Today* 19: 395–404

12 Loftus C, Huseby E, Gopau P, Beeson C, Goverman J (1999) Highly cross-reactive T-cell responses to myelin basic protein epitopes reveal a non-predictable form of TCR degeneracy. *J Immunol* 162: 6451–6457

13 Hiemstra HS, Van Veelen PA, Willemen SJ, Benckhuijsen WE, Geluk A, De Vries RR, Roep BO, Drijfhout JW (1999) Quantitative determination of TCR cross-reactivity using peptide libraries and protein databases. *Eur J Immunol* 29: 2385–2391

14 Miller SD, McRae BL, Vanderlugt CL, Nikcevich KM, Pope L, Karpus WJ (1995) Evolution of the T-cell repertoire during the course of experimental immune-mediated demyelinating disease. *Immunol Rev* 144: 225–244

15 McRae BL, Vanderlugt CL, Dal Canto MC, Miller SD (1995) Functional evidence for epitope spreading in the relapsing pathology of experimental allergic encephalomyelitis. *J Exp Med* 182: 75–82

16 Tuohy VK, Yu M, Yin L, Lkawcak JA, Kinkel RP (1999) Spontaneous regression of primary auto-reactivity during chronic progression of experimental autoimmune encephalomyelitis and multiple sclerosis. *J Exp Med* 189: 1033–1042

17 Yu M, Sohnson JM, Tuohy VK (1996) A predictable sequential determinant spreading cascade invariably accompanies progression of experimental autoimmune ence-

phalomyelitis: a basis for peptide-specific therapy after onset of clinical disease. *J Exp Med* 183: 1777–1788

18 Behan PO, Chaudhuri A, Roep BO (2002) The pathogenesis of multiple sclerosis revisited. *J R Coll Physicians Edinb* 32: 244–265

19 Van Noort JM, Van Sechel AC, Bajramovic JJ, El-Ouagmiri M, Polman CH, Lassmann H, Ravid R (1995) The small stress protein alpha B-crystallin as candidate autoantigen in multiple sclerosis. *Nature* 375: 798–801

20 Chabas D, Baranzini SE, Mitchell D, Bernard CC, Rittling SR, Denhardt DT, Sobel RA, Lock C, Karpuj M, Pedotti R et al (2001) The influence of the proinflammatory cytokine osteopontin in autoimmune demyelinating disease. *Science* 294: 1731–1735

21 Graw J (1997) The crystallins: genes, proteins and diseases. *J Biol Chem* 378: 1331–1348

22 Derham BK, Harding JJ (1999) Alpha-crystallin as a molecular chaperone. *Prog Ret Eye Res* 18: 463–509

23 Bajramovic JJ, Lassmann H, Van Noort JM (1997) Expression of alpha B-crystallin in glia cells during lesional development in multiple sclerosis. *J Neuroimmunol* 78: 143–151

24 Bajramovic JJ, Plomp AC, Van Der Goes A, Koevoets C, Newcombe J, Cuzner ML, Van Noort JM (2000) Presentation of alpha B-crystallin to T cells in active multiple sclerosis lesions: an early event following inflammatory demyelination. *J Immunol* 164: 4359–5366

25 Van Stipdonk MJB, Willems AA, Plomp AC, Van Noort JM, Boog CJP (1999) Tolerance controls encephalitogenicity of alphaB-crystallin in the Lewis rat. *J Neuroimmunol* 103: 103–111

26 Van Stipdonk MJB, Willems AA, Verbeek R, Boog CJP, Van Noort JM (2000) T- and B-cell non-responsiveness to self alpha B-crystallin in SJL mice prevents the induction of experimental allergic encephalomyelitis. *Cell Immunol* 204: 128–134

27 Thoua N, Van Noort JM, Baker D, Bose A, van Seche, AC, Van Stipdonk MJB, Travers PJ, Amor S (1999) Encephalitogenic and immunological potential of the stress protein alpha B-crystallin (peptide 1-16) in Biozzi ABH (H-2Adq1) mice. *J Neuroimmunol* 104: 47–57

28 Van Sechel AC, Bajramovic JJ, Van Stipdonk MJB, Persoon-Deen C, Geutskens SB, Van Noort JM (1999) EBV-induced expression and HLA-DR-restricted presentation by human B cells of alpha B-crystallin, a candidate autoantigen in multiple sclerosis. *J Immunol* 162: 129–135

29 Van Noort JM, Bajramovic JJ, Plomp AC, Van Stipdonk (2000) Mistaken self: a novel model that links microbial infections with myelin-directed autoimmunity in multiple sclerosis. *J Neuroimmunol* 105: 46–57

30 Picker LJ, Singh MK, Zdraveski Z, Treer JR, Waldrop SJ, Bergstresser PR, Maino VC (1995) Direct demonstration of cytokine synthesis heterogeneity among human memory/effector T cells by flow cytometry. *Blood* 86: 1408–1419

31 Waldrop SL, Pitcher CJ, Peterson D, Maino VC, Picker LJ (1997) Determination of anti-

gen-specific memory/effector CD4+ T cell frequencies by flow cytometry. *J Clin Invest* 99: 1739–1750

32 Goverman J, Woods A, Larson L, Weiner LP Hood L, Zaller DM (1993) Transgenic mice that express a myelin basic protein-specific T cell receptor develop spontaneous autoimmunity. *Cell* 26: 551–560

33 Matzinger P (1994) Tolerance, danger and the extended family. *Ann Rev Immunol* 12: 991–1045

34 Von Herrath MG, Guerder S, Lewicki H, Flavell RA, Oldstone MB (1995) Co-expression of B7-1 and viral ("self") transgenes in pancreatic beta cells can break peripheral ignorance and lead to spontaneous autoimmune diabetes. *Immunity* 3: 727–738

35 Li H, Cuzner ML, Newcombe J (1996) Microglia-derived macrophages in early multiple sclerosis plaques. *Neuropathol Appl Neurobiol* 22: 207–215

36 De Groot CJ, Bergers E, Kamphorst W, Ravid R, Polman CH, Barkhof F, Van der Valk P (2001) Post-mortem MRI-guided sampling of multiple sclerosis brain lesions: increasing yield of active demyelinating and (pr)reactive lesions. *Brain* 124: 1635–1645

37 Markovic-Plese S, McFarland HF (2001) Immunopathogenesis of the multiple sclerosis lesion. *Curr Neurol Neurosci Rep* 1: 257–262

38 Schwarzmann F, Jager M, Hornef M, Prang N, Wolf H (1998) Epstein-Barr viral gene expression in B-lymphocytes. *Leuk Lymphoma* 30: 123–129

39 Speck SH, Chatila T, Flemington E (1997) Reactivation of Epstein-Barr virus: regulation and function of the BZLF1 gene. *Trends Microbiol* 5: 399–405

HSP60-peptide interference with CD94/NKG2 receptors

Kalle Söderström

In Silico R&D, Entelos Inc., 110 Marsh Drive, Foster City, CA 94404, USA

Introduction

An efficient and sufficient inflammatory response mediated by cells belonging to both the innate and adaptive immune system, is crucial in order to clear infection and allow for tissue repair after damage. Natural killer (NK) cells are innate type of lymphocytes, which together with activated $CD8^+$ T-cells mediates immune responses against many pathogens and tumors. These lymphocytes can survey tissues looking for evidence that a cell has been altered and thereby prevent pathogen invasion and/or tumor growth and metastasis. Upon encountering an altered cell, they have the option to either kill the cell directly through release of lytic granules and/or to produce cytokines. Rapidly accumulating data show that certain MHC Class I specific cell-surface receptors, known previously to critically regulate NK cell functions, appear on large numbers of $CD8^+$ T-cells after activation [1–3].

In this chapter, I will focus on the role of CD94/NKG2 receptors, which are constitutively expressed on most resting and activated NK cells, and induced on subsets of activated T-cells. A description of a novel immune strategy for detection of stressed cells *via* peptide-dependent interference with CD94/NKG2-receptors will also be introduced. Potentially, these findings may result in novel peptide-based strategies to treat infection, cancer and autoimmune diseases.

MHC specific NK cell receptor families

The balanced action of positive and negative regulatory molecules determines the functional outcome of NK cell interaction with normal or "aberrant" cells. MHC Class I molecules are potent ligands for NK cell receptors (NKR) that negatively regulate their response towards normal cells. The loss – or down-regulation of one or several MHC Class I molecules, which is common during certain viral infections and neoplastic transformation – may result in NK cell activation, provided that the NK cells simultaneously receive sufficient stimulation. Such stimulation can be received

Heat Shock Proteins and Inflammation, edited by Willem van Eden

via activating NK cell receptors, cytokine receptors, adhesion molecules and sufficient interaction between activating receptors and their ligands expressed on the surface of tumors or pathogen-infected cells [4–6].

In humans, at least two structurally distinct groups of HLA-Class I specific receptors have been described [7, 8]. The first consists of C-type lectin-like receptors that include heterodimers composed of a common subunit (CD94) covalently associated with a distinct NKG2 chain, either inhibitory NKG2A, or activating forms NKG2C and NKG2E. The second group consists of immunoglobulin (Ig) domain-like receptors, including killer cell Ig-like receptors (KIRs), leukocyte Ig-like receptors (LIRs *a.k.a.* Ig-like transcripts, ILT) [9, 10]. In most cases, the ligands for the KIRs and LIRs are classical HLA Class I molecules, whereas the CD94/NKG2 receptors – which are conserved between human and mice – recognize non-classical HLA-E (or its murine ortholog Qa-1b) molecules when loaded with a nonapeptide derived from the leader sequence of certain HLA-A, -B, -C or -G heavy chains [11–13]. Likewise, mouse CD94/NKG2 receptors bind to Qa-1b presenting a similar peptide, termed qdm, which is derived from the leader sequence of H2D or H2L [14–16]. The CD94/NKG2 receptor approach to recognize HLA-E displaying a HLA Class I-leader peptide thereby allow NK cell subsets to survey a broader range of HLA-Class I molecules, whereas each KIR recognizes a group of HLA "allotypes" which each are capable of binding a large set of peptides. Thereby, KIR and CD94/NKG2 receptors complement each other in the surveillance of "aberrant" cells. It should be mentioned that although there are reports of peptide selective recognition by KIR the reason for this is still elusive [17–21].

CD94/NKG2A is expressed at bright levels on certain NK cell subsets

Up to 15% of all circulating lymphocytes in healthy individuals are NK cells. They can also be found in lymph nodes, bone marrow as well as in peripheral tissues such as decidua, liver, skin and intestine [22–24]. Freshly isolated NK cells can be broadly divided into two major subsets based on the cell surface levels of the CD56 marker, the $CD56^{dim}$ and the $CD56^{bright}$ populations [25–26]. The $CD56^{dim}$ population comprises approximately 90% of the NK cells in peripheral blood, whereas $CD56^{bright}$ cells seem to predominate in lymph nodes and peripheral tissues. Generally, the $CD56^{bright}$ population in peripheral blood expresses few and low levels KIRs and bright levels of CD94/NKG2A. In contrast, the $CD56^{dim}$ population expresses higher levels of KIRs, but lower levels of CD94/NKG2A [27]. The phenotypic division between these two NK subsets is associated with different effector functions. Resting $CD56^{dim}$ NK cells have a higher cytotoxic potential against most tumor targets. In contrast, the $CD56^{bright}$ population has the ability to produce impressive amounts of cytokines (IFN-γ, TNF-β, GM-CSF, IL-13) following direct stimulation by certain combinations of monocyte/DC derived cytokines (e.g. IL-1,

IL-12, IL-15, IL-18) or T-cell secreted IL-2 [26, 28, 29]. The production of cytokines by resting and/or activated NK cell subsets is also regulated by activating and inhibitory NKR signaling during cell-contact interactions, but there is much more to learn about how this is regulated and influenced by the local cytokine environment [30–39].

$CD56^{bright}$ NK cells might have unique migratory properties based on their expression of cell surface adhesion molecules and chemokine receptors [25, 40, 41], and the presence of this subset in human lymph nodes suggest that they play an active role in the skewing of adaptive immune responses [29]. As the $CD56^{bright}$ NK cell subset is also capable of producing IL-10 and TGF-β, these cells may also play an active part in the down-modulation of inflammatory responses [26]. Besides being present in lymph nodes, $CD56^{bright}$ NK cell also represents the major NK subset present in the synovial fluid of patients with arthritis [42, and our unpublished observation] where up to 20% of all lymphocytes are NK cells. These synovial-NK cells also express high levels of CD94/NKG2A and low levels of KIRs. In this respect, synovial-NK cells are phenotypically related to the small peripheral blood $CD56^{bright}$ population and the lymph nodes subset. Based on these findings it is likely that HLA-E play a central role in the functional regulation of lymph node-NK cells and synovial-NK cells. Besides its function to maintain tolerance in NK cells, recent data in mice show that expression of high levels of CD94/NKG2A on NK cells correlate with lower level of apoptosis [43], which may suggest that repetitive HLA-E ligation is necessary to maintain receptor expression and survival of these cells.

Like classical HLA Class I molecules, HLA-E is expressed in most tissues and on most cell types, including circulating leukocytes [13, 44]. It should be noted, however, that the expression of HLA-E on cells in the synovial tissue has not yet been assessed and the existence of this synovial-NK subset has so far only been reported in the synovial fluid.

CD94/NKG2 receptors are induced on $CD8^+$ CTLs and inhibit anti-viral and anti-tumoral responses

It is becoming increasingly apparent that upon activation conventional $CD8^+$ T-cells can express CD94/NKG2A, which regulates their response against viral infections and tumors [45–47]. Viral infections also up-regulate HLA-E mRNA in antigen-presenting cells [48, 49]. The CD94/NKG2A receptor can persist on T-cell subsets long after pathogen clearance, but stable expressions, however, seem to mainly depend on whether the pathogen is capable of establishing a persistent infection (e.g., polyoma virus). The CD94/NKG2A expression wanes among virus-specific memory T-cells in mice that are infected with virus that are acutely cleared, such as LCMV and influenza [45–47]. Perhaps repetitive antigenic stimulation by either viral or cross-reactive self-antigens may help to maintain CD94/NKG2A expression, which may be cytokine dependent and/or influenced by the particular cell type presenting the antigenic epitopes. Interestingly, in HTLV-1 infected patients clonally expanded $CD8^+$ T-cells expressing CD94/NKG2A receptors are not those with TCR-specificity against the immunodominant HTLV-1 antigen, suggesting that CD94/NKG2A appear mainly on T-cells with TCRs specificity for subdominant antigens [50]. Low frequency of $CD94/NKG2A^+$ T-cells in these patients was associated increased risk for inflammatory disorders associated with HTLV-1 infections suggesting that these cells may help to control immunopathologies associated with infections [50].

CD94/NKG2A expression is also found on a high proportion of $CD8^+$ T-cells infiltrating human tumors, such as melanoma and astrocytoma [51–54]. The receptor can restrain successful anti-tumor CTL responses, as shown *in vitro* by anti-CD94 antibody mediated masking on melanoma-specific T-cells [55].

Interesting new data show that only certain TCR specificities determines whether $CD8^+$ T-cell subsets are committed to express high levels of inhibitory CD94/NKG2A receptors after TCR engagement [56]. Most, but not all $NKG2A^+$ T-cells also expressed NKG2C mRNA, and cross-linking CD94 showed that the inhibitory NKG2A form was dominant when expressed. However, subsets of T-cells apparently express activating NKG2 mRNA in the absence of NKG2A, and anti-CD94 cross-linking on such cells provided a marked co-stimulation to TCR-mediated lysis [56, 57]. Furthermore, $CD94/NKG2C^+$ but NKG2A- T-cells were totally refractory for NKG2A induction [56]. Taken together, it appears that distinct $CD8^+$ T-cells subsets exist which are committed to either a dominant NKG2-inhibitory or NKG2-activating pathway, which in the case of NKG2A committed T-cells appear to be dictated by their TCR specificity. Finally, it should be noted that other lymphocyte subsets express CD94/NKG2 receptors. For example, constitutive expression is found on the majority of γδ T-cells [58], and conventional mouse and human $CD4^+$ T-cells can up-regulate CD94/NKG2 receptors after activation [59, 60].

HSP60 peptide interference with CD94/NKG2 receptors

Heat shock protein 60 (HSP60) is a highly conserved protein present in cells of virtually all living organisms where it performs important roles in various cellular processes [61, 62]. In eukaryotic cells, it is exported to the mitochondria where it assists in folding or refolding of proteins. Up to 90% of the total HSP60 pool is found within the mitochondria in healthy cells [63], and a mitochondrial targeting sequence is responsible for directing HSP60 to this location [64]. An increased production and altered distribution of HSP60 is observed when cells are subjected to various forms of stress such as metabolic changes, viral and bacterial infections, and increased temperatures [65]. Despite its strong conservation, HSP60 is known as one of the major antigens recognized during a wide variety of bacterial and parasitic diseases [66]. Even an inflammatory response itself without apparent infection, activate T- and B-cells that carry receptors for self-HSP60 derived epitopes [67]. Recent evidence shows a physiological role of self-HSP60 responses connected to immunoregulatory mechanisms, and it is becoming increasingly apparent that the functional properties of the T-cells activated against certain HSP60-peptides is as critical as their antigen specificity [68, 69]. For example, a strong proliferative T-cell response to self-HSP60 was shown to correlate with a good clinical prognosis in newly diagnosed patients with juvenile RA [70]. Moreover, prior HSP60 vaccination with either whole HSP60 protein, HSP60 DNA, or HSP60 derived peptides can either protect from induction of an experimental autoimmune disease [70–75], or halt the progression of Type I diabetes in human [76]. This may suggest that episodes of acute and sufficient HSP60 autoimmunity may efficiently activate regulatory adaptive immune mechanisms that can subsequently combat the effects of an otherwise destructive autoimmune process.

Twelve years ago, Imani and Soloski showed that the cell surface levels of Qa-1b, the mouse HLA-E homologue, could be up-regulated during heat stress [77]. Recently, the same group showed that Qa-1b is also able to present conserved HSP60 peptides of either mouse or bacterial HSP60 origin [78]. Based on these findings, we reasoned that stress-induced expression of endogenous HSP60 could result in the generation of peptides that gain access to the HLA-E presentation pathway. By searching the full-length human HSP60 sequence we identified four potential HLA-E binding peptides [79]. Interestingly, one of the peptides is located in the mitochondrial targeting sequence of human HSP60 (a similar peptide is found within the mouse HSP60), and this peptide as well as one of the other peptides was found to bind HLA-E. Our *in vitro* experiments demonstrated that the HSP60 signal peptide (HSP60sp) could gain access to HLA-E intracellularly, in particular when cells are subjected to increased cellular stress [79]. Although increased HLA-E cell surface levels on stressed cells parallels increased NK-cell cytotoxicity, direct binding assays demonstrated that soluble HLA-E/HSP60sp tetramers did not engage either the inhibitory CD94/NKG2A or the activating CD94/NKG2C receptors. Importantly, although increased susceptibility towards NK-cell mediated lysis was observed with stressed target cells, these could efficiently be rendered resistant by simply adding a protective HLA-E binding peptide to the assay medium. Moreover, HSP60sp could effectively compete for HLA-E binding with an HLA-Class I signal peptide on non-stressed target cells. Our *in vitro* findings therefore support the notion that HSP60sp can gain access to HLA-E *in vivo*, which raises fundamental and important questions about the role of mitochondria in stressed, but non-apoptotic cells, during activation of immune responses, and about the mechanisms responsible for protein translocation from mitochondria to other intracellular compartments.

A greater fraction of such HLA-E/HSP60sp complexes are likely to be expressed on stressed cells *in vivo*, and the balance between protective and non-protective HLA-E molecules would thereby allow CD94/NKG2A+ NK cells to gradually differentiate between "normal" and "abnormal" cells in a peptide-dependent manner (Fig. 1). This could be of particular importance not only for the $CD56^{bright}$ CD94/NKG2A+ NK cell subset present in lymph nodes and in the inflamed joint, but also for the subsets of activated T-cells that expresses this receptor. Hypothetically, HLA-E presenting stress-induced peptides may be involved in the fine-tuning of NK and T-cell-mediated responses during infections and inflammatory responses without necessarily involving a pathogen-induced cellular change. Efficient uncoupling of CD94/NKG2A receptors would most likely not only modulate NK cell-mediated responses, but also lower the threshold for TCR-mediated activation.

Recent data provided by Moser et al., demonstrate that effector CTLs against viral antigens became restrained through expression of CD94/NKG2A receptors, which inhibited both proliferation and lysis, with a dramatic influence of acute infection and oncogenesis by polyoma virus in mice [46]. The HLA-E loading of stress-peptides, such as HSP60sp, that is not only induced in stressed cells but also interferes with the protection normally conferred *via* CD94/NKGA receptor recognition of HLA-E, provides an explanation for the role of this receptor during the regulation of T-cell responses. The CD94/NKG2A receptor thereby complements the TCR pathway to help discrimination between healthy or sick cells, not only by surveillance of reduced proportion of MHC Class I molecules but also the increased accessibility to HLA-E of stress-induced peptides. This mechanism may thereby restore lysis by subsets of restrained CD94/NKG2A+ T-cells against low amounts of virus- or tumor derived peptide:MHC complexes and, in addition, potentially break T-cell tolerance against over-expressed tumor-antigens. Therefore a local increase of non-protective stress-peptides competing for HLA-E binding might allow for more efficient surveillance of persistent virus and tumor cells, without causing immunopathologies as the dominant inhibitory CD94/NKG2A pathway would ensure tolerance to healthy cells expressing sufficient levels of protective HLA-E in the surrounding. Further studies to address this issue should first test whether CD94/NKG2A expressing T-cells can be influenced by stressed induced changes in target cells.

IFN-γ is a key cytokine in the response against virus and tumors. A central role played by IFN-γ is its capacity to up-regulate MHC and thereby enhance antigen presentation to T-cells. IFN-γ induces not only classical HLA Class I molecules but also HLA-E itself [80], which will have an abundant intracellular supply of HLA-leader peptides, likely leading to increased cell surface levels of protective HLA-E complexes. Therefore, HLA-E up-regulation induced by IFN-γ could play an important feed back mechanism during infection by restraining excessive NK and CTL effector functions. The fine-tuning of early IFN-γ release may critically determine whether a virus will be either successfully cleared, or if it may establish persistence *via* premature dampening of anti-viral responses. This possibility is supported by the

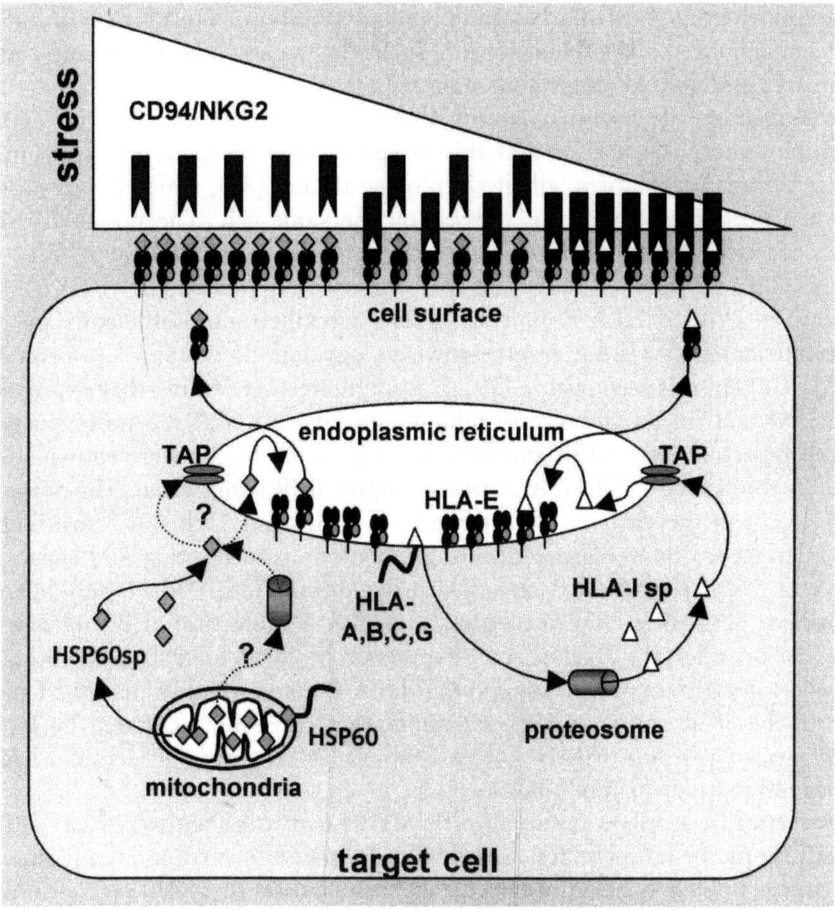

Figure 1
Inside the endoplasmic reticulum during non-stressful conditions nascent HLA-E molecules refold around HLA-leader peptides and then migrate to the cell surface for presentation to CD94/NKG2 receptors [11] (right part of the figure). During increased cellular stress, competitive peptide replacement in HLA-E can occur via *increased access to HSP60 signal peptides [79] (left part). High levels of HLA-E loaded with HSP60 signal peptides appear on the cell surface of stressed cells, which interfere with CD94/NKG2 receptor recognition.*

finding that IFN-γ can facilitate hCMV persistence [81], perhaps partially due to IFN-γ mediated up-regulation of HLA-E loaded with a peptide derived from hCMV-UL40 as a mechanism responsible for evasion from NK cell attack [80, 82]. In addition, IFN-γ may assist ovarian tumor cells to evade CTL detection [83]. It was shown that short term ovarian tumor lines can express HLA-G, which carries a leader peptide with a particularly strong binding affinity for HLA-E [84] and IFN-γ treatment enhanced HLA-E and HLA-G levels dampening autologous anti-tumor CTLs in a CD94/NKG2A dependent manner [83].

The biological role of the activating CD94/NKG2 receptors is not yet clearly defined. However, it is known that this receptor binds the same HLA-E/peptide ligand as the inhibitory form, albeit with lower affinity [84], which ensures tolerance when both receptors are expressed on the same cell. The possibility that HLA-E presenting stress-induced peptides binds with higher affinity to CD94/NKG2C, or another activating receptor, as compared to CD94/NKG2A is appealing, as stressed HLA-E expressing cells are killed more efficiently by NK cells despite increased HLA-E levels. However, our data do not support a role for CD94/NKG2C in this recognition [79]. It is an interesting finding that expression of CD94/NKG2C in the absence of inhibitory CD94/NKG2A receptors exists in the T-cell repertoire in healthy individuals [56]. These T-cells were shown to be completely refractory to NKG2A expression upon TCR stimulation. The presence of such T-cells in the synovial fluid of patients with acute forms of arthritis that are often triggered or exacerbated during infections is interesting and suggest a unique role for CD94/NKG2C during such conditions [85]. As the only ligands described for activating CD94 receptors are HLA-E molecules in complex with HLA leader peptides, the T-cell surface expression of CD94 activating receptors in the absence of inhibitory ones, suggest that HLA-E may provide a potent co-stimulation to the TCR-pathway. Such co-stimulation could be facilitated by IFN-γ dependent up-regulation of MHC Class I molecules providing an abundant supply of leader peptides for HLA-E.

Hypothetically, T-cells committed to the CD94-activating pathway carry TCRs with low affinity for self-peptides, as these T-cells otherwise may induce widespread tissue damage upon activation. Alternatively, these subsets carry self-reactive TCRs but are restricted to only certain tissues [e.g., gut-IEL], in which the particular self-peptides may not be expressed, or efficiently presented on APC in the tissue environment. However, recent data shows that CD94/NKG2C$^+$, but NKG2A-, T-cell clones can be derived from peripheral blood of healthy individuals [56]. *In vitro* experiments established a potent CD94-dependent co-stimulation to TCR-mediated cytotoxicity. Therefore, a plausible explanation for the existence of such T-cells is that they carry TCRs against self-antigens that are not normally displayed in the periphery. In this respect, it is interesting to note that during acute measles virus infection, activation of IFN-γ producing T-cells specific for abundant – but not normally displayed – self-peptides presented by HLA-A*0201 can be detected [86]. These cells were detected at low levels in the T-cell repertoire, but were functionally silent in the absence of infection. The precise mechanism how these peptides gain access to HLA-A*0201 and displayed on the cell surface upon infection is unclear. In light of the emerging evidence that CD94/NKG2 receptors appear on CD8$^+$ T-cells subsets early after infection, it is tempting to speculate that a distinct anti-viral response is driven by the T-cell population that is committed to the activating-CD94 pathway. Early IFN-γ release during virus infection, leading to enhanced HLA-E cell surface expression on infected- and bystander cells, could provide potent co-stimulation to TCRs expressed on $CD94/NKG2C^+$ but $CD94/NKG2A^-$ T-cells. Massive display of normally hidden self-peptides on polymorphic HLA Class I facilitated by IFN-γ could, however, result in immunopathology.

In summary, it is becoming increasingly apparent that not only IFN-γ induction of polymorphic HLA molecules, but also non-classical HLA-E molecules may play a fundamental role in the orchestration of immune responses against virus and tumors. Our recent data show that endogenous HSP60sp and mutated HLA Class I-leader peptides can efficiently compete with protective HLA-E peptides and gradually decrease the threshold for NK cell activation *in vitro* [79]. Possibly, such peptides may similarly act to decrease the threshold for $CD94/NKG2A^+$ T-cell activation. In contrast, similar peptides may instead gradually increase the threshold for activation by cells expressing activating CD94/NKG2 receptors in the absence of inhibitory receptors, which could limit immunopathologies associated with inflammatory diseases that are triggered or exacerbated by infections. It should, however, be clarified whether HLA-E in complex with HLA-leader peptides, or IFN-γ stimulation, may act to further enhance responses mediated by CD94/NKG2C expressing T-cells, and whether competition by HSP60sp in stressed cells may restrain their activation. Taken together, peptide presentation by HLA-E may critically fine tune inflammatory responses mediated by subsets of NK and T-cells, and a dysregulated balance between various HLA-E:peptide complexes during infection may act to either prematurely restrain or alternatively, unleash CD94/NKG2-dependent immune responses perhaps ultimately leading to a persistent disease.

Despite limited polymorphism in the HLA-E locus, Type I diabetes appears to be associated with a certain HLA-E allele [87]. In addition, a subset of HLA-DRB1* 0404 haplotypes (a known genetic risk factor for RA) contain a common set of mini-haplotypes in the Class I region, which confers an additional risk independent of the HLA-DRB1*0404 allele. This region contains ~12 functional genes, and includes HLA-E [88]. In the arthritic joint of most patients, there is a drastic accumulation of CD94/NKG2A expressing NK cells, but the presence of NK cells expressing a similar phenotype in the pancreas of Type I diabetic patients has not been reported.

These findings mentioned above raise further interesting questions. Can persistent and/or acute viral infections, tumors, and autoimmune diseases be treated with distinct HLA-E binding peptides? In any event, the potential to competitively displace HLA-E bound peptides *in vitro* exist and could lead to new therapeutics to treat a wide array of human diseases. Experiments in animal models will hopefully shed light on how CD94/NKG2 receptors bridge the link between innate and adaptive immune responses in health and disease.

Acknowledgements
The authors' previous address was Microbiology Tumorbiology Center, Karolinska Institutet, Box 280, S-171 77 Stockholm, Sweden. This work was performed at Karolinska Institutet supported by grants from the Swedish Cancer Society, Gustav V 80th Jubilee fund, Ulla and Gustaf af Ugglas' fund. The author is grateful for helpful discussions with Jakob Michaelsson, Cristina Teixeira de Matos, Adnane Achour, Lewis Lanier and Klas Kärre.

References

1 Raulet DH, Vance RE, McMahon CW (2001) Regulation of the natural killer cell receptor repertoire. *Annu Rev Immunol* 19: 291–330
2 Moser JM, Byers AM, Lukacher AE (2002) NK cell receptors in antiviral immunity. *Curr Opin Immunol* 14: 509–516
3 Anfossi N, Pascal V, Vivier E, Ugolini S (2001) Biology of T memory type 1 cells. *Immunol Rev* 181: 269–278
4 Ljunggren HG, Kärre K (1990) In search of the "missing self": MHC molecules and NK cell recognition. *Immunol Today* 11: 237–244
5 Cerwenka A, Lanier LL (2001) Natural killer cells, viruses and cancer. *Nat Rev Immunol* 1: 41–49
6 Biassoni R, Cantoni C, Pende D, Sivori S, Parolini S, Vitale M, Bottino C, Moretta A (2001) Human natural killer cell receptors and co-receptors. *Immunol Rev* 181: 203–214
7 Lanier LL (1998) NK cell receptors. *Annu Rev Immunol* 16: 359–393
8 Boyington JC, Brooks AG, Sun PD (2001) Structure of killer cell immunoglobulin-like receptors and their recognition of the class I MHC molecules. *Immunol Rev* 181: 66–78
9 Colonna M, Nakajima H, Navarro F, Lopez-Botet M (1999) A novel family of Ig-like receptors for HLA class I molecules that modulate function of lymphoid and myeloid cells. *J Leukoc Biol* 66: 375–381
10 Cosman D, Fanger N, Borges L, Kubin M, Chin W, Peterson L, Hsu ML (1997) A novel immunoglobulin superfamily receptor for cellular and viral MHC class I molecules. *Immunity* 7: 273–282
11 Braud VM, Allan DS, O'Callaghan CA, Söderström K, D'Andrea A, Ogg GS, Lazetic S, Young NT, Bell JI, Phillips JH et al (1998) HLA-E binds to natural killer cell receptors CD94/NKG2A, B and C. *Nature* 391: 795–799
12 Borrego F, Ulbrecht M, Weiss EH, Coligan JE, Brooks AG (1998) Recognition of human histocompatibility leukocyte antigen (HLA)-E complexed with HLA class I signal sequence-derived peptides by CD94/NKG2 confers protection from natural killer cell-mediated lysis. *J Exp Med* 2: 187: 813–818
13 Lee N, Llano M, Carretero M, Ishitani A, Navarro F, Lopez-Botet M, Geraghty DE

(1998) HLA-E is a major ligand for the natural killer inhibitory receptor CD94/NKG2A. *Proc Natl Acad Sci USA* 95: 5199–5204
14 DeCloux A, Woods AS, Cotter RJ, Soloski MJ, Forman J (1997) Dominance of a single peptide bound to the class I(B) molecule, Qa-1b. *J Immunol* 158: 2183–2191
15 Vance RE, Kraft JR, Altman JD, Jensen PE, Raulet DH (1998) Mouse CD94/NKG2A is a natural killer cell receptor for the non-classical major histocompatibility complex (MHC) class I molecule Qa-1(b). *J Exp Med* 188: 1841–1848
16 Kraft JR, Vance RE, Pohl J, Martin AM, Raulet DH, Jensen PE (2000) Analysis of Qa-1(b) peptide binding specificity and the capacity of CD94/NKG2A to discriminate between Qa-1-peptide complexes. *J Exp Med* 192: 613–624
17 Peruzzi M, Parker KC, Long EO, Malnati MS (1996) Peptide sequence requirements for the recognition of HLA-B*2705 by specific natural killer cells. *J Immunol* 157: 3350–3356
18 Peruzzi M, Wagtmann N, Long EO (1996) A p70 killer cell inhibitory receptor specific for several HLA-B allotypes discriminates among peptides bound to HLA-B*2705. *J Exp Med* 184: 1585–1590
19 Malnati MS, Peruzzi M, Parker KC, Biddison WE, Ciccone E, Moretta A, Long EO (1995) Peptide specificity in the recognition of MHC class I by natural killer cell clones. *Science* 267: 1016–1018
20 Rajagopalan S, Long EO (1997) The direct binding of a p58 killer cell inhibitory receptor to human histocompatibility leukocyte antigen (HLA)-Cw4 exhibits peptide selectivity. *J Exp Med* 185: 1523–1528
21 Mandelboim O, Wilson SB, Vales-Gomez M, Reyburn HT, Strominger JL (1997) Self and viral peptides can initiate lysis by autologous natural killer cells. *Proc Natl Acad Sci USA* 94: 4604–4609
22 King A, Loke YW (1991) On the nature and function of human uterine granular lymphocytes. *Immunol Today* 12: 432–435
23 Hata K, Zhang XR, Iwatsuki S, Van Thiel DH, Herberman RB, Whiteside TL (1990) Isolation, phenotyping, and functional analysis of lymphocytes from human liver. *Clin Immunol Immunopathol* 56: 401–419
24 Buentke E, Heffler LC, Wilson JL, Wallin RP, Lofman C, Chambers BJ, Ljunggren HG, Scheynius A (2002) Natural killer and dendritic cell contact in lesional atopic dermatitis skin – malassezia-influenced cell interaction. *J Invest Dermatol* 119: 850–857
25 Sedlmayr P, Schallhammer L, Hammer A, Wilders-Truschnig M, Wintersteiger R, Dohr G (1996) Differential phenotypic properties of human peripheral blood CD56dim+ and CD56bright+ natural killer cell subpopulations. *Int Arch Allergy Immunol* 110: 308–313
26 Cooper MA, Fehniger TA, Turner SC, Chen KS, Ghaheri BA, Ghayur T, Carson WE, Caligiuri MA (2001) Human natural killer cells: a unique innate immunoregulatory role for the CD56(bright) subset. *Blood* 97: 3146–3151
27 Jacobs R, Hintzen G, Kemper A, Beul K, Kempf S, Behrens G, Sykora KW, Schmidt RE

(2001) CD56 bright cells differ in their KIR repertoire and cytotoxic features from CD56 dim NK cells. *Eur J Immunol* 31: 3121–3127
28 Cooper MA, Fehniger TA, Ponnappan A, Mehta V, Wewers MD, Caligiuri MA (2001) Interleukin-1beta co-stimulates interferon-gamma production by human natural killer cells. *Eur J Immunol* 31: 792–801
29 Fehniger TA, Cooper MA, Nuovo GJ, Cella M, Facchetti F, Colonna M, Caligiuri MA (2003) CD56 bright natural killer cells are present in human lymph nodes and are activated by T cell derived IL-2: a potential new link between adaptive and innate immunity. *Blood* 101: 3052–3057
30 Mandelboim O, Kent S, Davis DM, Wilson SB, Okazaki T, Jackson R, Hafler D, Strominger JL (1998) Natural killer activating receptors trigger interferon gamma secretion from T cells and natural killer cells. *Proc Natl Acad Sci USA* 95: 3798–3803
31 Voss SD, Daley J, Ritz J, Robertson MJ (1998) Participation of the CD94 receptor complex in costimulation of human natural killer cells. *J Immunol* 160: 1618–1626
32 Perez-Villar JJ, Melero I, Rodriguez A, Carretero M, Aramburu J, Sivori S, Orengo AM, Moretta A, Lopez-Botet M (1995) Functional ambivalence of the Kp43 (CD94) NK cell-associated surface antigen. *J Immunol* 154: 5779–5788
33 Aramburu J, Balboa MA, Rodriguez A, Melero I, Alonso M, Alonso JL, Lopez-Botet M (1993) Stimulation of IL-2-activated natural killer cells through the Kp43 surface antigen up-regulates TNF-alpha production involving the LFA-1 integrin. *J Immunol* 151: 3420–3429
34 Takahashi K, Miyake S, Kondo T, Terao K, Hatakenaka M, Hashimoto S, Yamamura T (2001) Natural killer type 2 bias in remission of multiple sclerosis. *J Clin Invest* 107: R23–29
35 Warren HS, Kinnear BF, Phillips JH, Lanier LL (1995) Production of IL-5 by human NK cells and regulation of IL-5 secretion by IL-4, IL-10, and IL-12. *J Immunol* 154: 5144–5152
36 Rajagopalan S, Fu J, Long EO (2001) Cutting edge: induction of IFN-gamma production but not cytotoxicity by the killer cell Ig-like receptor KIR2DL4 (CD158d) in resting NK cells. *J Immunol* 167: 1877–1881
37 Piccioli D, Sbrana S, Melandri E, Valiante NM (2002) Contact-dependent stimulation and inhibition of dendritic cells by natural killer cells. *J Exp Med* 195: 335–341
38 Gerosa F, Baldani-Guerra B, Nisii C, Marchesini V, Carra G, Trinchieri G (2002) Reciprocal activating interaction between natural killer cells and dendritic cells. *J Exp Med* 195: 327–333
39 Ferlazzo G, Tsang ML, Moretta L, Melioli G, Steinman RM, Munz C (2002) Human dendritic cells activate resting natural killer (NK) cells and are recognized *via* the NKp30 receptor by activated NK cells. *J Exp Med* 195: 343–351
40 Campbell JJ, Qin S, Unutmaz D, Soler D, Murphy KE, Hodge MR, Wu L, Butcher EC (2001) Unique subpopulations of $CD56^+$ NK and NK-T peripheral blood lymphocytes identified by chemokine receptor expression repertoire. *J Immunol* 166: 6477–6482
41 Frey M, Packianathan NB, Fehniger TA, Ross ME, Wang WC, Stewart CC, Caligiuri

MA, Evans SS (1998) Differential expression and function of L-selectin on CD56^{bright} and CD56^{dim} natural killer cell subsets. *J Immunol* 161: 400–408
42 Dalbeth N, Callan MF (2002) A subset of natural killer cells is greatly expanded within inflamed joints. *Arthritis Rheum* 46: 1763–1772
43 Gunturi A, Berg RE, Forman J (2003) Preferential survival of CD8 T and NK Cells expressing high levels of CD94. *J Immunol* 170: 1737–1745
44 Koller BH, Geraghty DE, Shimizu Y, DeMars R, Orr HT (1988) HLA-E. A novel HLA class I gene expressed in resting T lymphocytes. *J Immunol* 141: 897–904
45 Moser JM, Byers AM, Lukacher AE (2002) NK cell receptors in antiviral immunity. *Curr Opin Immunol* 14: 509–516
46 Moser JM, Gibbs J, Jensen PE, Lukacher AE (2002) CD94-NKG2A receptors regulate antiviral CD8(+) T cell responses. *Nat Immunol* 3: 189–195
47 McMahon CW, Zajac AJ, Jamieson AM, Corral L, Hammer GE, Ahmed R, Raulet DH (2002) Viral and bacterial infections induce expression of multiple NK cell receptors in responding CD8(+) T cells. *J Immunol* 169: 1444–1452
48 Zhu H, Cong JP, Mamtora G, Gingeras T, Shenk T (1998) Cellular gene expression altered by human cytomegalovirus: global monitoring with oligonucleotide arrays. *Proc Natl Acad Sci USA* 95: 14470–14475
49 Huang Q, Liu D, Majewski P, Schulte LC, Korn JM, Young RA, Lander ES, Hacohen N (2001) The plasticity of dendritic cell responses to pathogens and their components. *Science* 294: 870–875
50 Saito M, Braud VM, Goon P, Hanon E, Taylor GP, Saito A, Eiraku N, Tanaka Y, Usuku K, Weber JN et al (2003) Low frequency of CD94/NKG2A-positive T lymphocytes in HTLV-1 associated myelopathy/tropical spastic paraparesis patients but not in asymptomatic carriers. *Blood* 102: 577–584
51 Pedersen LO, Vetter CS, Mingari MC, Andersen MH, Thor Straten P, Brocker EB, Becker JC (2002) Differential expression of inhibitory or activating CD94/NKG2 subtypes on MART-1-reactive T cells in vitiligo *versus* melanoma: a case report. *J Invest Dermatol* 118: 595–599
52 Becker JC, Vetter CS, Schrama D, Brocker EB, Thor Straten P (2000) Differential expression of CD28 and CD94/NKG2 on T cells with identical TCR beta variable regions in primary melanoma and sentinel lymph node. *Eur J Immunol* 30: 3699–3706
53 Vetter CS, Straten PT, Terheyden P, Zeuthen J, Brocker EB, Becker JC (2000) Expression of CD94/NKG2 subtypes on tumor-infiltrating lymphocytes in primary and metastatic melanoma. *J Invest Dermatol* 114: 941–947
54 Perrin G, Speiser D, Porret A, Quiquerez AL, Walker PR, Dietrich PY (2001) Sister cytotoxic CD8$^+$ T cell clones differing in natural killer inhibitory receptor expression in human astrocytoma. *Immunol Lett* 81: 125–132
55 Speiser DE, Pittet MJ, Valmori D, Dunbar R, Rimoldi D, Lienard D, MacDonald HR, Cerottini JC, Cerundolo V, Romero P (1999) *In vivo* expression of natural killer cell inhibitory receptors by human melanoma-specific cytolytic T lymphocytes. *J Exp Med* 190: 775–782

56 Jabri B, Selby JM, Negulescu H, Lee L, Roberts AI, Beavis A, Lopez-Botet M, Ebert EC, Winchester RJ (2002) TCR specificity dictates CD94/NKG2A expression by human CTL. *Immunity* 17: 487–499
57 Bellon T, Heredia AB, Llano M, Minguela A, Rodriguez A, Lopez-Botet M, Aparicio P (1999) Triggering of effector functions on a CD8⁺ T cell clone upon the aggregation of an activatory CD94/kp39 heterodimer. *J Immunol* 162: 3996–4002
58 Aramburu J, Balboa MA, Ramirez A, Silva A, Acevedo A, Sanchez-Madrid F, De Landazuri MO, Lopez-Botet M (1990) A novel functional cell surface dimer (Kp43) expressed by natural killer cells and T cell receptor-gamma/delta⁺ T lymphocytes. I. Inhibition of the IL-2-dependent proliferation by anti-Kp43 monoclonal antibody. *J Immunol* 144: 3238–3247
59 Romero P, Ortega C, Palma A, Molina IJ, Pena J, Santamaria M (2001) Expression of CD94 and NKG2 molecules on human CD4(+) T cells in response to CD3-mediated stimulation. *J Leukoc Biol* 70: 219–224
60 Meyers JH, Ryu A, Monney L, Nguyen K, Greenfield EA, Freeman GJ, Kuchroo VK (2002) Cutting edge: CD94/NKG2 is expressed on Th1 but not Th2 cells and costimulates Th1 effector functions. *J Immunol* 169: 5382–5386
61 Lindquist S, Craig EA (1988) The heat-shock proteins. *Annu Rev Genet* 22: 631–677
62 Bukau B, Horwich AL (1998) The Hsp70 and Hsp60 chaperone machines. *Cell* 92: 351–366
63 Soltys BJ, Gupta RS (1996) Immunoelectron microscopic localization of the 60 kDa heat shock chaperonin protein (Hsp60) in mammalian cells. *Exp Cell Res* 222: 16–27
64 Singh B, Patel HV, Ridley RG, Freeman KB, Gupta RS (1990) Mitochondrial import of the human chaperonin (HSP60) protein. *Biochem Biophys Res Commun* 169: 391–396
65 Parsell DA, Lindquist S (1993) The function of heat-shock proteins in stress tolerance: degradation and reactivation of damaged proteins. *Annu Rev Genet* 27: 437–496
66 Cohen IR, Young DB (1991) Autoimmunity, microbial immunity and the immunological homunculus. *Immunol Today* 12: 105–110
67 Anderton SM, van der Zee R, Goodacre JA (1993) Inflammation activates self hsp60-specific T cells. *Eur J Immunol* 23: 33–38
68 van Eden W (1991) Heat-shock proteins as immunogenic bacterial antigens with the potential to induce and regulate autoimmune arthritis. *Immunol Rev* 121: 5–28
69 Paul AG, van Kooten PJ, van Eden W, van der Zee R (2000) Highly autoproliferative T cells specific for 60 kDa heat shock protein produce IL-4/IL-10 and IFN-gamma and are protective in adjuvant arthritis. *J Immunol* 165: 7270–7277
70 Prakken AB, van Eden W, Rijkers GT, Kuis W, Toebes EA, de Graeff-Meeder ER, van der Zee R, Zegers BJ (1996) Autoreactivity to human heat-shock protein 60 predicts disease remission in oligoarticular juvenile rheumatoid arthritis. *Arthritis Rheum* 39: 1826–1832
71 Prakken B, Wauben M, van Kooten P, Anderton S, van der Zee R, Kuis W, van Eden W (1998) Nasal administration of arthritis-related T cell epitopes of heat shock protein 60 as a promising way for immunotherapy in chronic arthritis. *Biotherapy* 10: 205–211

72 van der Zee R, Anderton SM, Prakken AB, Liesbeth Paul AG, van Eden W (1998) T cell responses to conserved bacterial heat-shock-protein epitopes induce resistance in experimental autoimmunity. *Semin Immunol* 10: 35–41
73 Anderton SM, van der Zee R, Prakken B, Noordzij A, van Eden W (1995) Activation of T cells recognizing self 60 kDa heat shock protein can protect against experimental arthritis. *J Exp Med* 181: 943–952
74 Quintana FJ, Carmi P, Mor F, Cohen IR (2002) Inhibition of adjuvant arthritis by a DNA vaccine encoding human heat shock protein 60. *J Immunol* 169: 3422–3428
75 Billingham ME, Carney S, Butler R, Colston MJ (1990) A mycobacterial 65 kD heat shock protein induces antigen-specific suppression of adjuvant arthritis, but is not itself arthritogenic. *J Exp Med* 171: 339–344
76 Raz I, Elias D, Avron A, Tamir M, Metzger M, Cohen IR (2001) Beta-cell function in new-onset type 1 diabetes and immuno-modulation with a heat-shock protein peptide (DiaPep277): a randomised, double-blind, phase II trial. *Lancet* 358: 1749–1753
77 Imani F, Soloski MJ (1991) Heat shock proteins can regulate expression of the Tla region-encoded class Ib molecule Qa-1. *Proc Natl Acad Sci USA* 88: 10475–10479
78 Lo WF, Woods AS, DeCloux A, Cotter RJ, Metcalf ES, Soloski MJ (2000) Molecular mimicry mediated by MHC class Ib molecules after infection with gram-negative pathogens. *Nat Med* 6: 215–218
79 Michaelsson J, Teixeira de Matos C, Achour A, Lanier LL, Kärre K, Söderström K (2002) A signal peptide derived from hsp60 binds HLA-E and interferes with CD94/NKG2A recognition. *J Exp Med* 196: 1403–1414
80 Cerboni C, Mousavi-Jazi M, Wakiguchi H, Carbone E, Kärre K, Söderström K (2001) Synergistic effect of IFN-gamma and human cytomegalovirus protein UL40 in the HLA-E-dependent protection from NK cell-mediated cytotoxicity. *Eur J Immunol* 31: 2926–2935
81 Soderberg-Naucler C, Fish KN, Nelson JA (1997) Interferon-gamma and tumor necrosis factor-alpha specifically induce formation of cytomegalovirus-permissive monocyte-derived macrophages that are refractory to the antiviral activity of these cytokines. *J Clin Invest* 100: 3154–3163
82 Wang EC, McSharry B, Retiere C, Tomasec P, Williams S, Borysiewicz LK, Braud VM, Wilkinson GW (2002) UL40-mediated NK evasion during productive infection with human cytomegalovirus. *Proc Natl Acad Sci USA* 99: 7570–7575
83 Malmberg KJ, Levitsky V, Norell H, de Matos CT, Carlsten M, Schedvins K, Rabbani H, Moretta A, Söderström K, Levitskaya J et al (2002) IFN-gamma protects short-term ovarian carcinoma cell lines from CTL lysis *via* a CD94/NKG2A-dependent mechanism. *J Clin Invest* 110: 1515–1523
84 Vales-Gomez M, Reyburn HT, Erskine RA, Lopez-Botet M, Strominger JL (1999) Kinetics and peptide dependency of the binding of the inhibitory NK receptor CD94/NKG2-A and the activating receptor CD94/NKG2-C to HLA-E. *EMBO J* 18: 4250–4260
85 Dulphy N, Rabian C, Douay C, Flinois O, Laoussadi S, Kuipers J, Tamouza R, Charron D, Toubert A (2002) Functional modulation of expanded $CD8^+$ synovial fluid T cells by

NK cell receptor expression in HLA-B27-associated reactive arthritis. *Int Immunol* 14: 471–479

86 Herberts CA, van Gaans-van den Brink J, van der Heeft E, van Wijk M, Hoekman J, Jaye A, Poelen MC, Boog CJ, Roholl PJ, Whittle H et al (2003) Autoreactivity against induced or up-regulated abundant self-peptides in HLA-A*0201 following measles virus infection. *Hum Immunol* 64: 44–55

87 Hodgkinson AD, Millward BA, Demaine AG (2000) The HLA-E locus is associated with age at onset and susceptibility to type 1 diabetes mellitus. *Hum Immunol* 61: 290–295

88 Jawaheer D, Li W, Graham RR, Chen W, Damle A, Xiao X, Monteiro J, Khalili H, Lee A, Lundsten R et al (2002) Dissecting the genetic complexity of the association between human leukocyte antigens and rheumatoid arthritis. *Am J Hum Genet* 71: 585–594

Index

The PIR-Series
Progress in Inflammation Research

Homepage: http://www.birkhauser.ch

Up-to-date information on the latest developments in the pathology, mechanisms and therapy of inflammatory disease are provided in this monograph series. Areas covered include vascular responses, skin inflammation, pain, neuroinflammation, arthritis cartilage and bone, airways inflammation and asthma, allergy, cytokines and inflammatory mediators, cell signalling, and recent advances in drug therapy. Each volume is edited by acknowledged experts providing succinct overviews on specific topics intended to inform and explain. The series is of interest to academic and industrial biomedical researchers, drug development personnel and rheumatologists, allergists, pathologists, dermatologists and other clinicians requiring regular scientific updates.

Available volumes:

T Cells in Arthritis, P. Miossec, W. van den Berg, G. Firestein (Editors), 1998
Chemokines and Skin, E. Kownatzki, J. Norgauer (Editors), 1998
Medicinal Fatty Acids, J. Kremer (Editor), 1998
Inducible Enzymes in the Inflammatory Response,
D.A. Willoughby, A. Tomlinson (Editors), 1999
Cytokines in Severe Sepsis and Septic Shock, H. Redl, G. Schlag (Editors), 1999
Fatty Acids and Inflammatory Skin Diseases, J.-M. Schröder (Editor), 1999
Immunomodulatory Agents from Plants, H. Wagner (Editor), 1999
Cytokines and Pain, L. Watkins, S. Maier (Editors), 1999
In Vivo *Models of Inflammation*, D. Morgan, L. Marshall (Editors), 1999
Pain and Neurogenic Inflammation, S.D. Brain, P. Moore (Editors), 1999
Anti-Inflammatory Drugs in Asthma, A.P. Sampson, M.K. Church (Editors), 1999
Novel Inhibitors of Leukotrienes, G. Folco, B. Samuelsson, R.C. Murphy (Editors), 1999
Vascular Adhesion Molecules and Inflammation, J.D. Pearson (Editor), 1999
Metalloproteinases as Targets for Anti-Inflammatory Drugs,
K.M.K. Bottomley, D. Bradshaw, J.S. Nixon (Editors), 1999
Free Radicals and Inflammation, P.G. Winyard, D.R. Blake, C.H. Evans (Editors), 1999
Gene Therapy in Inflammatory Diseases, C.H. Evans, P. Robbins (Editors), 2000
New Cytokines as Potential Drugs, S. K. Narula, R. Coffmann (Editors), 2000
High Throughput Screening for Novel Anti-inflammatories, M. Kahn (Editor), 2000
Immunology and Drug Therapy of Atopic Skin Diseases,
C.A.F. Bruijnzeel-Komen, E.F. Knol (Editors), 2000
Novel Cytokine Inhibitors, G.A. Higgs, B. Henderson (Editors), 2000
Inflammatory Processes. Molecular Mechanisms and Therapeutic Opportunities,
L.G. Letts, D.W. Morgan (Editors), 2000

MIX
Papier aus verantwortungsvollen Quellen
Paper from responsible sources
FSC® C105338

If you have any concerns about our products,
you can contact us on
ProductSafety@springernature.com

In case Publisher is established outside the EU,
the EU authorized representative is:
Springer Nature Customer Service Center GmbH
Europaplatz 3, 69115 Heidelberg, Germany

Printed by Libri Plureos GmbH
in Hamburg, Germany